ENTANGLED WORLDS

TRANSDISCIPLINARY THEOLOGICAL COLLOQUIA

Theology has hovered for two millennia between scriptural metaphor and philosophical thinking; it takes flesh in its symbolic, communal, and ethical practices. With the gift of this history and in the spirit of its unrealized potential, the Transdisciplinary Theological Colloquia intensify movement between and beyond the fields of religion. A multivocal discourse of theology takes place in the interstices, at once self-deconstructive in its pluralism and constructive in its affirmations.

Hosted annually by Drew University's Theological School, the colloquia provide a matrix for such conversations, while Fordham University Press serves as the midwife for their publication. Committed to the slow transformation of religio-cultural symbolism, the colloquia continue Drew's long history of engaging historical, biblical, and philosophical hermeneutics, practices of social justice, and experiments in theopoetics.

Catherine Keller, *Director*

ENTANGLED WORLDS

Religion, Science,
and New Materialisms

CATHERINE KELLER AND
MARY-JANE RUBENSTEIN,
EDITORS

FORDHAM UNIVERSITY PRESS ⚭ NEW YORK ⚭ 2017

Copyright © 2017 Fordham University Press

All rights reserved. No part of this publication may be reproduced, stored in a retrieval system, or transmitted in any form or by any means—electronic, mechanical, photocopy, recording, or any other—except for brief quotations in printed reviews, without the prior permission of the publisher.

Fordham University Press has no responsibility for the persistence or accuracy of URLs for external or third-party Internet websites referred to in this publication and does not guarantee that any content on such websites is, or will remain, accurate or appropriate.

Fordham University Press also publishes its books in a variety of electronic formats. Some content that appears in print may not be available in electronic books.

Visit us online at www.fordhampress.com.

Library of Congress Cataloging-in-Publication Data

Names: Keller, Catherine, 1953– editor.
Title: Entangled worlds : religion, science, and new materialisms / Catherine Keller and Mary-Jane Rubenstein, editors.
Description: First edition. | New York, NY : Fordham University Press, 2017. | Series: Transdisciplinary theological colloquia | Includes bibliographical references and index.
Identifiers: LCCN 2016058786 | ISBN 9780823276219 (cloth : alk. paper) | ISBN 9780823276226 (pbk. : alk. paper)
Subjects: LCSH: Materialism. | Materialism—Religious aspects. | Religion and science.
Classification: LCC B825 .E58 2017 | DDC 201/.61—dc23
LC record available at https://lccn.loc.gov/2016058786

Printed in the United States of America
19 18 17   5 4 3 2 1
First edition

CONTENTS

Introduction: Tangled Matters | *Catherine Keller and Mary-Jane Rubenstein* 1

MATTER, ANEW

What Flashes Up: Theological-Political-Scientific Fragments | *Karen Barad* 21

Vegetal Life and Onto-Sympathy | *Jane Bennett* 89

Tingles of Matter, Tangles of Theology | *Catherine Keller* 111

Agents Matter and Matter Agents: Interpretation and Value from Cells to Gaia | *Philip Clayton and Elizabeth Singleton* 136

THE MATTER OF RELIGION

The Matter with Pantheism: On Shepherds and Goat-Gods and Mountains and Monsters | *Mary-Jane Rubenstein* 157

Material Subjects, Immaterial Bodies: Abhinavagupta's Panentheist Matter | *Loriliai Biernacki* 182

Theophanic Materiality: Political Ecology, Inhuman Touch, and the Art of Andy Goldsworthy | *Jacob J. Erickson* 203

Interdisciplinary Ethics: From Astro-Theology to Cosmo-Liberation Theology | *Theodore Walker Jr.* 221

Vascularizing the Study of Religion: Multi-Agent Figurations and Cosmopolitics | *Manuel A. Vásquez* 228

ETHICOPOLITICAL ENTANGLEMENTS

Stubborn Materiality: African American Religious Naturalism
and Becoming Our Humanity | *Carol Wayne White*   251

Grace in Intra-action: Complementarity and the Noncircular
Gift | *Terra S. Rowe*   274

The Door of No Return: An Africana Reading
of Complexity | *Elías Ortega-Aponte*   299

The Trouble with Commonality: Theology, Evolutionary
Theory, and Creaturely Kinship | *Beatrice Marovich*   317

List of Contributors   331

◈ Introduction: Tangled Matters

CATHERINE KELLER AND
MARY-JANE RUBENSTEIN

It is not just that we are entangled in matter—we subjects who read, write, and ruminate on what "we" are. We are materializations entangled in other materializations; we happen in our mattering. What matters in our ethics, our politics, our worlds entangles us in and as new materializations. And at this juncture, it entangles scholarship in retrievals and rethinkings of matter itself. Even disciplines that struggle with long histories of disembodied transcendence are registering the effects.

The "new materialisms" currently coursing through cultural, feminist, political, and queer theories seek to displace human privilege by attending to the agency of matter itself. Far from being passive or inert, they argue, matter acts, creates, destroys, and transforms—and, thus, is more of a process than a thing. "One could conclude," write Diana Coole and Samantha Frost, "that 'matter becomes,' rather than that matter is."[1] Calling as they do on the insights of quantum mechanics, general relativity, complexity theory, and nonlinear biology to theorize matter as matter*ing*, these thinkers work against much of what is often denigrated as "mere" materialism. Of course, the word will never make a safe slogan. It stimulates the whole modern array of old familiar materialisms, along with the dualist reactions against it, presuming pretty much the same inertly predictable stuff. A different materialism can therefore be introduced only in the company of such a caveat as Jane Bennett's: "American materialism, which requires buying ever-increasing numbers of products purchased in ever-shorter cycles, is antimateriality."[2] It junks old stuff so that we will buy buy buy new, and so conceals "the vitality of matter." Moreover, it destroys it, producing the waste-monsters that are deadening the atmosphere and the oceans.

Taking cues from Whitehead, Deleuze and Guattari, Stengers and Prigogine, and Margulis and Sagan, the new materialists mobilize a revivified materiality against such toxic materialisms. They accomplish this in part by rejecting traditional ontological hierarchies—especially those that seek clear distinctions between spirit and matter, life and nonlife, or sentience and nonsentience. They do not, however, embrace a straight reductionism, whether to cells, particles, or any purported fundament. Nor do they appeal to the strictly "flat ontology" of the speculative realists, according to whom "everything exists equally—plumbers, cotton, bonobos, DVD players, and sandstone, for example."[3] Like the new materialists, the speculative realists (especially those on the "object-oriented" branch) seek to unsettle philosophy's traditional privilege of the human but reject what they call "process relationalism" insofar as it privileges "alliances" over entities, "couplings" over objects, and motion over rest.[4] Demanding a theory of "sharp, specific units," these thinkers proclaim the ontological equality of every discrete *thing*.[5]

For the new materialists, by contrast, ontology is "monolithic but multiply tiered"; in other words, things are not simply "equal" because things are not "things" in the first place.[6] Rather, things—actual entities—are multiplicities, assemblages, hybrids, resonance machines, sonority clusters, intra-actions, complexities, and viscous porosities[7]—all terms that variously express the insight that each cell, organism, vegetable, and photon is irreducibly composed of what Karen Barad would call an "intra-active" host of others.[8] This constitutive alterity holds for humans as much as for anything else; as Donna Haraway reminds us, "human genomes can be found in only about 10 percent of all the cells that occupy the mundane space I call my body; the other 90 percent of the cells are filled with the genomes of bacteria, fungi, protists, and such.... I become an adult human being in company with these tiny messmates. To be one is always to *become with* many."[9] We could perhaps call such an ontology *folded*, even fractal, but not flat. And it is certainly not "hierarchical"; as Deleuze and Guattari caution, "It is already going too far to postulate an order descending from the animal to the vegetable, then to molecules, to particles. Each multiplicity is symbiotic; its becoming ties together animals, plants, microorganisms, mad particles, a whole galaxy."[10]

Given that all things are constantly becoming (and coming undone) in relation to constitutive multiplicities, it becomes difficult to say what the compendium of "all things"—or a "world"—might be. Faced with this difficulty and concerned about the ecological inaction that world-language purportedly instills, speculative realists Ian Bogost and Timothy Morton suggest we think of "world" as "just another being among the muskmelons and the lip balms," and that we resist at all costs the effort to gather such beings into a coherent (Heideggerian) *"world picture."*[11] Uncomfortable with totalization on the one hand and fragmentation on the other, we would like to propose a third option. As this volume employs the term, "world" is not just another *thing*, but neither does it constitute a single totality; rather, with the political philosopher William Connolly, we call upon "world" to name a provisional set of "multiple, interacting open systems of different viscosity morphing at different speeds."[12]

Accordingly, the volume itself imposes no homogeneous concept, not even of "the new materialism," let alone *the* world." Rather, it assembles a multiplicity of experimental perspectives on mattering, and so a multiplicity of worlds, which is to say a multiplicity of multiplicities. Whatever a "world" is, the motivation for its entanglement in the present multiplicity will neither reduce it to a philosophical or scientific interest, nor inflate it to a religious whole. The converging velocities of these transdisciplinary engagements do not rest with the description of a world. They charge the materiality of their cosmos with ethical intensity, with the spirit of specifically mattering bodies and with the chance of a cosmopolitics of the earth.

The politics that collects itself heterogeneously under the banner of new materialism comes energized already by its feminist, queer, and anti-racist orientations. It always already engages dialectical materialism and the range of counter-capitalist economies. But then as Pheng Cheah elucidates, the question is less that of the "relevance of these new materialisms to political thought" and concrete politics than of how they "put into question the fundamental categories of political theory including the political itself."[13] For this reason, Cheah surfaces the "urgency of rethinking the ontological bases" of the words and worlds of the political. Such ontological rethinking rides the force field of a world vastly exceeding the human, but tendering us no escape from the nonhuman. We are not enlisting troops for the "inhuman" or the "posthuman," though certainly all the antihumanisms help break down anthropocentric delusion. Rather, it is a matter of being differently, perchance responsibly, human—and so, as Connolly puts it, "organized by a host of nonhuman processes."[14] The focus, as he elucidates in *The Fragility of Things*, "is on our entanglements with heterogeneous entities and processes in a world in which humanity matters immensely."[15]

Humanity matters for good and for ill, from the perspective of the vast variety of nonhuman beings composing the life of the planet. And so the entangling and entangled worlds of this rhizome of writers get thought and rethought at every scale. For "when the smallest parts of matter are found to be capable of exploding deeply entrenched ideas and large cities," matter itself comes into question and brings us all with it.[16] The quantum has unsettled our pretensions toward objectivity and individuality; its apocalyptic bomb-power has rerouted history; and yet it has not yet shattered commonsense notions of the elements of matter. "Perhaps this is why contemporary physics makes the inescapable entanglement of matters of being, knowing, and doing, of ontology, epistemology, and ethics, of fact and value, so tangible, so poignant."[17]

These allusions to the multilayered, many-genred, polyvocal ecology of our entanglements are not casual. The term "entanglement" was coined in English and German (*Verschrankung*) by Erwin Schrödinger in a 1935 letter to Einstein, in the heat and shock of the emergent quantum mechanics.[18] He considered entanglement "not one but the characteristic trait of quantum mechanics, the one that reinforces its entire departure from classical lines of thought."[19] The observer and the observed, the subject and the object, enter into a dance barely glimpsed before in modern science. And the instruments of observation make the entanglement visible even as they join in to complicate it. The relation of entanglement, known also as nonlocality or nonseparability, so disturbed the common sense—of physics, yes, but of modernity itself—that Einstein called it (with dread) "spooky action at a distance."[20] He was haunted for the rest of his life by his failure to disprove it. The spookiness begins with two particles, entangled in a laboratory, flying off in opposite directions. If then you measure twin A, it will react, as quanta do, unpredictably. The spooky thing is that twin B will (as you measure it simultaneously) react exactly at the same moment, in exactly the same way.[21] No matter how far apart: the width of a grain of sand or of a galaxy. Physicists tend to reject the idea that a signal is being propagated at some mind-boggling speed, vastly exceeding the velocity of light; what seems likelier is that the particles are not sending a signal at all, but somehow remain nonseparable and so in touch, persistently entangled. At any distance.

Here we can only note that what the physicist Brian Greene calls this "earth-shattering result" has tested out both theoretically and physically for three decades, before which physicists (such as David Bohm) who tried in the United States to investigate the phenomenon were treated as heretics, accused of "juvenile deviationism."[22] In this century it is above all Karen Barad who from the scientific side has released the profound philosophical reverberations of quantum mechanics, discerning not only its use but its *meaning*: "We are of the universe," she reminds us; "there is no inside, no outside."[23] Rethinking the bottomless entanglement of observer and observed, matter and discourse, *cosmos* and *ethos*, this stuff and that, Barad teases not only a fundamental ontology but, more radically, "a tissue of ethicality," out of our basest matter: "There is only intra-acting from within and as part of the world in its becoming."[24] If that's the case, we'd better get our intra-actions right.

The entanglements visibly and invisibly composing this volume mean to provoke all manner of deviant materializations—across a plane of ontological interrelation in which humans are remembering our inextricable interembodiments—before we really do shatter the earth. Among these constitutive nonhumans are, to be sure, minerals, vegetables, and animals. But those of us assembled here also leap with the quantum to another set of spooky actants: spirits, gods, and other subtle bodies. We attend with a variety of lenses to a variety religious traditions, where embodiment thins, speeds, or slow into the extremities of the world. From within a religious perspective, any becoming comes wrapped up in meanings and materialities that exceed and include it. But of course, the major religious traditions also carry long normative routines of disembodiment. Christianity in particular has long suffered from a condition of we could call "materiaphobia." Whether in ascetic practice, Pauline theology, or casual presumption, the spirit got pitted against the flesh, or the soul against the body, in league with a God who remains condescendingly above and apart from the world that "He" created. The world is "good" (yes, sure, of course), but it stepped out of line; the soul yielded to the flesh, and the flesh to sinful seduction. Matter, especially as the theological tradition digested its Platonism, was deemed inferior to and outside of what matters.

Not only theology as a constructive practice, but the study of religion as a descriptive discipline may also operate materiaphobically. Religious studies in its incipience conveyed a bemused fascination with the "animism" of "primitive" non-Europeans, according to which "what we call inanimate objects—rivers, stones, trees, weapons, and so forth—are treated as intelligent living beings."[25] When this spiritually materialist "animism" edged into "fetishism," however, the fascination turned to repulsion. As E. B. Tylor explains, "Fetishism will be taken as including the worship of 'stocks and stones,' and thence it passes by an imperceptible gradation into Idolatry."[26] While no longer likely to pass such smugly theological judgments, the study of religion has borne like all other human and social sciences the Cartesianism that imported theological dualisms into modernity. Secularisms may proclaim one or another materialism, but may also remain hostile to the intra-actively vital versions engaged here. Such materialists however as Barad, Bennett, Connolly, and Haraway break down the very subject-object, observer-observed binary that constitutes "religion" as a mere object, and which renders any form of religious practice objectionable.[27] And behind all those binaries lurks that of mind versus matter, as incoherent when it takes theological form as it is in scientific reductionism.

Alongside the theological degradations of materiality, however, there have all along been the voices of a positively material theology—even in the simultaneously materiaphobic traditions of Judaism and Christianity—voices pointing to the biblical tradition in which there is little textual support for such disembodiment. There is not only the exuberant affirmation of the universe as the creative effect of God ("O Lord how manifold are your works! In wisdom you have made them all. The earth is full of your creatures" [Psalm 104:24].) There is also the scriptural absence of any concept of a changeless, unaffected One merely affecting the world from outside; and there is only salvation—

salvus, healing, shalom—by way of material justice, embodied love, revealing incarnation. There are no immortal souls, only resurrections of *bodies*, and in the meantime communal bodies, anticipatory of a great cosmic reembodiment, earthly and political, called the New Jerusalem or New Heaven and Earth. These voices point out that even Christian orthodoxy was formed in reaction to a Gnostic vilification of materiality.[28] But at the same time, and not quite coherently, that same orthodoxy—as though in Oedipal transcendence of a Judaism marked as materialist, literalist, fleshly—formed deadly patterns. In other words, the old multiplicity of multiplicities gathered as "Christianity" has fomented toward its matter both love and hate, prophetic respect and phobic disdain.

Among students of Christian theology over the past few generations, there have been intensive resurgences of the ancient materiality of scripture. In liberation, feminist, ecological, and queer theologies emphasis on social justice at a planetary pitch has revived the prophetic legacy.[29] Philosophically, these material theologies have drawn on the sources of socialism but also on Whitehead's quantum-inflected cosmology in the ever-unfolding forms of process theology. These materializations are not one, and they are indeed resistant to the background "logic of the One."[30] They resist the specific modes of oppression, particularly patriarchal, whereby the materiaphobia of the tradition finds its roots in an ancient *matri*phobia. Religious studies, too, has increasingly turned toward the *stuff* of religion, too long overlooked by its unconsciously Protestant and overwhelmingly male forebears. These contemporary historians, anthropologists, and sociologists of religion attend variously to "folk" practices, healing waters, ritual objects, enchanted machines, sacred spaces, salvific commodities, musical instruments, videotapes, telegraphs, relics, and bread.[31]

Although both theology and religious studies have long engaged numerous materialities and theories thereof, this volume sets forth the first multivocal conversation between religious studies, theological scholars, and the body of "the new materialism."[32] Although Jane Bennett and Karen Barad, like many others gathered under that name, resist the label's tendency to overstate the "new,"[33] we retain it here to mark our engagement with the thinkers who have been so gathered.

Will we then speak of a religious materialism, a new religious materialism, or even—given the specific cases in this collection—a Jewish, a Hindu, a Christian materialism? As to the answer to that question, there must remain here some suspense. We do, however, want to hint at the polyvalent intensities released by the essays in this volume: startling oscillations between micro and macro processes, between virtual infinities and queer intimacies, between the inhumans that surround and populate animals, vegetables, humans. There emerge the distinct voices of vibrantly variegated race-refractions, sexual dis/orientations, religious metamorphoses. They crisscross the economies and ecologies that are shifting the geologies, species, and force fields of the earth. In particular, the present conversation transverses several fields of academic discourse normally kept hygienically separate. Here such disciplines touch, transgress, and contaminate each other across their several carefully specified contexts. What Barad says of the virtual particles in a quantum field lends a parable for this volume: "*All touching entails an infinite alterity, so that touching the other is touching all others, including the 'self,' and touching the 'self' entails touching the strangers within.*"[34] And in the responsiveness of this mutual touching of science, religion, philosophy and theology, indeed also of the social sciences and of the arts, the growing complexity of our entanglements takes on a consistent ethical texture of urgency.

This volume moves in three constructively interfering waves. The first sets forth a variety of perspectives on the agency of matter, from the quantum to the vegetal to the theopolitical to the cosmic. The second brings these perspectives to bear on a range of theologies, each of which animates materiality as the site of divine unfolding. And the third examines the ethical and political work that material theologies might do to unsettle the racist, colonial, and ecocidal legacies of their materiaphobic counterparts.

What Karen Barad calls the "entanglement of matter and meaning" here takes on new resonances as it attracts religion-focused and theological response. Of course, there is a robust, century-long tradition of the dialogue between science and religion. Indeed, Whitehead's work early advanced such an exchange.[35] But often the science/religion dialogue has assumed rather standardized forms of interaction between two well-formed, if comically imbalanced, modern orthodoxies. In the practice of transdisciplinary theology in general and in this assemblage in particular, we hope for something more like the intra-action that does not presume the established, preexistent representatives. We all bring vast histories and habits, but we bring them for the sake of an entangled becoming. The "new" attention to matter—even if the label has already gotten old for some—touches on the deep tissues of meaning, the layers so densely sedimented in the religious ways of the world. For good and for ill. For cliché or for new matterings.

MATTER, ANEW

In "What Flashes Up," Karen Barad exposes a startling new sense of matter. The "agential realist" interpretation of quantum physics in her monumental *Meeting the Universe Halfway* had already brought the indeterminacy and relationality—the "intra-activity"—of quantum ontology into resonance with human ethics: All beings compose and partake in the responsive structure of the world. "Intra-acting responsibly as part of the world means taking account of the entangled phenomena that are intrinsic to the world's vitality and being responsive to the possibilities that might help us flourish."[36] In the present discussion, Barad draws Walter Benjamin's messianic "now-time" via Judith Butler and quantum field theory into a deep meditation on the matter of time, a time that breaks from the scientific and political modernisms of purportedly linear progress. Here, in an exploration bursting with transdisciplinary theological possibility, Barad reads matter itself, by virtue of its entanglement with a persistently not-nothing quantum void, as messianically inflected, kabbalistically encrypted, politically explosive. And so the "differential entanglements" that join us in our inextricable alterities echo "the mystical depth of Isaac Luria's cosmology, with the iterative rematerialization of the world based on ongoing intra-actions whereby ... the world is re-created anew in each moment." Neither human figure nor final closure, the messianic lightning flash that is the possibility of justice *now* breaks science open to its constitutively social, political, and theological alterities. It is not that the messianic can be simply "naturalized," but that in the "thick now" no bit of matter can any longer be purged of ethical meaning or indeed of revolutionary possibility. *"The flashing up of the messianic is written into the very structure of ... matter-time-being itself."*

Turning our gaze to the macroscopic register of the world's constitutive inhumanity, Jane Bennett attends to the spectrum of "Vegetal Life and Onto-Sympathy." Making a move startlingly congruent with Barad's other-than-human "feeling-with," Bennett calls her version "onto-sympathy." Tailing Darwin, Bergson, and particularly Thoreau in their adventures into the nonhuman, she magnifies the latter's insight that "the vegetality of humans and the vegetality of plants can somehow *communicate*. Why should I not, asks Thoreau, 'have intelligence with the earth? Am I not partly leaves and vegetable mould myself?'" Here Bennett finds a pre-echo of queer theory, particularly of Bersani's "failed subject," its outside that is "mine without belonging to me." Dropping out of conventional selfhood, Bennett sketches Thoreau's "psychedelic claim that each of the different materials—clay, water, sand, vines, human flesh—are variations on the theme of the droplet or 'moist thick *lobe.*'" In the interest of hallucinogenic diffraction, we might mention the "tiny droplets ... mingling and interacting with one another, expanding and conjoining until a sap pours forth from them" observed by Moses, according to the thirteenth-century text of the kabbalistic Iyyun Circle.[37]

Like Whitehead and Barad at the level of the quantum, Bennett destabilizes at the level of the vegetal any borderline between the organic and the inorganic—let alone of the living and lifeless, the touching-feeling and inert. These mobilities render uninteresting the sort of boundaries that "Man" has erected beneath himself, hoping thereby to secure proximity to the transcendent above. In their teeming fields of more-than-humanity, such new materialist reflections detect and amplify a new transspecies vitality, an intra-agential solidarity issuing forth for ethics and politics. Might these new multiplicities of multiplicities, inhuman but not dispassionate, teach us here something we ignore at our own manifold peril? Something beyond the vertical sovereignty modeled on the God classically forbidden to feel-with, the One who could only move and never be moved?

Reflecting on matter and its intimate ultimacy, Catherine Keller wonders about uses of the phrase "Christian materialism." She finds some shocking incongruity in the deployment of the phrase, and proceeds to locate it within a long and complicated history of religious materializations—in particular within Christian variegations of the "body of Christ." She takes the opportunity to stage an exchange between Barad and Whitehead, who reject the concept of "matter" for something very like the reason Barad nearly a century later embraces it. The vibratory becomings, from the quantum up, that drive both of them to a relational ontology come into potential solidarity here with a vibrant theological collective. Indeed the quantum mystery of entanglement, in which differences do not wash out interdependence, may find itself amplified in the subtle body of a Supreme Entanglement. But there will be many nicknames for the space—in its "Tingles of Matter, Tangles of Theology"—in which, for example, process, ecofeminist, and queer Christianities have been teaming up with ancient relics of holy matter and new diffractions of our teeming bodies.

Zooming out to the biospheric and in to the microscopic, Philip Clayton and Elizabeth Singleton affirm in "Agents Matter and Matter Agents" the agency of "every living being," from "the humble eukaryotic cell" to Gaia as an open and evolving whole. Beginning from what they call "the first agents," which is to say the simplest biochemical systems in which we can discern "teleology," the authors reveal each living thing as a network of intentions, environments, adaptations, communications, toxins, nutrients, and porosities. In dynamic, differential resonance with Barad, they thereby reveal (living) matter's primordial entanglement with meaning: "When a single-celled organism spins its flagellum in order to move up a glucose gradient and obtain more nourishment," Clayton and Singleton explain, "it *interprets* the higher glucose concentration as a *good* and acts in order to ingest more of it" (emphasis added). Moreover, this pan-agential biosemiotics entails—and necessarily so—an "ethic of embodied responsibility." After all, to recognize living beings as "valuing" is to recognize them as "valuable"—not simply "in themselves," but rather as "participants in the entire system of life . . . held together by networks of bodies and objects" and collectively composing the fragile, interdependent biosphere in, through, with, and as which anything that lives, lives.

THE MATTER OF RELIGION

In the meantime, amid all this mattering and multiplying of the agencies with whom we come entangled, God-talk comes to matter again. Relieved of impassive transcendence and immateriality, what sort of body might "God" signify? And freed of sovereign omnipotence, what sort of liberating agency? When religious figurations reappear, provoking unfamiliar forms, they demand the ministry of new language. But perhaps even more disturbingly than philosophy, theology comes haunted by deep pasts. What is "to come" comes superposed with what has not been allowed to become, indeed with possibilities blocked by religious authority. So in this part, each of the five authors reads divinity as emphatically and emancipatively materializing. At the same time, they each expose specific historical tangles of religious traditions whose past gifts of novelty and strangeness, lost in the margins or actively hereticized, have hardly yet been received.

For instance, what *is* "The Matter with Pantheism"? Mary-Jane Rubenstein asks a question that the Abrahamic traditions have with deafening success managed to silence for centuries. Baffled by this legacy of "name-calling," she wonders what is really at stake. It was precisely for the "error" of granting God a body—not just one brief, Galilean one but the magnificent matter of the whole universe—that Spinoza was vilified. Even the trope of "the body of God," when it surfaces as pan*en*theism through scientifically engaged process and ecofeminist theologies, hesitates to display its Spinozan ancestry. Pantheism has served as such a foil for orthodoxy, such a settled no-no, that even liberal and progressive theologies—willing to risk an impressive host of heresies—quickly reassure readers that no, at least we are not *that*. So Rubenstein takes us on a journey back to the earlier accounts of the "pan," replete with "Shepherds and Goat-Gods and Mountains and Monsters." Her goal is not to propose a new materialist pantheism, but rather to consider what has been lost in the name-calling. What "intra-carnational" possibilities, what ecologies of planetary enfleshment, and what queerly intraspecies sites of divinity have been, to our peril, repressed along with pantheism?

Far and free from this Abrahamic force field, Loriliai Biernacki considers "Material Subjects, Immaterial Bodies" in the work of the medieval Indian Tantric philosopher and mystic Abhinavagupta. Bringing to light a thousand year-old and all-too-timely spiritual philosophy, Biernacki writes that "Abhinavagupta inherits an already sophisticated model of body as both physical and nonmaterial," a "subtle body" that belies both Western and Asian tendencies to dualism and monism. Biernacki teases open this alternative materiality, nonseparable from consciousness, by drawing it into an engagement with the distinct quantum models of Karen Barad and Henry Stapp. Of course, there are significant differences between the tantra and the quanta; in particular, Biernacki notes Abhinavagupta's concern to establish an "overarching subjectivity" whereas Barad has an intra-determinate relationality. Nevertheless, Biernacki finds materialist promise in the over-arch, channeling into current theoretical vibrancy Abhinavagupta's extraordinary, panentheist teaching that "the highest principle is indeed the body of all things." In his superposed interferences with quantum theory, this medieval Indian thinker thereby offers our new materialisms some old "options for thinking about matter and the body" as an actively becoming, "material-discursive occurrence."

The next two essays consider the ways God matters in Christianity. In "Theophanic Materiality," Jake Erickson has us following a trail marked by little piles of stones, cairns along a path in the Himalayas, and back through the "land art" of Goldsworthy. Along the way, Barad's and Bennett's materialisms not only set the stones into theory (and motion), but enable a startling rereading of the ancient Eastern Christian traditions of theophany. Erickson's retrieval also of the suppressed ninth-century thought of John Scottus Eriugena—for whom God is manifest "through bodies in bodies"—glows with the dark brilliance of the path not taken. Ultimately, Erickson's concept of "theophanic materiality," in which "divine energy is entangled in the performance of indeterminate material agencies," goes so far as to "create a conceptual possibility for the queer intimacy of divinity and earth." This matters beyond cosmological mysticism: the uncommodifiable ordinariness of Goldsworthy's materials, the mobilization of counter-capitalist pleasures, and disciplines of attention to things, he imagines, "might effect a reimagined response to our current ecological crises."

The urgencies of contemporary theological practice burst into an extraterrestrial register in Theodore Walker's passage "From Astro-Theology to Cosmo-Liberation Theology." Drawing on the English natural philosopher and cleric William Derham, whose 1715 *Astro-Theology* performs an early modern integration of science and religion, Walker initiates a development of his previously proposed "constructive postmodern astro-theology." Working with a Whiteheadian metaphysical realism, African American liberation theology, and Barbara Holmes's originary synthesis of the latter with physical cosmology, Walker solicits an interdisciplinary cosmology to "inspire more liberating struggles." He argues that a robust and ultimately panentheistic realism, embodying a biblical prophetic ethics and a natural scientific universe, is a needed antidote to cynicism. Does the consonance of "astro-" and "Afro-" seems to hint here at what in the world does matter? (As we write, "Black Lives Matter" reverberates through the land.) The feedback loops between the farthest flung matter and race-class-gender-sex-ecological entanglements register, for instance, in the concreteness of an African American/Africana dimension that has been on the whole underdeveloped in the new materialist literature. We are composed, Walker writes, "of microscopic, cellular, subcellular, molecular, atomic, subatomic, and quantum parts-particles-wavicles and events." For Walker, it is this "cosmic context" that demands and emboldens the work of what he now calls "cosmo-liberation theology."

Attentive to the movement—in fact the flows—that interrupt the singular integrity of any purported subject, Manuel Vásquez mobilizes religious studies beyond its human fixation. His essay, "Vascularizing the Study of Religion," attends to the entanglement of "religious phenomena" with "their sociocultural and environmental contexts" as themselves the entangled products of "multiple types of agencies." With and beyond Actor Network Theory, and in conversation with both Bennett and Barad, Vásquez examines the Mexican-diasporic Light of the World Church to reveal its sites and celebrations as effects of complex flows of bodies, information, images, capital, and affect across various mediations of time, borders, satellite links, and screens. For Vásquez, such "transnational and electronic networks lead to the unsettling of established, territorialized cultural and religious habits and to the articulation of hybrid practices, identities, landscapes, and institutional arrangements." It is this vital, de- and re-territorialized hybridity of people, places, and stuff—material, virtual, ideological, and legislative—that comes to light in his "vascularized" methodology. Here "religion" itself is revealed as irreducibly "polycentric" and "multidirectional," a "lumpy landscape . . . of varying intensity, extensity, morphology, and durability," whose bodies exceed, resist, and reconfigure the strictures of the *anthropos* and nation-state alike.

ETHICOPOLITICAL ENTANGLEMENTS

Although all the essays in this volume gesture toward the ethical and political stakes of their variegated re-materializations, those gathered into this final part offer concrete visions of collective responsibility. In "Stubborn Materiality," Carol Wayne White augments and complicates our collective effort to unsettle "the human" by focusing on African American "subjects whose full humanity was often questioned or denied." In the harrowing wake of enslavement, black religions in North America have consistently proclaimed African Americans' full humanity by means of dynamic, evolving "theological tactics . . . ontological justification[s,] and ethical reasoning." Following the paths charted by Anna Julia Cooper, W. E. B. Du Bois, and James Baldwin, White traces here a discourse of "human desire" that unsettles nearly everything we associate with "humanism"—such as human autonomy, human exceptionalism, ontological hierarchies, and "environmental" instrumentalism. The task, she acknowledges, is "complex yet important": to recognize at once the full humanity of "black (and other marginalized) subjects" while also celebrating "the interrelatedness of all natural processes. Whatever conceivable notion of black humanity I claim in this essay will be ontologically enmeshed and entangled with other forms of natural life." Calling upon the "religious naturalism" of Ursula Goodenough, Donald Crosby, and Loyal Rue, White calls this ontological enmeshing and entangling, this more-than-humanism that speaks from and to the experience of African American *becoming*, "sacred humanity."

Like White, Terra Rowe contests in "Grace in Intra-Action" the narrow, often all-too-theological humanisms that have set "us" apart from "nature," thereby justifying its exploitation. "The inconvenient truth," she writes, "is that climate change only lends further evidence to the theory that we live in what Karen Barad calls an "intra-active universe." Plugging into the history of Christian eschatologies, Rowe locates the denial of this intra-action in unilateral theologies of grace, according to which God alone is active, and humans do nothing but receive. "Particularly from an ecological and climate-concerned perspective," she explains, the problem is that "the 'free' gift is not only free of obligation, but also of the intra-dependent relations that we now understand to make up the fabric of . . . existence." At the other end of the spectrometer, the danger of an obligatory gift is that it binds the recipient in an agonistic, escalating indebtedness. Mobilizing Jacques Derrida's and John Milbank's stubbornly opposed—even mutually exclusive—readings of Marcel Mauss, Rowe heralds the former's "unconditional grace" and the latter's "exchangist (or intra-active) universe" as *complementary* in the Bohrian sense. For the sake of the planet we inhabit and intra-actively compose, what we mean by "the gift"—and thus by grace—must mean, irreducibly and irreconcilably, "*both* an other-oriented, unconditional gift and a relational ontology of . . . exchange" (emphasis added).

In the essay of Elías Ortega-Aponte, the reader is "being touched and glimpsed" by the ghosts of an alternative spatiality. We feel the rhythms of Puerto Rican bomba begin to materialize "the sense of an alternate spiritual space." In the process, a vibrational shift in the social study of religion becomes possible. With Dionne Brand and Édouard Glissant, he offers "haunting" as an analytical strategy "to reflect the contemporary situation and increased complexity of life in 'late modernity,'"—but a life that "may itself be haunted by the specter of the transatlantic slave trade and its aftermath." Ortega-Aponte's "The Door of No Return: An Africana Reading of Complexity," engages social theoretical incorporations of complexity theory to render space "relational, unfolding as network, and with continually emerging properties." Indeed, in this altered space, affect and materialist perspectives may be "catching up," he quips, with the insights emerging from the Afro-Diaspora experience. "If haunting reveals repressed possibilities and alternate arrangements, then . . . let ghosts loose to haunt every life, institution, and social arrangement," but also their theorization. For example, he argues, the figurations of the "Door of No Return, the abyss, and the plantation are more than sociocultural happenings that shape history, economics, and cultural worlds." Rather, with their traumatic afterlives they are "deeper, implicating the nature of reality itself." For the haunting, in particular in its altered and altering Africana spatiality, is also drumming up—by way of its embodied memory and its layered networks—a greater material complexity.

Finally, in "The Trouble with Commonality," Beatrice Marovich offers a meditation on the category of "creature," which along with "mortal" has the distinction of being one of "the few words to express the fact that humans are earthlings, bound in a kind of kinship with other creatures on the planet." Like "mortal," she argues, "creature" carries with it an ineradicable theological past; in fact, even an anti-theologian like Richard Dawkins uses the term to signal the radical dependence and vulnerability of earthly life-forms under the regime of natural selection; for him as for contemporary theologians of "kinship," the term "creature" comes "umbilically connected to a sovereign power over life." For Marovich, such theological and scientific constructions of creatureliness locate our interspecies commonalities "in our incapacities," neglecting both "the plurality of creaturely life and the differentiated nature of creaturely powers and capacities." Concerned to attend both to bonds and distinctions, Marovich deploys Barad's "diffractive" methodology to configure creaturely kinship as one of "connective distinction." Although she concedes with Derrida that all creatures are radically different from one another, she also insists that, by means of this alterity, human and "other-than-human" life-forms are radically "intra-dependent." In this connective distinction, Marovich finds a bond of "strange kinship," which neither assimilates planetary others to the (always human) same nor writes them off as insignificant to our "own" flourishing.

• • •

Should this volume be read then as a companion to *New Materialisms*? The offering of a new theological materialism or a new materiality of religion? Or might it rather be something allied but other (a bit alien?), connectively distinct, pulsing across wide and wild frequencies, from the quantum physical to the materially religious to the polydoxically theological? Amid the turbulent cosmopolitics of race and sex, of religion and irreligion, of species and planet, these strategies of materialization insist in each case on mattering ethically. In gratitude for each of our authors, we turn to them now, in the hope that together we stir rather than satisfy interest in the entangled worlds we so precariously inhabit.

NOTES

1. Diana Coole and Samantha Frost, "Introducing the New Materialisms," in *New Materialisms: Ontology, Agency, and Politics*, ed. Diana Coole and Samantha Frost (Durham, N.C.: Duke University Press, 2010), 10.
2. Jane Bennett, *Vibrant Matter: A Political Ecology of Things* (Durham, N.C.: Duke University Press, 2010), 5.
3. Ian Bogost, *Alien Phenomenology: Or, What It's Like to Be a Thing* (Minneapolis: University of Minnesota Press, 2012), 6. The term "flat ontology" was coined by Manuel Delanda, and has been adopted broadly by the speculative realists, including Graham Harman, Levi Bryant, Ian Bogost, and Timothy Morton. See Manuel Delanda, *Intensive Science and Virtual Philosophy* (London: Bloomsbury Academic, 2013); and Graham Harman, *Towards Speculative Realism* (Winchester, UK: Zero Books, 2010).
4. Timothy Morton, *Hyperobjects: Philosophy and Ecology after the End of the World* (Minneapolis: University of Minnesota Press, 2013), 119; Bogost, *Alien Phenomenology*, 7.
5. Morton, *Hyperobjects*, 119–20. For a roundup of Object-Oriented Ontology's rejection of relation, as well as the terminology of "materialism," see Graham Harman, "I Am Also of the Opinion That Materialism Must Be Destroyed," *Environment and Planning D: Society and Space* 28 (2010): 772–90.
6. Coole and Frost, "Introducing the New Materialisms," 10. For a new materialist rejoinder to Harman's above-cited essay, see Jane Bennett, "Systems and Things: On Vital Materialism and Object-Oriented Philosophy," in *The Nonhuman Turn*, ed. Richard Grusin (Minneapolis: University of Minnesota Press, 2015).
7. The terms "multiplicity" and "assemblage" can be found in Gilles Deleuze and Félix Guattari, *A Thousand Plateaus*, trans. Brian Massumi (Minneapolis: University of Minnesota Press, 1987). The rest of these terms can be found in the following sources, respectively: Donna Haraway, "A Cyborg Manifesto: Science, Technology, and Socialist-Feminism in the Late Twentieth Century," in *Simians, Cyborgs, and Women: The Reinvention of Nature* (New York: Routledge, 1991); William Connolly, "The Evangelical-Capitalist Resonance Machine," *Political Theology* 33, no. 6 (December 2005), 869–86; Jane Bennett, *Vibrant Matter: A Political Ecology of Things* (Durham, N.C.: Duke University Press, 2010), 23; Karen Barad, *Meeting the Universe Halfway: Quantum Physics and the Entanglement of Matter and Meaning* (Durham, N.C.: Duke University Press, 2007); Diana Coole and Samantha Frost, "Introducing the New Materialisms," in *New Materialisms: Ontology, Agency, and Politics*, ed. Diana Coole and Samantha Frost (Durham, N.C.: Duke University Press, 2010), 29; Nancy Tuana, "Viscous Porosity: Witnessing Katrina," in *Feminist Materialisms*, ed. Stacy Alaimo and Susan Hekman (Bloomington: Indiana University Press, 2008).
8. Barad, *Meeting the Universe Halfway*, 36.
9. Donna Haraway, *When Species Meet* (Minneapolis: University of Minnesota Press, 2007), 3–4.
10. Deleuze and Guattari, *Thousand Plateaus*, 250.

11. Bogost, *Alien Phenomenology*, 12; Morton, *Hyperobjects*, 127.
12. William Connolly, *A World of Becoming* (Durham, N.C.: Duke University Press, 2011), 39.
13. Pheng Cheah, "Nondialectical Materialisms," in *New Materialisms: Ontology, Agency, and Politics*, ed. Diana Coole and Samantha Frost (Durham, N.C.: Duke University Press, 2010), 89.
14. William Connolly, *The Fragility of Things: Self-Organizing Processes, Neoliberal Fantasies, and Democratic Activism* (Durham, N.C.: Duke University Press, 2013), 49.
15. Ibid., 50.
16. Barad, *Meeting the Universe Halfway*, 3.
17. Ibid.
18. Schrödinger wrote the letter in response to Einstein's coauthored paper, which infamously charged quantum mechanics of incompletion: Albert Einstein, B. Podolsky, and N. Rosen, "Can Quantum-Mechanical Descriptions of Reality Be Considered Complete?," *Physical Review* 47 (1935): 777–80. Schrödinger's first formal treatment of entanglement can be found in Erwin Schrödinger, "Discussion of Probability Relations between Separated Systems," *Proceedings of the Cambridge Philosophical Society* 31, no. 4 (1935): 555–63.
19. Schrödinger, "Discussion of Probability Relations between Separated Systems," 555.
20. Bruce Rosenblum and Fred Kuttner, *Quantum Enigma: Physics, Encounters Consciousness* (Oxford: Oxford University Press, 2007), 188.
21. By "same," we mean anti-correlated, which is to say if one particle's spin along a particular axis is determined to be "up," its partner will always spin "down." For a popular introduction to the physics of entanglement, see Brian Greene, *The Fabric of the Cosmos: Space, Time, and the Texture of Reality* (New York: Vintage, 2005), 77–126.
22. The mathematician John Nash (subject of the movie *A Beautiful Mind*) tried to stand up for him. Decades later, Nash blamed his plunge into insanity on his own attempt—triggered by Oppenheimer's making that accusation against Bohm—"to resolve the contradictions in quantum theory" (Louisa Gilder, *The Age of Entanglement: When Quantum Physics Was Reborn* [New York: Vintage, 2009], 222). See also David I. Kaiser, *How the Hippies Saved Physics: Science, Counterculture and the Quantum Revival* (New York: W. W. Norton, 2012), 26–27.
23. Barad, *Meeting the Universe Halfway*, 396.
24. Ibid.
25. Edward Burnett Tylor, *Religion in Primitive Culture*, vol. 2 (New York: Harper and Row, 1958), 61.
26. Ibid., 230. On the Victorian fetishism of "fetishism," see Peter Melville Logan, *Victorian Fetishism: Intellectuals and Primitives*, Studies in the Long Nineteenth Century (Albany: State University of New York Press, 2010).
27. See William Connolly, *Why I Am Not a Secularist* (Minneapolis: University of Minnesota Press, 2010). Identifying himself neither with "secularism" nor "postsecularism," Connolly labors to strengthen with the help of progressive religious commu-

nities an alternative "resonance machine" to the resentful "Evangelical-Capitalist" assemblage. See also Connolly, "Evangelical-Capitalist Resonance Machine." On Haraway's similar inability to become "a secular humanist," see Nicholas Gane and Donna Haraway, "When We Have Never Been Human, What Is to Be Done? Interview with Donna Haraway," *Theory, Culture & Society* 23, no. 7–8 (2006): 137.

28. Jürgen Moltmann, *Theology of Hope* (Minneapolis: Fortress Press, 1993); Elizabeth A. Johnson, *Quest for the Living God: Mapping Frontiers in the Theology of God* (New York: Bloomsbury Academic, 2011); Caroline Walker Bynum, *Christian Materiality: An Essay on Religion in Late Medieval Europe* (New York: Zone Books, 2011).

29. See, for example, James Cone, "Whose Earth Is It Anyway?" *Cross Currents* (2000): 36–46; Barbara A. Holmes, *Race and the Cosmos: An Invitation to View the World Differently* (London: Bloomsbury T&T Clark, 2002); Stephen H. Moore and Mayra Rivera, eds., *Planetary Loves: Spivak, Postcoloniality, and Theology*, Transdisciplinary Theological Colloquia (New York: Fordham University Press, 2010); Mayra Rivera, *The Touch of Transcendence: A Postcolonial Theology of God* (Louisville, KY: Westminster John Knox Press, 2007); Whitney Bauman, *Theology, Creation, and Environmental Ethics: From Creatio Ex Nihilo to Terra Nullius*, Routledge Studies in Religion (New York: Routledge, 2009); Whitney Bauman, Richard R. Bohannon, and Kevin O'Brien, eds., *Grounding Religion: A Field Guide to the Study of Religion and Ecology* (New York: Routledge, 2010).

30. Laurel Schneider, *Beyond Monotheism: A Theology of Multiplicity* (New York: Routledge, 2008), 1.

31. Winnifred Sullivan, *The Impossibility of Religious Freedom* (Princeton, N.J.: Princeton University Press, 2007); Robert Orsi, *The Madonna of 115th Street: Faith and Community in Italian Harlem, 1880–1950*, 3rd ed. (New Haven, Conn.: Yale University Press, 2010); Jeremy Stolow, ed., *Deus in Machina: Religion, Technology, and the Things in Between* (New York: Fordham University Press, 2012); Jonathan Z. Smith, *To Take Place: Toward Theory in Ritual* (Chicago: University of Chicago Press, 1992); Kathryn Lofton, *Oprah: The Gospel of an Icon* (Los Angeles: University of California Press, 2010); Elizabeth A. McAlister, *Rara! Vodou, Power, and Performance in Haiti and Its Diaspora* (Los Angeles: University of California Press, 2002); Manuel A Vásquez, *More Than Belief: A Materialist Theory of Religion* (Oxford: Oxford University Press, 2011); Jenna Supp-Montgomerie, "World Wide Wire: Religion, Technology, and the Quest for Global Unity" (manuscript in progress); Bynum, *Christian Materiality: An Essay on Religion in Late Medieval Europe*; S. Brent Plate, *A History of Religion in 5 1/2 Objects: Bringing the Spiritual to Its Senses* (Boston: Beacon, 2014).

32. For a political-theological deployment of this strand of thinking, see Clayton Crockett and Jeffrey Robbins, *Religion, Politics, and the Earth: The New Materialism* (New York: Palgrave, 2012). The very compelling *Religious Experience and New Materialism: Movement Matters*, ed. Joerg Rieger and Edward Waggoner (New York: Palgrave Macmillan, 2015) was published after this book went into production.

33. For a critical treatment of the "newness" of this body of work with respect to feminist theories, and of its relative nonengagement of dialectical materialisms, see Sara Ahmed, "Imaginary Prohibitions: Some Preliminary Remarks on the Founding

Gestures of the 'New Materialism,'" *European Journal of Women's Studies* 15, no. 23 (2008): 23–39.

34. Karen Barad, "Deep Calls unto Deep: Queer Inhumanism and Matters of Justice-to-Come," conference paper delivered at the Transdisciplinary Theological Colloquium, Drew University, Madison, N.J., March 29, 2014.

35. Alfred North Whitehead, *Science and the Modern World* (New York: Free Press, 1997). The process theology movement has fostered multiple collaborations in the past half century between science and theology, as in David Ray Griffin, ed., *Physics and the Ultimate Significance of Time* (Albany: State University of New York Press, 1985). David Griffin has summarized much of this conversation in *Religion and Scientific Naturalism: Overcoming the Conflicts* (Albany: State University of New York Press, 2000).

36. Barad, *Meeting the Universe Halfway*, 396.

37. Cited in Catherine Keller, *Face of the Deep: A Theology of Becoming* (New York: Routledge, 2003), 236.

∾ Matter, Anew

❦ What Flashes Up: Theological-Political-Scientific Fragments

KAREN BARAD

Nature is Messianic by reason of its eternal and total passing away.
—WALTER BENJAMIN, "Theological-Political Fragments"

I set forth how this project—as in the method of smashing an atom—releases the enormous energy of history that lies bound in the "once upon a time" of classical historical narrative.
—WALTER BENJAMIN

THEOLOGY-POLITICS-SCIENCE

Even if the profound economic, social, and ecological crises of the twenty-first century are not the same as those of the first half of the previous century, the global political atmosphere is highly charged and fascism is on the rise once again. Conditions of gross economic disparity, war, loss of habitat and homeland, demonization and racialization of the dispossessed and most precariously positioned, and the militarization of national police forces designed to crush the seeds of revolutionary change that persist in pushing up through the cracks in the seeming totality of neoliberal capitalism, are at hand. At this moment, Walter Benjamin's electrifying insights intensify in relevance, crackle with energy, with the potential to break down the ambient ideology of perceived progress and break through the continuum of history, forming conducting paths arcing across spacetimes, constellations of glowing images condensed and crystallized inside the thick-now of the present. For Benjamin, questions of time and justice are inseparable. If the messianic played a role in Benjamin's later thinking, perhaps it was *because* of his political commitment to justice. A key insight of Benjamin's that continues to flash up before us is that the potential for justice exists in the thick-now of the present moment—what Benjamin calls "now-time" (*Jetztzeit*)—and not in pinning political hope on some future time.[1] How can we understand this thick-now that is shot through with the eternal—time outside of time, other times that are not the this-time of the present—flashing up in each moment? What is the nature of the relationship between the eternal and transitory, the infinite and the finite? Benjamin contends that "a revolutionary chance in the fight for the oppressed past" exists in this very structure of the thick-now, which the historical materialist "recognizes as a sign of a messianic cessation of happening"—a rupture in the continuum of time—a break from the unilinear conception of temporality as the continuous unfolding of the past into the future.² The radical political potential that exists in the thick-now of this moment requires thinking time anew—diffracting the past through the present moment, like the play of light inside a crystal.

This is an essay on theology, politics, and science. It engages with the work of the twentieth-century German Jewish philosopher Walter Benjamin and the philosopher Judith Butler in her writings on Benjamin. And rather than offering a straight narrative, a linear unfolding of a particular storyline, it experiments with montage and fragmentary writing, diffractively reading insights through one another, allowing the reader to explore various crystalline structures that solidify, if only momentarily in the breaking of continuity.

If the combination of Kabbalah and Marxism in Benjamin's work has already been a source of consternation for a number of scholars of his work, then adding quantum physics to the mixture is no doubt to risk an explosion, to say nothing of compounding the misunderstandings. But perhaps this is appropriate, not only because risk is unavoidable in the pursuit of justice, but also because Benjamin makes considered use of the metaphor of explosion, indicative of something essential about his methodology and his politics. On this point Butler writes:

> Something flashes up, but something also flashes through a historical continuum, understood as the "historical progress of mankind" that has instituted and even naturalized time as "homogenous and empty." Sometimes it seems that this flash comes from an explosive device, as when he remarks upon "the awareness [of the revolutionary classes at the moment of their action] that they are about to make the continuum of history explode."³

Indeed, Benjamin speaks of the need to "blast open the continuum of history."⁴ How are we to understand Benjamin's reliance on the metaphor of explosion, particularly given its destructive connotations? And if there is in Benjamin's thought the acknowledgment of a certain kind of "destruction" as the condition of the possibility of "construction"—the latter being an essential ingredient in his notion of critique—what is the nature of this "destruction"?⁵

"Benjamin," writes Susan Buck-Morss, "compares his own activity in 'constructing' dialectical images to that of an engineer, who 'blasts' things in the process of building them."[6] In addition to being a crucial methodological step, for Benjamin, "destruction"—the breaking open of the continuum of history—is a political act, a material de(con)struction of the continuum of history that is the condition of possibility for bringing the energetics of the past into the present and vice versa.[7]

Judith Butler's insightful reading of the nature of what Benjamin calls "divine violence" as a disruption of ongoing state-sanctioned violence is helpful in understanding what is at stake for Benjamin in positing "destruction" as construction's condition of possibility. In Butler's reading of Benjamin's famous Thesis IX on the Paul Klee painting *Angelus Novus*, she wrestles with a perplexing phrase in the thesis that indicates the origins of the famous storm holding the Angel of History back from acting on its desire to redeem the wreckage of the past: The "storm is blowing from Paradise."[8] In addition to this specification of the storm's origins—in Paradise—we know something else about it that seems to fly in the face of such an innocent seeming origin: the fact that "this storm is what we call progress,"[9] which is explicitly identified by Benjamin as a destructive force that leaves much ruin in its wake. How can these seemingly contradictory facts about the storm be reconciled?

Before proceeding further with Butler's analysis, let's recall this famous scene in Benjamin's "Theses on the Philosophy of History" in which he describes the Angel of History with his face "turned toward the past":

> Where we perceive a chain of events, he sees one single catastrophe which keeps piling wreckage upon wreckage and hurls it in front of his feet. The angel would like to stay, awaken the dead, and make whole what has been smashed. But a storm is blowing from Paradise; it has got caught in his wings with such violence that the angel can no longer close them. This storm irresistibly propels him into the future to which his back is turned, while the pile of debris before him grows skyward. This storm is what we call progress.[10]

Benjamin here offers us two different conceptions of history: The historicist sees a set of events connected by linear causality moving progressively forward into the future in the unfolding of events in "homogeneous, empty time," whereas, the Angel of History (the historical materialist) "blasts open the continuum of history" enabling a vision of history outside of the temporality of progress.[11] As such, the piling up of the wreckage of history appears as a single catastrophe—a past that must be confronted—rather than a chain of events in which time keeps moving on. The angel would like to redeem the past, awaken the dead, much like the legendary Messiah of traditional Judaism, but a storm called "progress" propels him into the future, preventing him from making reparations. (Benjamin's story is reminiscent of the Talmudic story about the baby messiah—who was born immediately after the Temple was destroyed and it lay in ruins—being blown away in a storm.)[12] The ruins of the past, what progress leaves in its wake, pile up in this crystallization of history into *now-time*—a "Messianic cessation of happening" that has "a revolutionary chance in the fight for the oppressed past."[13] The storm that is propelling the angel into the future is so violent the angel can't close its wings and stay where it is and attend to the past. But the angel, even as it is driven into the future, nonetheless continues to face the past, which continues to lay claim on the present: "The past carries with it a temporal index by which it is referred to redemption."[14] But where is redemption—a term traditionally associated with the messianic endpoint, reflecting the beginning point labeled "Paradise"—to be found in this scene of the past's wreckage?

Butler begins her explanation of the destructive nature of the storm with a question—"If something is being destroyed, is it perhaps forward movement itself?"—that she ultimately answers in the affirmative. "Indeed, if the figure of the storm is the means through which Benjamin introduces a particular notion of the messianic, we will be right to think that the messianic is not the same as progress, and whatever destruction it wreaks will be *of something that is itself destruction*."[15] A critique of the notion of progress is a central theme of Benjamin's "Theses," including a certain faith in progress among those on the Left who would invoke it in the fight against fascism. But for Benjamin, progress is powerless to act against the destructive force of fascism: "No unfolding historical development will overcome fascism, only a state of emergency that breaks with a certain faith in historical development."[16] As Benjamin says, "One reason why fascism has a chance is that in the name of progress its opponents treat it as a historical norm."[17] If both fascism and protests against it function not according to some exceptional mode of operation, but through the usual way things get done, through "democratic" elections and state-sanctioned forms of violence, then resistance to fascism requires a rupture of the continuum of history, the bringing about of a "real state of emergency."[18] Butler writes: "It is this belief [in progress] that is now wrecked, and that wreckage is what the angel clearly sees."[19] Crucially then, as Butler reads it, redemption for Benjamin is "a disruption of teleological history and an opening to a convergent and interruptive set of temporalities. This is a messianism, perhaps secularized, that affirms the scattering of light, the exilic condition, as the nonteleological form that redemption now takes. This is a redemption *from* teleological history."[20] Similarly, when Benjamin urges: "It is our task to bring about a real state of emergency" in the face of the perpetuation of state-sanctioned violence, it is not that he is advocating violence, a kind of violence that perpetrates bodily harm, but on the contrary, he is calling for *a disruption of the ongoing violence of the state*.[21] The messianic is therefore not figured in human form, as Butler compellingly argues, but rather, *the messianic is a break in the continuum of history*.

Importantly, this break in the continuum of history is a highly energetic event. As Benjamin writes in describing the *Arcades Project* to Ernst Bloch in 1935: *"I set forth how this project—as in the method of smashing an atom—releases the enormous energy of history that lies bound in the 'once upon a time' of classical historical narrative."*[22] The enormous energy released from the core of the atom, its nucleus, a mere speck of matter, made headlines around the globe in 1933, based on discoveries by the physicist Ernest Rutherford and his students at the Cavendish Laboratory: From a transformation of elements to the transformation of the world, the highly energetic possibilities were described by Rutherford as an "explosive violence" unleashed from a mere "fragment of an atom" leaving in its wake a profound transformation—the transformation of that very nature of nature that was believed to be fixed and given. From such events new elements are born out of old ones.[23] Little did the world know in 1935, when Benjamin wrote this line, the magnitude of the explosive force that would be unleashed from these discoveries a decade later. But even before the world would be shaken off its axis by the massive destructive forces unleashed on the cities of Hiroshima and Nagasaki, something about immensity or even infinity being contained in the smallest bits had spoken to Benjamin, one might say that this point was at the very core of his theological-political theory, if not his being or, for that matter, all being. Bound up in this tiniest bit of matter, this mere fragment of an atom, is a material force so enormous that its release could reconfigure the material conditions of possibility, the very fabric of *spacetimemattering*, producing a revolutionary chance in the present, in this moment of now-time.[24]

As Benjamin writes in the "Theses on the Philosophy of History": "Thus [the historical materialist] establishes a conception of the present as the 'time of the now' which is shot through with chips of Messianic time."[25] The "time of now" is not an infinitely thin slice of time called the present moment, but rather a thick-now that is a crystallization of the past diffracted through the present. Hence, reading against the grain of a historicist account of history in homogeneous empty time, Benjamin's historical materialist account entails a break in the continuum of history in specific material forms that are *exothermic*—that is, entail the release of energy, whether in the form of the reconfiguring of the nucleus of an atom, the discharge of a massive buildup of electrical potential through lightning flashes, or the reconfiguring of a constellation of atoms through a sudden process of crystallization. References to these kinds of physical phenomena occur at key moments in Benjamin's writing, and the release of energy sufficient to effect transformation marks the enormous revolutionary potential condensed into a single point: the thick-now of the present.

In an important sense, then, Benjamin's work is already shot through with the scientific, and the conjunction theological-political-scientific does not mark the addition of the third term so much as give it an explicit place in the conversation. In any case, questions of the nature of nature, time, and causality, are surely scientific, as well as political, ones. The point of making science an explicit part of the conversation is not to provide a scientific foundation that validates political and theological frameworks; on the contrary, the point is: to explode the temporality of the modernist conception of science, understood as a progressive process of knowledge accumulation, one embraced not only as the epitome of progress but as its very ideal; to blast apart the notion that science is an independent field of thought driven only by empirical findings devoid of any metaphysical, theological, or political commitments; and to destabilize the high authority accorded to science over all other ways of knowing. The point is also to understand how science deconstructs its own authority: how its own findings undermine the very modernist conceptions on which this progressive narrative rests, such as the scientific belief in the immutability of matter, the Newtonian conception of time as an external parameter that marches forward without interruption, determinism, the nature/culture dualism, and human exceptionalism. (The deconstruction of human exceptionalism is not about not caring about humans but rather about including within the analysis an understanding of how the "human" is constituted and against what constitutive outside.)[26] What is at stake, among other things, is a political commitment to undo the inflated authority of Western science, to put science explicitly in the loop of Benjamin's critique of progress and state-sanctioned violence, to rework the practices of technosciences in ways that hold them accountable and responsible for thinking about questions of justice "at the lab bench" (that is, *within* the very practices of science) while leaving open questions about what constitutes justice and for whom, to decolonize mainstream technoscientific practices while acknowledging and granting authority to alternative and indigenous practices for their scientific contributions, to understand the specific nature of the entanglements of political, economic, technoscientific, and ecological displacements and injustices, to think more carefully about the meta/physical understandings of nature that are always already built into our thinking about the theological-political, and, like Benjamin, to allow critique its constructive element.[27] In this essay, I highlight an important Benjaminian feature of critique (that is structurally reflexive or, rather, diffractive, of his ontoepistemology), by attending to (or at least hinting at) not only the "destructive" and "constructive" dimensions of quantum physics, one of the sciences featured herein, but also the very dynamics of annihilation-creation written into the core of the theory itself: In particular, what is at stake is blasting open the continuum of history, making evident state-sanctioned forms of violence that quantum theory—especially (but not exclusively) in its specific entanglements with the making of the atom bomb—has been a part of, while also breaking open its core to release the energy of its radical political potential.[28] This is not to say that all these issues will be fully addressed in this essay, far from it (indeed, this is but a fragment of my ongoing work on this larger project), but rather, the modest contribution here is to begin to open up for further investigation some of the important elements of Benjamin's work in relation to science.[29]

Making the scientific explicit in the conjunction "theological-political" also provides an opportunity for asking questions about whether new insights might thus be gained concerning long-standing questions regarding the nature of the relationship between the political and the theological. In other words, does working with this theological-political-scientific "mixture" open up thinking not only about the scientific in (intra-active) relationship to the political and theological but also the other elements—political and theological and their conjunction—in new ways? We might ask, for example, given the ongoing debate among Benjamin scholars as to whether the Kabbalistic and Marxist elements of his theory do, or could even possibly, cohere, whether highlighting the scientific dimensions of Benjamin's work lends any further insights. To take one specific tack, we might ask about the following: Given the profound troubling of the traditional notion of time and causality that quantum theory offers, do these shifts in thinking produce a reconstellating of theological-political-cosmological patternings that enable a reassessment of the configurations as previously understood? Which also raises questions of methodology and whether the scientific terms Benjamin calls on might be reconstellated and diffracted back through his methodology.[30]

Furthermore, in alliance with Butler's radical reclamation and resituating of the Jewish tradition in a way that provides resources from within Judaism for opposing Israeli state-sanctioned violence against Palestinians in all its various forms, while refusing claims of exceptionalism (that only Judaism can provide the ethical and political theory needed to interrupt such violence), this essay seeks to open up further possibilities for rethinking Judaism and in particular realigning it with its stated commitment to justice.[31] Scientific issues are surely relevant here as well, not only with respect to metaphysical questions but also practical issues, including those of ecological injustice (for example, water injustice—including issues of differential access, pollution, and diversion of water sources in the region), and other issues, which are not separable from political, economic, and social forms of injustice.[32]

ON METHODOLOGY: CRYSTALS, CONSTELLATIONS, AND LIGHTNING

Walter Benjamin was a philosopher of fragments and constellations. Discontinuity and juxtaposition played strongly in his works, over and against continuity and linear succession. This was not an arbitrary, nor particularly aesthetic, choice. As Benjamin understood it, crucial to the pursuit of justice is the disruption of the temporality of progress—the continuous flow of time that leaves the past behind while moving inexorably toward the future. In other words, it is the very notion of time itself as a progression—as a continuous flow of homogeneous, empty moments—that must be interrupted. In Thesis XIII of "Theses on the Philosophy of History" he makes this point directly:

> The concept of historical progress of mankind cannot be sundered from the concept of its progression through homogeneous, empty time. A critique of the concept of such a progression must be the basis of any criticism of the concept of progress itself.[33]

According to Benjamin, it is the "stubborn faith in progress" (Thesis X) that is undermining Marxism's revolutionary potential: "Nothing has corrupted the German working class so much as the notion that it was moving with the current" (Thesis XI). For Benjamin, the notion of progress, together with its affiliates—determinism, teleology, development, evolution, and positivism—stands in the way of urgently needed historical materialist analyses that "brush history against the grain" of historicist accounts (Thesis VII), which have written into them a certain faith in a brighter future that has yet to unfold. In the "Theses," Benjamin's passionate plea to the political Left is to purge itself of the idea of progress and the developmental conception of history that was inherent in German Idealism and interpolated into Marxism through Marx's "inversion of Hegel."

In the "Theses," Benjamin identifies this progressivist temporality as "homogeneous, empty time," the continuous flow of time as it marches forward without regard to any external forces. Or as Newton wrote in the *Principia*: "Absolute, true and mathematical time, of itself, and from its own nature flows equably without regard to anything external, and by another name is called duration."[34] "Homogeneous, empty time" is clock time, the time of modernity, capitalism, colonialism, imperialism, industrialism, militarism.

Train lines, assembly lines, communication lines, time lines. "Now" is the thinnest slice of time: an empty speck. Each moment is the same as all others. Moments lined up like so many beads of a rosary (Thesis XVIII A), forming an unbroken continuous chain of bits of momentariness—an infinitely long chain of infinitesimals. On the assembly line of moments each now is the same as all others, too thin to be of any substance. History is what happens *in* time, but time itself is independent of history. Time is universal, continuous, and unstoppable, moving relentlessly into the future. The past has passed, the future has yet to arrive, and all we have before us is the thinnest slice of a continuum, an ungraspable undifferentiated empty tick of the clock. There is no break; there are no breaks in this continuum of time. Moments slip through our fingers, but progress—the unilinear advance of time—is our insurance, assurance: The next moment, a new moment, will always arrive (just in time) to replace the last. Inevitability drives the assembly line of time. The passage of time unhaltingly moves into the future. Time advances; it progresses; it is what progress is.

Benjamin's efforts to inspire and contribute to forming a robust historical materialism are directly tied to a reconceptualization of the notion of history, not as something that happens *in* (the passive flow of) time, but rather as something the materialist is charged *to do* in *making time itself materialize* in ways that have "a revolutionary chance in the fight for the oppressed past."³⁵ It is worth quoting Benjamin at length on this important point:

> Historicism rightly culminates in universal history. Materialistic historiography differs from it as to method more clearly than any other kind. Universal history has no theoretical armature. Its method is additive; it musters a mass of data to fill the homogeneous, empty time. Materialistic historiography, on the other hand, is based on a constructive principle. Thinking involves not only the flow of thoughts, but their arrest as well. *Where thinking suddenly stops in a configuration pregnant with tensions, it gives that configuration a shock, by which it crystallizes into a monad.* A historical materialist approaches a historical subject only where he encounters it as a monad. *In this structure he recognizes the sign of a Messianic cessation of happening, or, put differently, a revolutionary chance in the fight for the oppressed past. He takes cognizance of it in order to blast a specific era out of the homogeneous course of history—blasting a specific life out of the era or a specific work out of the lifework.* As a result of this method the lifework is preserved in this work and at the same time canceled; in the lifework, the era; and in the era, the entire course of history. The nourishing fruit of the historically understood contains *time as a precious but tasteless seed.*³⁶

In this passage on methodology, Benjamin's materialist ontoepistemology shines through. The task of the historical materialist—much like a bench scientist—is not merely to rewrite history, but rather to do something material to history and time itself. The imagery here is packed with laboratory effects like crystallization, explosions, structures, configurations, extractions, and condensations.

Let's unpack some of what is going on in this dense thesis. There is, first of all, the matter of shocking a configuration pregnant with tensions by which it crystallizes into a monad. It is this very structure of the crystal/monad in which the historical materialist can recognize a revolutionary chance in the fight for the oppressed past. Hence, something of the past seems to be immanent in the crystal. Indeed, at the core of "the nourishing fruit of the historically understood" is *time* "as a precious but tasteless seed"—that is, revolutionary potential is immanent in this crystallization of time, blasted out of the continuum of history. As Benjamin repeatedly conjures monads and crystals as primary facets of his materialist methodology, two more quotes could be usefully juxtaposed here as we delve into this further:

> If the object of history is to be blasted out of the continuum of historical succession, that is because its monadological structure demands it. *This structure first comes to light in the extracted object itself.* And it does so in the form of the historical confrontation that makes up the interior . . . of the historical object, and *into which all the forces and interests of history enter on a reduced scale.*[37]

> In what way is it possible to conjoin a heightened graphicness [*Anschaulichkeit*] to the realization of the Marxist method? The first stage in this undertaking will be to carry over the principle of montage into history. That is, to assemble large-scale constructions out of the smallest and most precisely cut components. Indeed, to discover in the analysis of the small individual moment *the crystal of the total event*. And therefore, to break with vulgar historical naturalism. To grasp the construction of history as such.[38]

Here Benjamin speaks of historical forces being condensed on a reduced scale, making up the interior of the object. How is this crystal-like? Is there something of a monadological structure to crystals? How are we to understand this "crystal of the total event"?

Objects are not mere metaphors for Benjamin. They are instances of sensuous materiality. As such, let us take this object in hand and see if we can glimpse something about its structure. First let's recall some basic features of crystallization. Crystal formation can happen rapidly, or it can take hundreds or even thousands of years. Snowflakes form quickly during their descent to earth, whereas the enormous crystals (one as large as 39 feet in length and 13 feet in diameter, weighing 55 tons) in the Caves of Crystals in Naica, Chihuahua, Mexico, have been growing for half a million years. Whatever the scale, crystals are condensations of history, the frozen traces of forces acting through time.

Crystallization can also happen almost instantaneously, under the right circumstances. That is, there is a way to condense the duration of the crystallization process into a moment, as it were. In particular, crystallization can occur rapidly out of a "supersaturated solution"—a solution that contains more of a dissolved substance (e.g., a mineral) than the solution would normally hold. (For example, while at a higher temperature a solution can hold more of a dissolved substance, as the temperature is lowered the solution can become supersaturated if crystallization has yet to begin.) A supersaturated solution is an unstable configuration in which the crystallization is stalled. One way to get the process going is to "seed" the solution with a "seed crystal." A seed crystal is a small bit of the crystal that has the same structure as the crystal that is to result from the process of crystallization. Adding a seed crystal to a supersaturated solution initiates the process of precipitating out the crystals from solution. Once seeded, the crystallization process can proceed very rapidly; in some cases, the crystal forms (precipitates out of solution) in the blink of an eye. The entire "configuration pregnant with tensions" freezes into a crystal in a mere instant. (The seed crystal—the arrest of thinking—is that which "shocks" the configuration, crystallizing it into a monad, much like seeding a supersaturated solution.) The flow—the continuous movement of atoms in a fluid—is arrested at once, leaving a crystal—the condensation of historical forces—in its place. Crystallization is a discontinuous process. There are in fact multiple discontinuities that characterize this process, including the discontinuous change in structure and volume at the point of crystallization.

What else do we know about crystals? Crystallization is a process whereby atoms—parts of which are positively charged and parts of which are negatively charged—are slowed down as the temperature decreases and find very specific ways of configuring themselves according to the interatomic forces based on the specific polarization of the charged configuration. External forces also contribute to the shapes that form out of this "configuration pregnant with tensions." Crystals are patterned repetitions of specific organizations of atoms—they are specific patternings in (of) spacetime. Crystal structures can be discerned through a form of analysis called crystallography.

Imagine taking hold of a crystal and shining light through it. You would see the ray of light being dispersed by the crystal structure. The light rays bend when hitting the atoms of the crystal, and the deflected rays overlap with one another producing *diffraction patterns*. Diffraction is the physical phenomenon that is the basis of *crystallography*—the technique of using different kinds of waves to map the structure of crystals or even of molecules (Rosalind Franklin discovered the structure of DNA this way).[39] A range of different forces acting over time are condensed into this frozen configuration. Shining light on the crystal projects its internal structure outward, producing an image of this structure. In other words, "this structure comes to light in the extracted object itself," and shining light on the crystal makes a constellation of lights that illuminate the overall structural pattern. Some crystallographic images look just like a constellation of lights dotting the night sky (see Figure 1).[40]

Crystals are distinctive spatial structures, specific patternings, condensations of forces acting *through* time. And yet we are expecting from this, as per Benjamin's invocation, a disruption of process, a disjuncture not merely *in* but *of* time. The temptation is to read crystals metaphorically, switching spatial patterns for temporal patterns. But it is in fact possible to read the crystal-

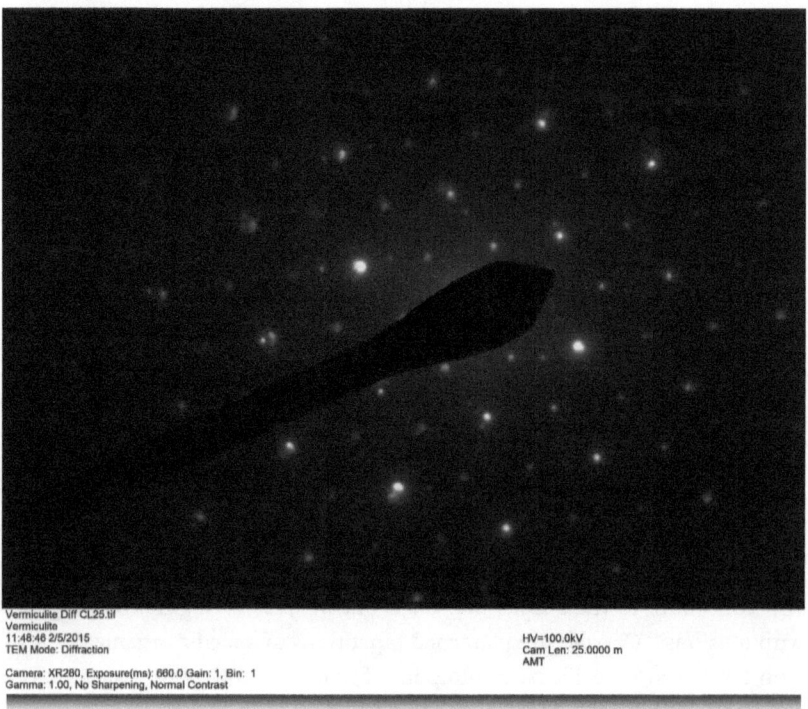

Figure 1. Electron diffraction pattern. (Used with the permission of Materials & Chemistry Laboratory, Inc.)

lization *of* time—the temporal patternings interior to this monad—from the material crystal we are already holding in our hands.

There's a marvelous way to think about crystals and the kinds of temporal patternings, condensations, and superpositions that Benjamin invokes by drawing on quantum physics' radical rethinking of time. One might object that this move would be anachronistic and couldn't possibly have been what Benjamin had in mind, but that would be to yoke time back into its usual formation, just when its disruption is precisely at issue here. If you'll permit me to turn the crystal in this way to examine some fascinating patterns for a bit, we can return shortly to an analysis that might seem closer to his text (how do we measure such a distance in any case when discontinuity is at issue?).

It was in the early years of the development of quantum theory, in 1927, in the process of investigating the properties of a material by bombarding it with electrons, that Clinton Davisson and Lester Germer found that the target the electrons had been impinging upon had changed its structure into a crystalline form, which had not been part of the plan of the experiment. The end result, steered as much by serendipity as human know-how, was that the physicists were able to demonstrate the wave nature of electrons by observing that an electron diffraction pattern was produced. The fact that electrons (small bits of matter) can (under the right circumstances) exhibit wave behavior is a well-known feature of quantum theory. Electron diffraction is evidence of the fact that an electron can be in a *superposition of states*—that is, *an electron* is not just in one place at a given time (like a particle), but in fact has an ontologically (*hauntologically*) indeterminate position, and exhibits a material ghostly non/presence in *multiple places at the same time*. It is a much lesser known fact that diffraction can happen not just in space but also in *time. Temporal diffraction* is indicative of *a superposition of times*—that is, *a single electron* can be in *a superposition of different times*, that is, materially coexist (*hauntologically* speaking) at multiple times (at once?).[41] Temporal diffraction would be a really rich way to think of Benjamin's notion of *Jetztzeit*, or now-time. *Jetztzeit* is a crystallization of times, of multiple temporalities, blasted out of the continuum of history: a superposition of times—moments from the past—existing in the thick-now of the present moment. And in fact, according to quantum physics the past is always open and can be reconfigured, but never in a way that loses track of (i.e., erases the trace of) all that has happened (all the various reconfigurings).[42] This seems key to the kind of intervention Benjamin would have the historical materialist make: *a reconfiguring of time (spacetimemattering) itself.* Quantum physics has flashed up in our discussion and will now quickly disappear, but only for the time being. Let's come back to Benjamin's discussion of a key temporal element in his methodology: the arrest of time or "dialectical standstill."

Benjamin writes about crystallization as coinciding with, or being precipitated out from, an arresting of time, a disruption of the flow of history, what he calls a "dialectical standstill." And in that standstill a crystalline constellation forms.

> Within the enlightened consciousness, political-theological categories arise. . . . And it is only within the purview of [political-theological] categories, which bring the flow of events to a standstill, that history forms, at the interior of this flow, as crystalline constellation.[43]

Constellations, like crystals, seem to be purely spatial arrangements, but Benjamin uses them in a temporal modality: In particular, if "standstill" indicates the arrest of time, the crystallization of history in that configuration indicates an array of times. How can we understand this? When we gaze up into the night sky and see specific spatial configurations of stars we call "constellations," the stars are not all the same distance from us. Some stars are farther away than others. And since the speed of light is a constant,[44] when we look at more distant objects we are looking deeper into the past. For example, when we look at our closest star, the sun, we are seeing the way it looked eight minutes ago—that is, we are watching in the present something that happened in the past. Staring at a constellation, we are witnessing multiple different pasts in the present, some more distant than others. Constellations are then images of a specific array of past events, a configuration of multiple temporalities, "a constellation in being."[45]

Now, in Benjamin's account, constellations aren't simply evident; rather, they are awakenings:

> It's not that the past casts its light on what is present, or what is present its light on what is past; rather, image is that wherein *what has been comes together in a flash with the now to form a constellation*. In other words, image is dialectics at a standstill. For while the relation of the present to the past is a purely temporal, continuous one, the relation of *what-has-been to the now is dialectical: is not progression but image, suddenly emergent.*—Only dialectical images are genuine images (that is, not archaic); and the place where one encounters them is language. ☐ Awakening ☐.[46]

> In the fields with which we are concerned, knowledge comes only in lightning flashes.[47]

According to Benjamin, an image is something that flashes up: "Image is that wherein what has been comes together in a flash with the now to form a constellation." Hence, Benjamin's lightning flash is not between spatial points but across times.

Let's stop here and touch on the phenomenon of lightning. Lightning is an energizing response to a highly charged and polarized field. Contrary to popular belief, lightning does not proceed in a continuous path from sky to ground. We know that lightning is prone to strike some objects, such as tall buildings, more than others. How can we account for that? That is, how does lightning *know* how to find such objects? Before a lightning bolt is born, the sky and the ground engage in a highly nonlinear form of communication with each other across large distances, vast disjunctures.

As I have written elsewhere,

> The lightning expert Martin Uman explains this strangely animated inanimate relating in this way: "What is important to note . . . is that the usual stepped leader [an electrical gesture that precedes a lightning strike] starts from the cloud without any 'knowledge' of what buildings or geography are present below. In fact, it is thought . . . that the stepped leader is 'unaware' of objects beneath it until it is some tens of yards from the eventual strike point. When 'awareness' occurs, a traveling spark is initiated from the point to be struck and propagates upward to meet the downward moving stepped leader, completing the path to ground."[48] What mechanism is at work in this communicative exchange between sky and ground when *awareness* lies at the crux of this strangely animated inanimate relating? And how does this exchange get ahead of itself, as it were? What kind of queer communication is at work here? What are we to make of a communication that has neither sender nor recipient until transmission has already occurred? That is, what are we to make of the fact that the existence of sender and receiver follows from this nonlocal relating rather than preceding it? What strange causality is effected?
>
> A lightning bolt is not a straightforward resolution of the buildup of a charge difference between the earth and a storm cloud: a lightning bolt does not simply proceed from storm cloud to the earth along a unidirectional (if somewhat erratic) path; rather, flirtations alight here and there and now and again as stepped leaders and positive streamers gesture toward possible forms of connection to come. The path that lightning takes not only is not predictable but does not make its way according to some continuous unidirectional path between sky and ground. Though far from microscopic in scale, it seems that we are witnessing a quantum form of communication—a process of iterative intra-activity.[49]

All of these strangely nonlocal communicative "gestures" happen before the lightning bolt forms, and even when that visible luminous flow of energy takes place, it does not proceed in a continuous or progressive path: "The part of the channel nearest the ground will [flash first], then successively higher parts, and finally the charge from the cloud itself [drains and makes visible flashes of light]."[50] Lightning flashes have no truck with traditional conceptions of causality or a unilinear progressive notion of temporality. An arcing dis/juncture, lightning is a connective thread, a *luminous entanglement*—and, I would argue, the way to understand this *entanglement* is in the technical sense that quantum mechanics understands the term—not as a connection between two preexisting objects at different places and times, but rather an intra-action through which "this" and "that," "here" and "there," "now" and "then" are formed. *Lightning is a jagged dis/continuous "moving toward" with innumerable interruptions.*[51]

Benjamin understands what-has-been diffracted through the now-time-being in a "dialectical reversal"; that is, the past remains open to what it might yet have been: "What has been is to become.... The facts become something that just now first happened to us."[52] Rolf Tiedemann, the editor/assembler of the posthumously published Benjamin work, *Passagen-Werk* (translated as *The Arcades Project*), further explains:

> The object of history goes on changing; it becomes "historical" (in this word's emphatic sense) only when it becomes topical in a later period. Continuous relationships in time, with which history deals, are superseded in Benjamin's thought by constellations in which the past coincides with the present to such an extent that the past achieves a "Now" of its "recognizability." Benjamin developed this "Now of Recognizability," which he sometimes referred to as his theory of knowledge, from a double frontal position against both idealism and positivistic historicism.... That the lineaments of the past are first detectable after a certain period is not due to the historian's whim; *it bespeaks an objective historical constellation.*[53]

As Benjamin puts it:

> History is the object of a construct whose site is not homogeneous, empty time, but the time filled by now-time [*Jetztzeit*]....To write history... means to cite history.[54]

Quantum physics has wrangled its way in again, this time "flashing up" through a markedly macroscopic phenomenon—lightning flashes—which play a key role in Benjamin's theory. Quantum physics radically disrupts the notion of "homogeneous, empty time." Time takes on an entirely different, and much more active, set of features. These include possibilities such as temporal discontinuity, temporal diffraction, temporal entanglement (relevant both to lightning and the "dialectical reversal" in Benjamin), and the "condensation" of time into an instant (a quantum field theory phenomenon related to temporal indeterminacy)[55] that seem to go to the core of Benjamin's notion of the dialectical image. (But now, appropriately enough, in introducing quantum physics, we have already gotten ahead of ourselves. More on this in the following sections.)

Lightning, constellations, and crystals play crucial roles in helping Benjamin articulate his notion of a dialectical image, which is the centerpiece of his methodology and (onto)epistemology.[56] For Benjamin, crystallization emerges with/in the temporal discontinuity, in the breaking of time's continuum. That is, it is the halting of time—the introduction of an essential discontinuity—that constitutes (and is coincident with) the seeding of the crystallization of multiple temporalities into a temporal pattern/ing outside of the flow of time. The past is not left behind, but rather, is diffracted through/in *Jetztzeit*, the now-time of the present moment. Energy is released in this crystallization, this reconfiguring of time, that takes place in the messianic cessation of happening. And as light diffracts through this crystal structure, "the genuine historical image" "flares up briefly," offering the historical materialist a revolutionary chance in the fight for the oppressed past.[57] That is, in/through this arresting of time, history—blasted out of a linear narrative of happenings in homogenous, empty time—becomes a materialization *of* time (spacetimemattering), a dialectical reconfiguring of time itself: The materialization/reconstellation/crystallization of time and history is made possible in the flashing up of images, luminous points of light, sparks released from the small individual moment of the crystal of the total event, if only momentarily. Crucially then, Benjamin's methodology constitutes a material intervention into the making of time and history.

Benjamin's is first and foremost a materialist methodology. According to Rolf Tiedemann, "Benjamin's intention was to bring together theory and materials, quotations and interpretation, in a new constellation compared to contemporary methods of representation."[58] Benjamin sought in the method of montage an alternative to representationalism and a robust historical materialism that breaks with the temporality of progress. The material fragments are neither to be read as causally linked (in the sense of linear causality) nor merely analogous. Rather, they are to be picked up like crystals, and turned around and around allowing the light to diffract through them, seeing the overall pattern that is already inside each fragment but also watching entire constellations of insights flash up, if only momentarily. How else might one have a chance of grasping the complex and intricate play of different forces that are condensed and concentrated into each object?

What I have discussed so far is one way of putting together some of the scattered fragments of Benjamin's discussion of dialectical images. It is a way that tries to make sense of the specifically material dimensions of his method, by fleshing out some important scientific and sensuous features of the particular objects that form the core of his method. My agential realist reading of these scientific phenomena (and my diffractive methodology)[59] joins with Benjamin in taking neither time nor being as a given, but rather understands time-being as something that materializes, as a specifically political and material practice, an iterative practice, a chance for intervention that is attuned to the possibilities for liberation, for justice, in the now of the present moment.

I am not proposing this as a historically situated reconstruction of how Benjamin understood things. I am not suggesting that he knew all details of the physics (let alone the quantum physics that is hinted at just a bit here, although much of the construction of the quantum theory took place in Europe in the 1920s and 1930s), or that he was somehow prescient about physics yet to come. There would be something ironic about making an argument of this kind and offering a seamless narrative that (presumes to) unfold(s) what Benjamin (already) had in mind, given that this kind of historical reconstruction must assume the very notions of history and time that Benjamin is criticizing. At least it would be an odd sort of thing to do without honoring the dis/junctions and dis/continuities.[60]

In what follows, and doesn't follow, in a break that is about to happen—a dis/continuity in form, a break in the flow of the essay—I invite you to participate with me in a mini-montage/diffraction experiment, where no linear narrative will be offered. I will place before us some fragments, in the style of Benjamin's *Arcades Project*, including insights from Kabbalah, quantum physics, and Benjaminian philosophy (drawing heavily on Butler's commentary) and let us see what constellations form and what flashes up. But first some preliminary remarks.

ON THE FRAGMENTS THAT FOLLOW

For someone who sought to blast open the continuum of history, an assemblage of fragments seems quite apt, and speaks not only of the ruins but also of innumerable possibilities for re-membering the past and conceptualizing history anew. Benjamin's *Arcades Project* is a collection of fragments, including reflections of various lengths and individual passages from various texts copied down as individual notes and assembled in constellations suggesting multiple reverberations or diffraction patternings among the fragments.

The next section of this essay begins the montage/diffraction experiment and contains theological, political, and scientific fragments, including passages from Walter Benjamin, Butler's readings of Benjamin, Kabbalah, Marx, and quantum physics. It is crucial to understand that just as there is no singular Benjaminian philosophy, or Marxism, or Kabbalah, there is no singular interpretation of quantum theory. There are in fact a host of different interpretations. In part because of the elevated authority accorded to science, there is a general tendency to accept anything that physicists have said about quantum theory as a statement of fact and to believe that it is fair game to select these different "facts" about quantum physics and paste any subset of them together as if this constitutes a coherent interpretation of the theory. To make matters worse, physicists are not always upfront about the fact that there are multiple and competing interpretations of quantum theory, and they often do not specify which interpretation they are using when they make pronouncements about quantum physics. In fact, physicists sometimes fail to appreciate difference in interpretational stances in their own presentations and thinking about these matters (both in their popular writings and their research papers). (The fact that physicists and philosophers of physics often speak of the work of Niels Bohr, Werner Heisenberg, and Erwin Schrödinger as making contributions toward a single interpretation—"the Copenhagen interpretation"—is a case in point, since these founding fathers of quantum physics fundamentally disagree about crucial interpretative issues.)[61] An antiphilosophical culture (which can be characterized by the widely known adage familiar to physics majors trained in the United States: "Shut up and calculate"), has predominated since the center of physics shifted across the Atlantic in the mid-twentieth century, and it has not served physics well. As for the stance I am using here, all the quantum physics fragments that I share here are based on my agential realist interpretation of quantum physics. Significantly, this particular interpretation of quantum physics is intentionally constructed in conversation with critical social and political theories, including Marxist, feminist, queer, critical race, poststructuralist, deconstructionist, postcolonial, decolonial, and science studies theories.[62] This already underlines the important point that none of the three arms of the theological-political-scientific conjunction are separable from the others, which is not to say that they are identical; rather, each is entangled with, shot through with, the other.[63]

The fragments from Benjamin and Butler's reading of Benjamin constellate around the theme of the nature of the relationship between the transcendent and the immanent, the divine and the profane, the eternal and the transient, the infinite and finite beings. The center of gravity is Judith Butler's remarkable essay "One Time Traverses Another: Benjamin's 'Theological-Political Fragment.'"[64] Butler begins this essay with the question of "whether Benjamin understands the divine as a purely immanent feature of the world," and responds with a rather complex suggestion: "The eternal . . . *traverses* the transient without exactly becoming transient and losing its status as the eternal."[65] She also puts it this way: "The immanent is itself broken up or *traversed* by what is eternal, which means that the eternal is both a feature of immanence and yet irreducible to it—it can be in it or of it without being fully encompassed by it."[66] The key point then is to "understand how one time *traverses* another without precisely that first time being absorbed by and contained in or by the time traversed."[67] The explication of the nature of this "traversal" is what occupies her in the rest of the essay. In order to address this, Butler carefully traces Benjamin's notion of the messianic, especially as he expresses it in the "Theses on the Philosophy of History" (the final work he penned). Butler writes: "In my reading of Benjamin, the messianic does not, and cannot, follow the trajectory of revelation. Rather, it is figured time and again in Benjamin's work as the traversal of one temporal modality in and through another."[68] I will not be offering a full presentation of Butler's ideas here, nor, for that matter, any of the other subject matters that this essay engages with, including quantum physics, agential realism, and Kabbalah.

The theological fragments that appear in the next section are drawn from a range of Jewish theological texts. There is no one book that constitutes the sum total of the Jewish tradition, and "the Book" is, if anything, an ongoing deconstruction and dispersal of itself. As Marc-Alain Ouaknin explains in *The Burnt Book: Reading the Talmud*, the Jewish textual tradition is not one, and is never settled: "The Book . . . is not a 'compilation,' a 'manual.' It is not the scene of a gathering together of signs; it is not a system. The Book is the scene of the 'impossible simultaneity of meaning,' of the 'nonassemblage,' and the unsynchronizable." It is a dissemination of meanings, the blasting open of sedimented ways of thinking. Ouaknin writes:

> The relation of the Book and time appears, then, as a fundamental relation by the fact that the Book is—or should be—the breaking up of the synchronizable, that is to say, of the recollectable. . . .
>
> The Book will always be the future book, the "book yet to come" [*à-venir*] or, simply the future [*avenir*]. The Book, by its impossibility of settling down in the "now," helps us to attain discontinuity and time-as-discontinuity. The Book introduces us into a time that "adds something new to being, something absolutely new."[69]

The question of temporality is at the forefront of that which constitutes the tradition, not as preservation and recollection, but on the contrary, as that which is iteratively reconfigured and never settles.

Interpretation cannot be repetition. The creation of meaning is a creation-production of time. We could be brought around to define a new time that is not the measurable time of the "watch or the calendar," but that of the creativity of interpretation. We could talk about "Ḥidush [new] time" or about "Talmudic time." In this way we would think, not about the change that is produced in time, but about time through change, and change through the speech of Ḥidush; "Ḥidush time" is the founding of History, of the World and Being. It is no longer a subjective or internal time, where time accelerates or slows down with the intensity of life, and of the events that fill existence. It is an ontological time. It is a time that "adds something new to being, something absolutely new." . . .

The importance of Ḥidush lies in the fact that it introduces a discontinuity into the rhythm of consciousness and into the process of being: it shatters the block of being.[70]

While Ouaknin does not mention Benjamin in *The Burnt Book*, the imagery that he uses is one of blasting, bursting open, and scattering the words and sentence to effect a complete reorganization of meaning.

Let us point out that what is important is not to rediscover [the original words] but to understand the scattering and bursting movement of the sentence and words and the way they are reorganized. . . . Reading becomes possible only by "leaps."[71]

There is even the invocation of this crystal-like image: "You must turn them over, turn them around again showing up all their facets in the hope that a gleam will burst forth."[72] And like Benjamin, Ouaknin also underlines the importance of a "destructive" element in the constructive methodology of reading according to Ḥidush time:

One must adopt a fundamentally critical attitude that one might call "destruction" in order to emphasize its radicality and importance. It is a matter of a totally positive act of destruction that will allow a man "to open himself up 'anew.'"[73]

This is not to say that Benjamin mimics or follows this particular practice or that there is something essentially Talmudic in his approach. Neither is it to argue for some exclusionary or essentialist practice unique to Judaism. On the contrary, it points to a de-essentializing, indeed, radical approach to the tradition that is already part of the tradition.

In this spirit, my presentations of Jewish theological ideas in what follows are always-already shot through with my readings of quantum physics, which are shot through with readings of Butler and Derrida, which are shot through with a political commitment to unsettling and blasting through the alluvium of injustices that block the flow of the lifeblood of the planet, including all its inhabitants, from circulating in ways that support mutual thriving and survival in our cohabitating intra-dependence.

FRAGMENTS, CONSTELLATIONS, AND DIFFRACTED ILLUMINATIONS

B'reishit. In the beginning of . . .[74]

* * *

The great medieval Torah commentator Rashi warns: Do not read the opening line of Genesis as "In the beginning God created the heavens and the earth . . ." because this is not what the Hebrew says. In fact, the grammatical construction of the opening sentence of Torah doesn't make sense. "*B'reishit*," the first word, is in "construct form," meaning that it should be the first of two nouns in a row, that would be translated: [the first noun] of [the second noun], as in *B'nei Yisrael*, "the children of Israel." "*B'reishit*" thus means "in [a/the] beginning of . . ." and there is no noun that directly follows. In fact, Rashi writes: "This verse says nothing but *darsheini*—'Expound me!'"[75] That is, according to Rashi, the opening word of the Torah is just that: an opening—an invitation to interpret; it is in effect an injunction against the very possibility of literal interpretations of the Bible. This invitation, which is integral to the Jewish tradition, has been taken up with great enthusiasm (*midrash*, hermeneutics, etc. etc.). Volumes have been written on the meaning of just this first word, *B'reishit* (e.g., *Tikkunei Zohar*). It could be said that *B'reishit*, the opening word of Torah, signals the indeterminacy of the whole: for the words of Torah lack vowels, which forecloses the possibility of the determinate specification of many of the words. The opening is nothing but an emphasis on the text's ongoing openness: on *the lack of determinacy*. Creation is an indeterminate matter. And the first word of the Torah, what seems to start out as the articulation of a beginning, uses the word beginning in such a way that it interrupts itself to question the very possibility of origin, linear temporality, determinism, and determinacy.

* * *

B'reishit. In the beginning of . . .

In the beginning ... *there was an interruption*, a rupture, *a break in the continuum of time* before time

a disruption, a hesitation, a pause, a stutter, a disjointure, a cut. In the beginning of ... before we even get to the beginning of ...

there is a break that disrupts the very possibility of origin and the unilinear unfolding of ....

In the beginning that was an originary dis/continuity that breaks open the continuum of time, before it gets started, before there is a beginning.

* * *

"This 'beginning,' like all beginnings, is always already threaded through with anticipation of where it is going but will never simply reach and of a past that has yet to come. It is not merely that the future and the past are not 'there' and never sit still, but that the present is not simply here-now. Multiply heterogeneous iterations all: past, present, and future, not in a relation of linear unfolding, but threaded through one another in a nonlinear enfolding of spacetimemattering, a topology that defies any suggestion of a smooth continuous manifold. Time is out of joint. Dispersed. Diffracted. Time is diffracted through itself. It is not only the nature of time in its disjointedness that is at stake, but also *disjointedness itself*. Indeed, the nature of 'dis' and 'jointedness,' of discontinuity and continuity, of difference and entanglement, and their im/possible interrelationships are at issue."[76]

* * *

"To Robespierre ancient Rome was a past charged with the time of now which he blasted out of the continuum of history. . . . It is a tiger's leap into the past. This jump, however, takes place in an arena where the ruling class gives commands. The same leap in the open air of history is the dialectical one, which is how Marx understood the revolution."[77]

* * *

Quantum leap—a discrete "jump" from here-now to there-then without being anywhere in between (original definition).

Quantum leap—a radical discontinuity, a breaking in continuity so profound that not even the boundary (discontinuity) between continuity and discontinuity can hold (according to agential realism).

"Quantum 'leaps' are not mere displacements in space through time, not from here-now to there-then, not when it is the rupture itself that helps constitute the here's and now's, and not once and for all. The point is not merely that something is here-now and then there-then without ever having been anywhere in between—that's bad enough, of course—but that here-now, there-then have become unmoored: there's no given place or time for them to be."[78]

Quantum—an originary dis/continuity, not in space through time, but in the iterative intra-active constitution/reconfiguring of spacetime(mattering). Dis/continuity—neither continuity nor discontinuity but rather *cutting together-apart* (one move). *Intra-actions* cut together-apart, differentiate-entangle. *Intra-action*, not interaction. Causality reworked: cause does not precede effect, no subject/object; "subject" and "object," "cause" and "effect" are mutually constituted in and through intra-action in "a 'holding together' of the disparate itself."[79] "*Agential cuts—intra-actions*—don't produce (absolute) separation, they engage in *agential separability*—differentiating and entangling (that's one move, not successive processes)."[80] *Quantum*—that which distinguishes Newtonian physics from quantum physics. There is no "quantum world"—a natural quarantining of quantum queerness, keeping it at a safe distance from the quotidian, the realm of "normal" events. The attempts to safeguard the everyday from the queerness of quantum phenomena is merely an instance of queerphobia, not an empirical fact.[81]

* * *

In the Passover story, liberation from slavery was a matter of a *dillug* (leap)—an interruption of history—a breaking through by *passing over*: a quantum leap. A new calendar, a break from the continuum of oppression. A splitting of the sea, a rupture in the continuum, and a journey across the desert to bring forth new possibilities (see quantum field theory, quantum vacuum, nothingness). *Dillug*— a quantum leap: not from here to there, now to then, but rather, an entanglement of here-there now-then, a material reconstellating of spacetime(mattering).[82]

* * *

"What characterizes revolutionary classes at their moment of action is the awareness that they are about to make the continuum of history explode. The Great Revolution introduced a new calendar. The initial day of a calendar presents history in time-lapse mode. And, basically, it is this same day that keeps recurring in the guise of holidays, which are days of remembrance. Thus, calendars do not measure time the way clocks do; they are monuments of a historical consciousness."[83]

* • *

Revolutionary actions require breaks in the continuum—temporal dis/continuities, a dis/jointure of time. (Derrida's *Specters*, the specters of Marx, surely haunt these temporal reimaginings/reconfigurings.) Revolution: an interruption, a break in historical continuity, a disruption/reconfiguring/reconstellating of the current material conditions of possibility, effecting an opening up of/to what-might-yet-have-been-and-might-yet-be. New calendar, messianic time—a flashing up of the covered-over calendars of the oppressed, monuments to historical consciousness.[84]

* • *

Rashi begins his commentary on the first word of Torah (*B'resheit*) by asking why the Torah doesn't start with the first commandment, rather than with the beginning which is not a beginning. The first commandment given to the Jewish people does not happen at Mount Sinai, but before that, just prior to their exodus from Egypt. The first commandment occurs in the middle of the Torah, in the third book of Moses, and yet it does mark an origin of sorts. Not a beginning in time, but *of* time. The first commandment in the Torah is to disrupt time—to make a new calendar, a new temporality: "This month shall be for you the beginning of months; it shall be for you the first of the months of the year" (Exodus [*Shemot*] 12:2). Rashi writes: "[God] showed [Moses] the moon in its renewal and said to him, 'When the moon renews itself it will be the beginning of months for you.'"[85] The Jewish people are commanded to count the months by the moon. In Hebrew the word "month" (*Ḥodesh*) is from the same root as "new" ("innovation") (*Ḥadash*): the new moon—which marks the beginning of the month—brings renewal, not mere repetition, but iteration. Unlike calendars that track the sun, thereby keeping the cycles the same, such as the Julian and Gregorian calendars, the lunar calendar has a different rhythm, a reminder that change is possible. Each new moon is a minor holiday. A new moon, a new month, the new month of *Nisan*—the new moon preceding the full moon of the Passover holiday, a new year; one of four new years in the Jewish calendar,[86] a calendar synchronized to temporal indeterminacy.

* * *

"The importance of Ḥidush (new) lies in the fact that it introduces a discontinuity into the rhythm of consciousness and into the process of being: it shatters the block of being."[87]

* * *

The commandment to count time by the moon appears right before the Exodus because counting by the moon is connected to liberation: "The sun has no sort of newness or change, it is always in a fixed position and never changes itself, but the moon is continually in a situation of change and self-renewal. It has small self-renewals every day and every hour, and at every time it is in a different and unique position, no day is similar to another; and it has a great self-renewal each month, in that a time comes when the moon makes itself small, until it reaches complete nothingness ... and then the light of the moon returns and renews itself.... Every hour in a [being's] life is a unique situation unto itself, and one is always able to renew oneself as a new creature."[88]

* * *

The Jewish calendar is attuned to change and to remembering, attuned to multiple rhythms at the same time—*one rhythm traversing another*: the phases of the moon and the movement of the sun in the sky. Celestial beings in motion, wheels within wheels, free will within determinism, a shattering of linear causality based on linear time. Syncopated to the rhythmic pull of two large forces: the moon and the sun. The number of months in a year is twelve, except when it's thirteen. The stars matter but are not determinative.[89]

* * *

Jewish holidays are days of remembrance—of re-memberings, of material entanglements to other times, to past moments that are alive in the present, not merely as idealized visions within the mind of individual selves, but brought into existence in the now-time of the present through iterative communal material acts. Not eternal return but rather eternal recurrence, each iteration with a difference. The holidays are portals in time, openings to re-membering, reconfigurings of time-being. The past remains open to what might yet have been. Not looking back at the past, but bringing the past into the present: a simultaneity (coexistence) of times—multiple times at the same time: *Jetztzeit-Jetztseins*. Time-lapse mode—history condensed into the present moment: a monad, a diffracted heterogeneous moment of time, now-time-being.[90]

* * *

"Physicists now claim to have empirical evidence that it is possible not only to change the past, but to change the very nature of being itself ... in the past."[91] "The point is that the past was never simply there to begin with and the future is not simply what will unfold; the 'past' and the 'future' are iteratively reworked and enfolded through the iterative practices of spacetimemattering."[92]

* * *

"What has been is to become.... The facts become something that just now first happened to us."[93]

* * *

Temporal indeterminacy, time diffraction, temporal entanglements. A given object can be in a superposition of positions: two or more places at the same time. This is an expression of spatial indeterminacy. Less well known, but also empirically verified, is the fact of *temporal indeterminacy and temporal diffraction*: the fact that an object can be in a superposition of *times*—multiple times at the same.... hauntologically coexisting multiple times in the here-now of time-being. Multiple temporalities, diffracted in, threaded through this moment.[94]

* * *

"The appearances of superposition, of overlap, ... come with hashish."[95]

* * *

"Remembrance is not about demonstrating or telling a history, and neither is it finally about the excavating and subsequent monumentalization of a past.... Remembrance works against history, undoes its seamless continuity ... [the] breaking forth of another temporality into one characterized by its uniformity and its progress."[96]

* * *

The past is not fixed, not given, but that isn't to say that the trace of all memories can simply be erased. Memory is not a mere property of individual subjects, but a material condition of the world. *"Memory—the pattern of sedimented enfoldings of iterative intra-activity—is written into the fabric of the world.* The world 'holds' the memory of all traces; or rather, the world *is* its memory (enfolded materialization)."⁹⁷

*   *   *

"The soothsayers who found out from time what it had in store certainly did not experience time as either homogeneous or empty. Anyone who keeps this in mind will perhaps get an idea of how past times were experienced in remembrance—namely, in just the same way. We know that the Jews were prohibited from investigating the future. The Torah and the prayers instruct them in remembrance, however. This stripped the future of its magic, to which all those succumb who turn to the soothsayers for enlightenment. This does not imply, however, that for the Jews the future turned into homogeneous, empty time. For every second of time was the strait gate through which the Messiah might enter."⁹⁸

*   *   *

"Everything is foreseen, and free will is given."⁹⁹

*   *   *

Every year at Passover, each Jew is obligated to see themselves leaving the Narrow Place (*Mitzrayim*).¹⁰⁰ Every year, each day, each moment. A syncopation of rhythms. Wheels within wheels. Each day, each Jew is commanded to remember, or rather, to re-member, that you too [here-today] are enslaved in the Narrow Place (*Mitzrayim*) and require a *dillug*, a break in the continuum of time.¹⁰¹ A stopping of time, to take in the fact that Israel is both oppressed and oppressor. An olive on the seder plate (to remember the oppression of Palestinians under Israeli occupation). A re-dedication to work for justice for all peoples, all beings.

*   *   *

Quantum "entanglements are not intertwinings of separate entities, but rather irreducible relations of responsibility. There is no fixed dividing line between 'self' and 'other,' 'past' and 'present' and 'future,' 'here' and 'now,' 'cause' and 'effect.' . . . Entanglements are not a name for the interconnectedness of all being as one, but rather specific material relations of the ongoing differentiating of the world. Entanglements are relations of obligation—being bound to the other—enfolded traces of othering. Othering, the constitution of an 'Other,' entails an indebtedness to the 'Other,' who is irreducibly and materially bound to, threaded through, the 'self'—a diffraction/dispersion of identity. 'Otherness' is an entangled relation of difference (*différance*). Ethicality entails noncoincidence with oneself."[102]

* * *

"An ethics of entanglement entails possibilities and obligations for reworking the material(ized) [configurations of] the past and the future. As the quantum eraser experiment shows, it is not the case that *the* past (a past that is [allegedly fixed and] given) can be changed (contrary to what some physicists have said), or that the effects of past actions can [thereby] be fully mended, but rather that the 'past' is always already open to change. There can never be complete redemption [in the sense of a full restoration of the way it was], but spacetimematter can be productively reconfigured, as im/possibilities are reworked. Reconfigurings don't erase marks on bodies—the sedimenting material effects of these very reconfigurings—memories/re-member-ings—are written into the flesh of the world. Our debt to those who are already dead and those not yet born cannot be disentangled from who we are. What if we were to recognize that differentiating is a material act that is not about [absolute] separation, but on the contrary, about making connections and commitments?"[103]

* * *

Time doesn't unfold as unilinear forward motion. Jewish clock time is surreal relative to the standard. A nighttime hour is not the same length as a daytime hour. Sunset to sunrise divided by twelve is the length of a nighttime hour; sunrise to sunset divided by twelve is the length of a daytime hour. Midnight (halfway through the hours of nighttime—midnight) is different from day to day. Indeed, every hour is different. Every day hours are different lengths. Imagine calibrating train schedules, capitalist expansion, and empire building to such clocks. Imagine a theory of relativity synchronized to such clocks. Einstein would never have left his work at the patent office. Shabbat is cut loose from time: 25 hours outside of time (from sunset until the stars come out (twilight)—belongs to both days).

* * *

The commandment to count days, to "count the Omer," between the Jewish holidays of Passover and Shavuot: spheres within spheres (worlds within worlds, wheels within wheels). All sefirot inside each sefirah: Different constellations of sefirot constitute different flows of energy, different forces. Shavuot—the giving of the Torah at Mount Sinai, thousands of years ago, a whole history crystallized into now-time: The break in the continuum of time produces new conditions of possibility, new Torah coming through. Tradition rewritten once again: History is only ever citational. "To write history ... means to cite history [anew]."[104] Judaism re-newing its founding commitment to justice. "Justice, justice shall you pursue."[105]

* * *

Kabbalah turns tradition on its head, which is, in fact, traditional.

* * *

"In every era the attempt must be made anew to wrest tradition away from a conformism that is about to overpower it."[106]

* * *

Shavuot—the coming through of Torah anew, at Mt. Sinai 3000 years ago, entangled with this moment, this now-time (of Shavuot each year). "The crystal of the total event" (Benjamin). Ben Bag Bag said of the Torah, "Turn it and turn it over again, for everything is in it" (*Pirkei Avot*, 5:25).[107]

* * *

"And the exposition [of Torah] is not what is essential, but rather action."[108]

* * *

"The philosophers have only *interpreted* the world, in various ways; the point, however, is to *change* it."[109]

* * *

Torah, the blueprint for the world, existed before the world.[110] Torah doesn't follow chronology; it can't be read as linear sequence. "There is no before and after in Torah."[111] The multiple "creation stories" do not synchronize and do not simply unfold, they are dis/jointed narratives of temporal indeterminacy tied to justice-to-come. The Kabbalists explain that the "creation story" is not a narrative of the unfolding of things in linear time; rather, it is a map of the cosmic structure and the ethical practices that follow from it.

* * *

B'reishit. In (a/the) beginning of . . .

The first letter of the first word *B'reishit*, the letter *bet*, can mean "with" as well as "in."

B'reishit: With (B') a point (reishit) . . .

So, the first three words—*B'reishit* [With a point] *bara* [created] *Elohim* [one of the names of God that is plural yet singular, Kabbalistically associated with *Gevurah*—boundary-making]—can be read:

With a point—an infinitesimal bit of the Infinite— . . . created Elohim

(Blasphemy within blasphemy—part of the tradition)

* • *

Lurianic Kabbalah (which is not one) takes this point seriously. The Kabbalistic account of Isaac Luria starts before the beginning—with the creation of the conditions of possibility of creation.[112] It starts with *Eyn Sof*—No-thing-ness. (*Eyn Sof*—literally "without end," infinite. *Eyn Sof* is *Ayin*, literally "nothingness," but also "spring" or "source").[113] Before the creation of light or anything else, there was light—the infinite light of *Eyn Sof*.

> Before the emanations were emanated and the creations created, a most supreme, simple light filled the whole of existence. There was no vacant place, no aspect of empty space or void, but everything was filled by that simple, infinite light . . . what is called the light of the Infinite (*or Eyn Sof*).[114]

Luria explains that the *Eyn Sof* (the Infinite) sets out to make a space for creation by contracting/concentrating its light within itself. This *tzimtzum* (contraction/concentration) of the light of the Infinite produces a space within which to create the world, not a world detached from or absent of *Eyn Sof* (see below), on the contrary, but with a chance for finitude.[115]

> You should know at the beginning of everything, the whole of existence was a simple light called the light of the Infinite. . . . There was no empty space or open space. . . . The Infinite contracted itself in the middle of its light, at its very central point, withdrawing to the circumference and the sides, leaving an open space in between.[116]

The emphasis on the center is noteworthy since it isn't clear what can be made of a notion of a center of something infinite. But a point is surely made about this point.

> The Infinite contracted/concentrated itself at its midpoint, in the exact center of its light.[117]

And at the same time, there is a related point: "The space created is [itself] a mere infinitesimal point in contrast with God's infinity. Nevertheless, ... it is the space in which all dimensions of existence are formed."[118]

* * *

In the beginning there was an involution, not evolution, a kind of self-touching,[119] such that creation does not simply start with, but materializes with/in a point, an infinitesimal bit of the infinite.

* * *

"Each letter is a world, each word, a universe."[120]

* * *

"This point is especially not a point, an indivisible place, a touchable limit, but a spacing—forever."[121]

* * *

The *tzimtzum* is not a subtraction or a negation in any ordinary sense. It is a contraction/concentration—a pulling back of an infinite light from a space, a mere point of the infinite, to leave an infinitesimally small space/place (*Ḥalal panui*) of infinitely less intensity. How are we to understand the "subtraction" of one infinity from another? Surely, this is not the same operation as the subtraction of finite numbers. But there are different sizes of infinity,[122] and so there is in fact a rather oblique way to do the nonstandard "subtraction." According to Lurianic Kabbalah, the result of the *tzimtzum* is not to produce a space that is fully empty—it is not empty, and it is also not not-empty. What remains is the mere trace of the Infinite (like the trace of wine that remains after the glass has been emptied)—the *reshimu*—the indeterminate empty/not-empty no-thing-ness fills-and-does-not-fill the space created inside an infinitesimal bit of the Infinite.[123]

* •   *

"From the point of view of classical [Newtonian] physics, the vacuum [is complete emptiness: it] has no matter and no energy. But the quantum principle of ontological indeterminacy calls the existence of such a zero-energy, zero-matter state into question, or rather, makes it into a question with no decidable answer. Not a settled matter, or rather, no matter. And if the energy of the vacuum is not determinately zero, it isn't determinately empty. In fact, this indeterminacy is responsible not only for the void not being nothing (while not being something), but it may in fact be the source of all that is, a womb that births existence."[124]

* •   *

After the *tzimtzum* (contraction/condensing) of the Infinite light, the *Eyn Sof* emitted a *kav*—an infinitely thin thread of light into the space. What remains at this point is not nothing, but rather a structured nothingness—the condition for the possibility of creation. The *kav-reshimu* intra-action starts off an emanation that after a long complicated set of further intra-actions results in a coarsening of the light on the way to making finite material beings. This emanation is not a one-way unfolding but an intra-action that depends on the actions of material beings in its co-creation.[125]

* •   *

Quantum field theory (QFT) is a combination of quantum physics, relativity, and field theory. QFT rewrites our understanding of the nature of space, time, and matter, or more precisely, *spacetimemattering*, and nothingness. According to QFT, the vacuum can't be determinately nothing because the indeterminacy principle allows for fluctuations *of* the quantum vacuum. "The quantum vacuum is more like an ongoing questioning of the nature of emptiness than anything like a lack. The ongoing questioning of itself (and *itself* and *it* and *self*) is what generates, or rather *is*, the structure of nothingness. The vacuum is no doubt doing its own experiments with non/being. In/determinacy is not the state of a thing, but an unending dynamism."[126]

* •   *

"When it comes to the quantum vacuum, as with all quantum phenomena, ontological indeterminacy is at the heart of (the) matter ... and no matter. Indeed, it is impossible to pin down a state of no matter or even of matter, for that matter. The crux of this strange (non)state of affairs is the so-called energy-time indeterminacy principle, but because energy and matter are equivalent [see Einstein, $E = mc^2$], it seems appropriate to call it the 'being-time' or 'time-being' indeterminacy principle."[127]

* * *

"Crucially, an indeterminacy in the energy of the vacuum translates into an indeterminacy in the number of particles associated with the vacuum, which means the vacuum isn't (determinately) empty, nor is it (determinately) not empty. These particles that are and aren't there as a result of the time-being indeterminacy relation, are called 'virtual particles.' *Virtual particles are quantized indeterminacies-in-action.* Virtual particles are not present (and not absent)—they are hauntological non/presences—that are material. Virtual particles do not traffic in a metaphysics of presence. They do not exist in space and time. They are ghostly non/existences that teeter on the edge of the infinitely fine blade between being and nonbeing. Admittedly, virtuality is difficult to grasp. Indeed, this is its very nature."[128]

* * *

"The Newtonian nature of space, time, matter, and the void are undone by quantum physics. In particular, quantum physics undoes the Newtonian assumptions of separability and metaphysical individualism. There are no individual entities running in the void. Matter is always already caught up with nothingness."[129]

* * *

The *kav* serves as a thread of connection between the *Eyn Sof* and creation. There is a gap, a dis/continuity between the *kav* and the surrounding light of the *Eyn Sof.* Joined and not joined, touching and not touching (*mati v'lo mati*), *separate and not separate* (agential separability).[130] The *reshimu*, which fills and doesn't fill the space, is at once the ultimate realm of limitation/manifestation and pure openness.[131] The *reshimu* is the recurrent play of in/determinacy of being/nonbeing that is the condition of possibility for existence. The nothingness (no-thingness) out of which creation arises is indeterminately empty/not-empty. Creation is and is not (or rather, is indeterminately) *ex nihilo*, and this is the point. The point out of which creation arose.

Everything changed in the run-up to WWII, during, and after: Time will never be the same. Being and time were together remade. No longer an independent parameter relentlessly marching forward, time is no longer continuous or one. There are no longer individual moments; there never were. "Time is diffracted, imploded/exploded in on itself: each moment made up of a superposition, a combination, of all moments (differently weighted and combined in their specific material entanglements). And directly linked to this indeterminacy of time is a shift in the nature of being and nothingness."[132]

"According to QFT, the vacuum can't be determinately nothing because the indeterminacy principle allows for fluctuations *of* the quantum vacuum. How can we understand 'vacuum fluctuations'?[133] If the physicist's conception of a field can be likened to a drumhead, with a zero-energy state being akin to a perfectly still drumhead, and a field with a finite energy being identified with a drumhead in one of its (quantized) vibrational modes (like the 3D analog of harmonics of a string), then while classical physics fashions the vacuum state as perfectly still, without any vibrations, according to quantum physics the vacuum state, although it has zero-energy, is *not* determinately still as a result of the energy-time indeterminacy principle. *Vacuum fluctuations are the indeterminate vibrations of the vacuum or zero-energy state. Indeed, the vacuum is far from empty, for it is filled with all possible indeterminate yearnings of time-being; or in this drum analogy, the vacuum is filled with the indeterminate murmurings of all possible sounds: it is a speaking silence.* What stories of creation and annihilation is the void telling? How might we approach the possibility of listening?"[134]

"If we are trying to figure out in what sense rhythm is eternal, then it cannot be that we can discover *this* or *that* rhythm, and can say that this one is eternal and the other one is not. We cannot really give an example without making a mess of what we are trying to say. In any case, we don't really have to, since it seems, at least for Benjamin, that not every rhythm is eternal—only the one that characterizes the rhythm of downfall; so we are left having to understand what that means."[135]

* * *

"The vacuum is flush with yearning, bursting with innumerable imaginings of what could-be/have-been. The quiet cacophony of different frequencies, pitches, tempos, melodies, noises, pentatonic scales, cries, blasts, sirens, sighs, syncopations, quarter tones, allegros, ragas, bebops, hiphops, whimpers, whines, screams are threaded through the silence, ready to erupt, but simultaneously crosscut by a disruption, dissipating, dispersing the would-be sound into non/being, an indeterminate symphony of voices.... A polyphony of emptiness."[136]

* * *

"Virtual particles are not in the void but *of* the void. They are on the razor edge of non/being. The void is a lively tension, a desiring orientation toward being/becoming. The void is flush with yearning, bursting with innumerable imaginings of what-could-be/might-yet-have-been."[137] That is, virtuality is the material wanderings/wonderings of nothingness. Vacuum fluctuations are ongoing virtual/thought experiments of the void: testing out every conceivable possibility, flashes of creation-annihilation, birth-death of every possible time-being. *Virtual transience* is what's up, or rather *the indeterminate play of what's up-what's down, creation-annihilation, downfall-renewal*. Multiple and varied rhythms diffracted through one another—the dance of the vacuum.

* * *

"If the recurrent character of transience is eternal, meaning that it has no beginning and no end, then the recurrence itself is not transient.... However, the particular form of recurrence that characterizes transience requires the multiplicity of transient beings or processes in duration, since without that multiplicity and duration, there would be no recurrence (or reiteration) and so no punctual condition for what Benjamin calls the rhythm of transience."[138]

* * *

Nothingness is not nothing. The vacuum fluctuations constitute the structure of this no-thingness: *The structure of nothingness is one of eternal virtual transience—the play of indeterminacy.*[139] Which is not to say that the quantum vacuum is always the same, it never is/was/will-be, but virtual transience is in any case its defining characteristic, and manifests as the eternal transience of matter. If (a sufficient amount of) energy is put into the vacuum it produces matter together with its anti-matter partner (since $E = mc^2$—energy and matter are equivalent), and all matter and antimatter have finite lives and inevitably find their downfall: They emerge from and fall back into nothingness. But what is the nature of matter such that it is governed by transience? In particular, what is matter's relationship to nothingness?

*   *   *

"Indeed, the problem with which we must come to terms is the following: How can the eternal characterize the recurrent character of transience while not being the same as the transience it characterizes? ... The eternal recurrence of the transient is itself not transient; even as we must understand this eternal recurrence as specifically characterizing the transience, it is neither absorbed nor defeated by the transience that it characterizes. The one remains irreducible to the other, and yet, the eternal recurrence at issue is less a Platonic form ... than a particular feature of musical, poetic, or metric patterns—closer perhaps to Pythagoras than to Plato.... Rhythm belongs to movement.... One hears a rhythm in the midst of a melody, or moves with a rhythm when one dances, and there might be a silent rhythm, but that means that it is punctuated by something noisy, or that it is itself punctuated by a field of sound."[140]

*   *   *

"Matter fell from grace during the twentieth century. It became mortal. Very soon after that it was murdered, exploded at its core, torn to shreds, blown to smithereens. The smallest of smallest bits, the heart of the atom, was broken apart with a violence that made the earth and the gods quake. In an instant, in a flash of light brighter than a thousand suns, the distance between Heaven and Earth was obliterated—not merely imaginatively crossed by Newton's natural theology-philosophy, but physically crossed out by a mushroom cloud reaching into the stratosphere. 'I am become death, the destroyer of worlds.'"[141]

*   *   *

"If transience, however, is what characterizes the time of *this* life, and of all life or living processes, then a principle or a movement that is eternal or nontransient traverses this time of the living. I use the term *traverse* to characterize the radical alterity of this eternity to transient life; eternity characterizes the recurrence of transience, and so informs all transient things."[142]

* • *

"Birth and death, it turns out, are not the sole prerogative of the animate world; 'inanimate' beings also have finite lives (troubling the inanimate/animate boundary and calling forth what the stakes are/have-been in maintaining it). 'Particles can be born and particles can die,' explains one physicist.[143] In fact, 'It is a matter of birth, life, and death that requires the development of a new subject in physics, that of quantum field theory.... Quantum field theory is a response to the ephemeral nature of life.'"[144]

* • *

How does quantum field theory understand the nature of matter?

"According to classical physics, a particle can stand on its own. We simply place a particle in the void—a Democritean delight. But according to QFT, a physical particle, even a (presumably) structureless point particle like an electron, does not simply reside in the vacuum as an independent entity, but rather is inseparable from the vacuum. The electron is a structureless point particle 'dressed' with its intra-actions with virtual particles: It intra-acts with itself (and with other particles) through the mediated exchange of virtual particles. (For example, an electron may intra-act with itself through the exchange of a virtual photon, or some other virtual particle, and that virtual particle may further engage in other virtual intra-actions, and so on.) Not every intra-action is possible, but the number of possibilities is infinite. In fact, some of the exchanges themselves constitute infinite contributions to the particle's energy-mass, and there are an infinite number of possible virtual intra-actions. This would constitute an infinite contribution to the mass of the electron. But how can this be when the mass of a physical electron is clearly finite (indeed, it's pretty darn small from our perspective)? The explanation physicists give is that the lone ("bare") point particle's contribution is infinite as well (infinitely negative due to the negative charge of the electron), and when the two infinities (that of the bare electron and that of the self-energy) are properly added together (or rather, subtracted from each other), the sum is a finite number, and not just any finite number but the one that matches the empirical value of the mass of the electron. In other words, an electron is not just "itself" but includes as part of its structure a "cloud" of an indeterminate number of virtual particles. All this may seem like a farfetched story, but it turns out that vacuum fluctuations have direct measurable consequences."[145]

Richard Feynman, one of the physicists who designed the "renormalization" procedure for consistently subtracting the infinities to produce a finite electron, quipped that the self-touching that the electron engages in that produces an infinity is "immoral." According to the renormalization procedure the "bare" electron (which is mathematically infinite) is "*dressed*" with the infinite contributions of the virtual particles of the vacuum such that, in the end, the physical electron is finite. (I'm using technical language here!) That is, what renormalization entails is the subtraction of two infinities to get something finite. This renormalization procedure necessarily entails taking into account the infinite possible intra-actions with all virtual particles in all possible ways, that is, *all possible virtual histories are condensed into each bit of matter.*[146]

* * *

In Lurianic Kabbalah, as part of the "chaining down" (*seder hishtalshelut*), the order of the emanation of the Infinite into the making of finitude, which is a highly elaborate and complex process, there are many levels where more rarefied lights become "*enclothed*" in coarser ones (as increasingly coarse vessels are made, until they are "crystallized out").[147] In a stage of emanation called the "World of Points" (*Olam Ha-Nekudot*), the lower *sefirot* (vessels made of lights) were not able to hold the light and the vessels shattered. The main parts of the broken vessels remained, but bits of the vessels in the form of shards and sparks fell. The shards that were struck by the sparks emerged as *kelippot (husks)*, which form the matter of the material world. The sparks live within the *kelippot* as their life force. *Bits of the infinite are threaded through the finite.* It is the task of humankind to liberate the sparks that exist in each bit of matter. "The world of points was like a sown field whose seeds could not bear fruit until they had first split open and rotted."[148]

* * *

"Where thinking suddenly stops in a configuration pregnant with tensions, it gives the configuration a shock, by which it crystallizes into a monad.... In this structure [the historical materialist] recognizes the sign of a Messianic cessation of happening, or, put differently, a revolutionary chance in the fight for the oppressed past. He takes cognizance of it in order to blast a specific era out of the homogeneous course of history.... The nourishing fruit of the historically understood contains time as a precious but tasteless seed."[149]

* • *

"History is the subject of a structure whose site is not homogeneous, empty time, but time filled by the presence of the now [Jetztzeit].... A historian who takes this as his point of departure stops telling the sequence of events like the beads of a rosary. Instead, he grasps the constellation which his own era has formed with a definite earlier one. Thus he establishes a conception of the present as the 'time of the now' which is shot through with chips of Messianic time."[150]

* • *

The breaking of the vessels (*shevirat ha-kelim*) is a major element of Lurianic Kabbalah. In Luria's account, the making of finite beings is not a unilinear developmental process of emanation that proceeds in a one-way teleological unfolding from the source to creation. Rather, creation is created in incomplete form, and it is the task of human beings (or perhaps all worldly beings) to engage in community-shared iterated material practices that liberate *the sparks, the tiny bits of light from the fallen shattered vessels (coarsened lights produced through elaborate sets of intra-actions) that are threaded through the material world.* The liberated sparks ascend and repair the *sefirot*, allowing the flow (*shefa*) of the Infinite to come down into the material world with fewer impediments. Also, written into this cosmology is the iterative rematerialization of the world based on ongoing intra-actions between the upper and lower worlds. On this account the world is not created once, but rather it is re-created/co-created anew in each moment: an iterative intra-active reconfiguring/reconstellating of the world.[151]

* • *

"Nothingness is not the background against which something appears, but an active constitutive part of every 'thing.' As such, *even the smallest bits of matter, electrons—infinitesimal point particles with no dimensions, no structure—are haunted by, indeed, constituted by, the indeterminate wanderings of an infinity of possible configurings of spacetimemattering in their specificity. Entire worlds inside each point; each specifically configured. Matter is spectral, haunted by all im/possible wanderings, an infinite multiplicity of histories present/absent in the indeterminacy of time-being.*"[152] Finitude is shot through with infinity, with an infinity of polyrhythmic vibrations of nothingness that flash up and fall away (eternal virtual transience) in the eternal iterative play—the dynamism—of indeterminacy.

* * *

"It is the history of the oppressed that flashes up within present time, disrupting the continuity and contesting its progressive claim."[153]

* * *

"The true picture of the past flits by. The past can be seized only as an image which flashes up at the instant when it can be recognized and is never seen again."[154]

* * *

"The messianic does not, and cannot, follow the trajectory of revelation. Rather, it is figured time and again in Benjamin's work as the traversal of one temporal modality in and through another."[155]

* * *

"The messianic, understood not as a human figure, becomes the recurrent and rhythmic downfall or going under that characterizes the transience of all living things."[156]

* * *

"A human centered temporal perspective that presumes and struggles against transience is supplanted by the rhythmic nature of an eternal recurrence of passing away.... One temporal perspective interrupts, converts, and follows from another. The sentence in which Benjamin relays this news works by ... shifting the perspective from that of human finitude to one that is decidedly not human centered. Something recurs, repeats in the sentence: the human-centered temporal perspective shifts into an eternal perspective on recurrent transience, one that requires the abandonment of the anthropocentric perspective itself."[157]

* * *

"Even the smallest bits of matter are an unfathomable multitude. Each 'individual' always already includes all possible intra-actions with 'itself' through all the virtual others, including those that are noncontemporaneous with 'itself.' That is, every finite being is always already threaded through with an infinite alterity diffracted through being and time. Indeterminacy is an un/doing of identity that unsettles the very foundations of non/being.... *Ontological indeterminacy, a radical openness, an infinity of possibilities, is at the core of mattering.*"[158]

* * *

"In an important sense, in a breathtakingly intimate sense, touching, sensing, is what matter does, or rather, what matter is: Matter is condensations of responses, of response-ability. Each bit of matter is constituted in response-ability; each is constituted as responsible for the other, as being in touch with the other. Matter is a matter of some intimacy, of cohabitating, of touching, of being in touch, of responses to yearnings."[159]

* * *

"Nothingness is not absence, but the infinite plentitude of openness. Infinities are not mere mathematical idealizations, but incarnate marks of in/determinacy. Infinities are a constitutive part of all material 'finities,' or perhaps more aptly, 'af/finities' (*affinities*, from the Latin, 'related to or bordering on; connection, relationship'). Representation has confessed its shortcomings throughout history: Unable to convey even the palest shadow of the Infinite, it has resigned itself to incompetence in dealing with the transcendent, cursing our finitude. But if we listen carefully, we can hear the whispered murmurings of infinity immanent in even the smallest details. Infinity is the ongoing material reconfiguring of nothingness; and finity/finitude is not its flattened and foreshortened projection on a cave wall, but an infinite richness. The idea of finitude as lack is lacking. The presumed lack of ability of the finite to hold the infinite in its finite manifestation seems empirically unfounded, and cuts short the infinite agential resources of undecidability/indeterminacy that are always already at play. Infinity and nothingness are not the termination points defining a line. Infinity and nothingness are infinitely threaded through one another so that every infinitesimal bit of one always already contains the other. *The possibilities for justice-to-come reside in every morsel of finitude* [including in the thick-now, the now-time, of the present moment]."[160]

* * *

"In reality, there is not one moment that does not carry its *own* revolutionary opportunity in itself.... [Crucially,] the power this moment has to open a very particular, heretofore closed chamber of the past. Entry into this chamber coincides exactly with political action."[161]

* * *

"For nature is Messianic by reason of its eternal and total passing away. To strive after such passing, even for those stages of man that are nature, is the task of world politics."[162]

WHAT FLASHES UP

Meditating on this constellation of theological-political-scientific insights, one point (among many others) that flashes up is this: *The messianic—the flashing up of the infinite, an infinity of other times within this time—is written into the very structure of matter-time-being itself.*[163]

The political import of such a claim is multiple. For one thing, this claim extends weak messianic power and eternal transience to *all material beings* and not merely those that are identified as "organic" or "living." Furthermore, it provides a way to link Benjamin's earlier and later works, the "Fragment" and the "Theses," by connecting rhythmic passing away, or eternal transience, a theme in the earlier work, to the seemingly episodic flashing up of the past, found in the latter piece. According to my agential realist materialist reading of QFT, all material beings are susceptible to downfall. This includes institutions, ideologies, imaginaries, and affective states, which (according to agential realism) are specific material configurations of worlding—and all of which are thereby susceptible to change and impermanence. Benjamin sought to interrupt a long-standing belief in progress; this was on behalf of his yearning for the destruction of fascism, for its downfall and the rebuilding of new institutions out its ruins.[164] Can we understand the messianic—the flashing up of the other times within this time—as linking up the transience of all material beings (e.g., the downfall of fascism) with the messianic as the flashing up of the past in a moment of danger? The rhythmic and the episodic seem to be at odds, but according to (my reading of) QFT, the past (and the future, indeed all possible temporalities and histories) flash up in the thick-now/now-time of the present. All rhythms are at play (virtually) in the vacuum/nothingness, and out of this can emerge large amplitude virtual expressions of past events (where events are neither singular nor locatable although they leave specific traces). And "seiz[ing] hold of a memory as it flashes up in a moment of danger"[165] is then not only a hopeful message about the impermanence of violent regimes but also an opportunity for taking up the charge of weak messianic power[166]—taking responsibility to rework the past in the present (which is not the same as denying the past, but on the contrary, of being present to it on behalf of the oppressed and their erased histories).[167]

At the same time, in the course of her subtle and nuanced reading of Benjamin, Butler offers some important cautionary remarks that seem, at least at first blush, to create substantial trouble for the claim above. It seems prudent to heed them or least take them seriously. In what follows, let's consider this particular claim that has flashed up in light of Butler's cautions. Here are Butler's cautionary remarks (rounding out a paragraph within which she clarifies what she means by *eternal*):

> As clarified earlier, the eternal no longer is posited exclusively in a spiritual domain considered as distinct from the worldly one, but becomes characteristic of transience or downfall that defines the worldly. At the same time, it would be a mistake to say that the eternal is now expressed in some final and readable form within the worldly or worldly terms, since we have already established that nothing in the historical or the worldly can adequately or properly refer to the messianic. So neither the eternal nor the messianic is expressed directly or instantiated in the transient and finite world—it enters in a more oblique way. One might suspect that the messianic is not all of what is meant by the eternal, but for Benjamin, it seems to be the name for "nature in its eternal and total transience." If we conclude that the messianic is nature, and it *includes* the eternal, we have missed the point that the messianic works in and through nature and its transience in a way that remains irreducible to it. The conclusion is not that the messianic belongs to another order, but only that it operates within this one as a constitutive alterity—breaking in, breaking out, flashing up, confounding without collapsing the spheres of this-worldly and the otherworldly.[168]

If we say that the flashing up of the eternal/infinite is written into the very structure of matter-time-being itself, have we not failed to heed Butler's cautionary remarks? Have we not naturalized the messianic, or even the eternal, collapsing the spheres of this-worldly and otherworldly, and making the rhythm of transience this-worldly, that is, itself transient (thereby bringing an end to the recurrence of transience)? The stakes, as Butler explains, are nothing less than whether violence and downfall (e.g., of "this or that regime or some other existing state of affairs") is "a certain action of the divine, the divine in action."[169] Crucially, her reading of Benjamin understands the messianic not as wholly immanent, but rather as a matter of one time traversing another, which short-circuits theocratic claims.

One response might be to say that my claim is about the nature of matter, and not nature per se, and that this claim arises within a materialist account that is not reducible to concerns about nature as separate from culture. Although such an objection sidesteps rather than responds to Butler's cautions, it does draw attention to an important point about the need to blast open terms such as "matter," "nature," and "ontology" that have been deeply sedimented in fixed and particular forms within Western philosophical discourses. This is surely part of what is at issue here, but let's not move too quickly, but rather take up Butler's cautions in detail. *Slow reading*—an arresting of thinking,[170] at least a slowing down, moving slowly through words and sentences carefully crafted, a practice of opening up to the possibilities of important insights flashing up—is an anticapitalist praxis. Not picking up a work and dismissing it or slamming it before it is given its due, before it is even understood and moving on to the next trendy theory. Critique is an indispensable practice, but there is nothing inherent in critique that makes it anticapitalist; critique too can be a handmaiden of capitalism, engaging in and enabling a continuing logic of disposability and training the mind to operate in the mode of progress, always looking to the next exciting idea, turning aside the old in favor of the new. The possibilities for countering an economy of disposability include composting ideas, turning them over, reading against the grain, reading through, aerating the encrusted soil to stimulate new growth.[171]

Before I address Butler's cautionary remarks that round out this paragraph, we'll need to unpack the first half of this dense paragraph. In the first sentence of the paragraph Butler summarizes one of her main points about the nature of the eternal in relation to the spiritual and worldly domains. Butler provisionally separates these two domains, and explains the nature of their inter- or perhaps intra-action: The eternal is not ("no longer") restricted to the spiritual domain, but rather becomes characteristic of transience that defines the worldly. In particular, Butler explains that the eternal "enters into" or "informs" the worldly through the relation of the eternal recurrence of transience without the eternal itself getting swallowed up by and conditioned by the temporality of worldly transience, which would belie its eternality. "Eternity," Butler writes, "characterizes the recurrence of transience, and so informs all transient things; and yet, it is not reducible to the transience of any of those things."[172]

This feeds directly into the next sentence of the paragraph we are considering, which cautions that it would be "a mistake to say that the eternal is now *expressed* in some final and readable form within the worldly or worldly terms." It may seem that the first clause of the statement already stands on its own, without the explanatory phrase that follows, since the nature of the traversal is such that eternity crosses over but is not encompassed by the worldly. That is, one would in any case expect that the eternal—that which is without beginning or end, the endless, the infinite—cannot be adequately or properly described in limited terms. But Butler follows this phrase with its justification (signaled by the connector "since") based on what she has argued is the ahistorical nature of the messianic: "*since* we have already established that nothing in the historical or worldly can adequately or properly refer to the messianic."[173] At first blush it might seem that Butler is conflating the eternal and the messianic here, which would be odd. It is one thing to claim that both the eternal and the messianic lack adequate or proper referentiality in worldly terms and quite another to equate them. The paragraph we are discussing is precisely the one in which Butler explicates what she means by the eternal (to the degree that anyone can presume to do this, as she herself remarks). So how are we to understand these key terms—the eternal and the messianic?

Remarking on the opening lines of Benjamin's "Theological-Political Fragment," Butler writes that Benjamin's formulation of the messianic "afflicts our very capacity to refer to the Messiah as a historically embodied human figure. Indeed, the Messiah or, rather, the messianic is the name for the break in referentiality that occurs when something historical seeks to relate itself to the messianic and necessarily fails."[174] Butler reads the messianic in Benjamin's "Theological-Political Fragment" as follows:

> The messianic figures the failure of the historical to signify the divine (which is one reason why the messiah cannot be a historical instantiation of the divine); the messianic, understood not as a human figure, becomes the recurrent and rhythmic downfall or going under that characterizes the transience of all living beings. The eternal thus traverses the transient without exactly becoming transient and losing its status as eternal.[175]

What remains to be answered is precisely the nature of this traversal. Let's leave this crucial matter aside for the moment and step back to consider what's been offered so far. Have we here entered a realm of abstraction that renders the messianic wholly unrecognizable from the perspective of its more usual renderings?

One might be willing to grant that a break in referentiality may be a feature or mark of the messianic, but it would seem that this does not exhaust the primary characteristics of this important term; surely, there are other kinds of breaking that are important in defining the messianic, such as a breaking with state-sanctioned violence, a breaking of the continuum of time, which is the condition for the possibility of redeeming the past, to name some dimensions of the messianic as they are expressed in Benjamin's "Theses."[176] In *Parting*

Ways, Butler offers a Benjaminian notion of the messianic that is significantly more resonant with (and at the same time also divergent from) traditional understandings of this term: "The messianic is a counterdoctrinal effort to break with temporal regimes that produce guilt, obedience, extend legal violence, and cover over the history of the oppressed."[177] She then explains: "The messianic emerges as a way of exploding that particular chronology and history in the name of recovering in scattered form those remnants of suffering's past that in indirect ways comport us to bring to an end regimes whose violence is ... physical."[178] At this point we might ask if we are being offered entirely different and disconnected, perhaps even inconsistent, accounts of the messianic? It here where Butler's brilliance shines through with crystalline clarity. She writes:

> In the "Theses on the Philosophy of History" (1940), it is the history of the oppressed that flashes up within present time, disrupting its continuity and contesting its progressive claim. In each case, we have to understand how one time traverses another, without precisely that first time becoming absorbed and contained in or by the time traversed. And although the focus changes in the twenty-one-year interval between these two pieces [the other being "Theological-Political Fragments"] from eternity to the history of the oppressed, both modalities are referred to as *messianic*, *meaning a traversal or a breaking through of one time in another*.[179]

In other words, there is a crucial feature of the messianic that underlies these two (or, indeed, the various) seemingly disparate takes: *The messianic is a traversal or breaking through of one time in another*. Continuing to leave the question of the nature of this traversal aside for a bit longer, let's return to the original set of questions about the paragraph we are considering, understanding that Butler does not collapse or equate the eternal and the messianic.

Can we now understand the first clause as logically following from the second? Butler claims that the messianic is "the recurrent and rhythmic downfall or going under that characterizes the transience of all living things," and that "the eternal thus traverses the transient without exactly becoming transient and losing its identity as eternal."[180] Hence, the messianic is a matter of the eternal traversing the this-worldly, and since nothing in the this-worldly can adequately or properly refer to the messianic, it follows that not even the eternal as it becomes characteristic of the transient can be expressed in some final and readable form within the worldly or worldly terms. In other words, the eternal is a matter of transience—of *endings* not unendingness—in its this-worldly form.

Which leads Butler to her next statement that "neither the eternal nor the messianic [which is not to conflate the two] is expressed directly or instantiated in the transient or finite world—it enters in a more oblique way." This more oblique way is clearly in the form of a "traversal" (yet to be discussed).

Butler next underlines the fact that while it would be wrong to conflate the messianic and the eternal—"One might suspect that the messianic is not all of what is meant by the eternal"—for Benjamin, the messianic is the name for "nature in its eternal and total transience," that is, it is, in particular, the eternal characteristic of transience that constitutes the messianic: "For nature is Messianic by reason of its eternal and total passing away."[181] This crucial point of Benjamin's is the very point Butler has set out to explicate.

Now we come to her primary cautionary remarks. The first one is this: "If we conclude that the messianic is nature, and it *includes* the eternal, we have missed the point that *the messianic works in and through nature* and its transience in a way that remains irreducible to it."[182] We need to pause at this juncture and first consider what kind of relationship between the messianic and nature is being figured here. If it is the messianic that is the sole actor, the one who acts, and acts unilaterally "in and through" nature, which remains a passive receptacle open to the eternal-yet-fleeting interruptions by this external force, then haven't we come to a place where nature is figured as the feminine awaiting the penetrating act of (yet another force coded as masculine, this time) the messianic? Does nature passively await the messianic to open it up to possibilities of redemption/revolution? That is, do we have here, in one form or another, a reinscription of the nature/culture dualism, or its interruption? Given Butler's marked rejection of the inscription model of social constructivism in *Bodies That Matter*, we might expect her to flag this in Benjamin if that were her reading of his position regarding the nature of nature in its eternal transience.[183] If, however, the nature of the mark of the eternal is not an inscription, then can we understand the eternal's traversal of nature as performative, as constitutive of nature itself? This goes to the point I raised at the outset: How is nature here understood? And can we understand nature itself as agentive when it comes to the messianic? We can no longer delay an encounter with the question of the nature of the *traversal*, which is key to understanding the nature of the messianic in its relation to nature.

Butler explains what she means by the verb "traverse" in this way: "I use the term *traverse* to characterize the radical alterity of this eternity to transient life; eternity characterizes the recurrence of transience, and so informs all transient things; and yet, it is not reducible to the transience of any of those things."[184] She continues: "Indeed, the problem with which we must come to terms is the following: how can the eternal characterize the recurrent character of transience while not being the same as the transience it characterizes"?[185] Butler goes on to explain that this cannot be answered by postulating a Platonic form for this eternal recurrence of transience and that the nature of this recurrence is rhythmic and therefore "closer perhaps to Pythagoras than to Plato."[186] She then offers some examples of rhythms punctuating one another, such as in contrapuntal music, in seeking to grasp something of this traversal. And yet she concludes this analogical exploration by saying that "it cannot be that we can discover this or that rhythm, and can say that this one is eternal and this other one is not."[187] At the same time, it is only the eternal rhythm that concerns us because it is the one that characterizes downfall, that is, the eternal rhythm of downfall. Butler continues:

> The eternal recurrence at play in this formulation is precisely not an immanent dimension of life, but an *entry* or *interruption of the eternal* within the transient. It is in transience without being of it, and this leads us to ask how one temporal modality (eternity) can *enter into and inform another* (transience) without ever becoming fully absorbed by the latter. To understand this *entry* of one temporal modality into another, one has to turn to Benjamin's notion of the messianic.[188]

Now we're coming to the crux of things. Butler names here a very particular conception of the messianic that goes directly to this question:

> Indeed, as I hope to show, the hyphen that links the *theological* with the *political* in the title of this fragment names a way that the messianic operates as the flashing up of one time within another or, in this passage, a timelessness within the domain of time. We have to understand how an atemporal mode breaks into another, or how an atemporal mode breaks out from with a temporal one. *Such traversals are possible only on the condition that finite and present temporality does not contain the other temporality that runs through it, flashes up within it, or breaks into or out of it.*[189]

The key point is that "to say that any given transient object or life is informed by an eternal recurrence of transience is not to say that that eternal recurrence is transience, will stop at some time, or be finished once and for all. For transience to be informed by the eternal means that it is interrupted or broken up by a temporal order that exceeds its frame, indeed, a temporal order that is not itself transient (and that strictly speaking, is not a temporal modality, but an atemporal one)."[190] In other words, *nature is constituted as transient in and through the breaking-in of an atemporal temporality.*

And this brings us to the concluding sentence of the paragraph we've been considering: "The conclusion is not that the messianic belongs to another order, but only that it operates within this one [the this-worldly] as a constitutive alterity—breaking in, breaking out, flashing up, confounding without collapsing the spheres of this-worldly and the otherworldly."

If the very nature of this "breaking in, breaking out, flashing up" is such that its relationship to nature (and it the very nature of this nature that is at stake and in question) is constitutive, rather than one that figures nature in a passive role awaiting the mark of the messianic, that is, if the nature of this "flashing up" is such that there is no nature that endures until it receives the mark of transience, but rather is constituted in and through eternal transience, does this action (intra-action) with the eternal somehow constitute nature as agentive? That is, *can we understand nature, which is eternally transient in its very constitution, as itself agentive and alive with possibilities?*[191] And how is this accounted for? For in reading Benjamin's explication of the messianic in the "Theses," it seems clear that the revolutionary Lurianic possibilities that are shot through the thick-now of the present are materially co-constituted; in other words, each moment, each bit of matter is materially shot through with the possibilities for change and transformation. And if the messianic, by Benjamin's lights, is surely not human, if his "formulation afflicts our very capacity to refer to the Messiah as a historically embodied figure,"[192] then the nature of nature must itself be agentive.

At this point, I want to return to the claim that opened this section, and consider it in light of what we have learned about Butler's understanding of Benjamin's notion of the messianic. Let's remind ourselves of this very point: *The messianic—the flashing up of the infinite, an infinity of other times within this time—is written into the very structure of matter-time-being itself.*

Although this claim relies heavily on my agential realist reading of quantum field theory (which also constitutes a further elaboration of agential realism), I would remind the reader, as I cautioned at the outset of this essay, that this is not to position the scientific (in particular, QFT) as the foundation for the theological-political, but on the contrary, to understand the scientific as always already shot through with the theological-political in a way that undermines the privileging of the former over the latter. In particular, the reader will do well to remember that my agential realist interpretation of quantum physics was crafted using a diffractive methodology of reading insights drawn from quantum physics through those of social and political theories (which, in particular, means that it is already shot through with some of Butler's insights).

A few notes and qualifications before I begin. First of all, as I mentioned previously, I will not be offering a systematic exposition of QFT here; rather, in what follows I rely on the montage, or shifting diffraction patterning, of fragments on QFT as diffractively read through other fragments, in the previous section of the essay. Second, if the "eternal" is usually understood as the temporal dimension of the infinite, which leaves the spatial to one side, I will speak about the "infinite" in what follows, since this term covers both spatial and temporal aspects of the infinite, and, in QFT, in particular, space and time do not exist as separate categories but rather are combined into spacetime (or rather, space, time, and matter become spacetimemattering).[193] I will, however, occasionally use "eternal" to specifically strike up resonances with Butler's analysis. Also, if nature, as Benjamin understands it, is that which is susceptible to downfall (often identified as "the living"), we will not be deviating from this in focusing on matter (since the conception of matter presented here calls into question the animate/inanimate binary).

As I have explained in detail elsewhere, according to (my agential realist reading of) QFT, there is an important sense in which the eternal/infinite does in fact operate with/in and through matter in its very constitution and as a constitutive alterity.

> The infinite touch of nothingness is threaded through all being/becoming, a tangible indeterminacy that goes to the heart of matter. Matter is not only iteratively reconstituted through its various intra-actions, it is also infinitely and infinitesimally shot through with alterity.[194]

What is the nature of this constitutive alterity and its operation with/in and through matter in its very constitution? It is the dynamism of the in/determinacy of the nothingness (the quantum vacuum) and its inseparability from matter that is the source and substance of this constitutive alterity. The quantum vacuum is an active polyrhythmic polyphonic play of no/thing-ness, the polyamorous multiply desiring material sensuous vibrancy of the virtual. The virtual in its materiality is the infinite play of the flashing up of an infinity of possible "histories," diffraction patternings of spacetimemattering in their ongoing reconfiguring, which derives from the indeterminacy (an ongoing dynamism) of time-being, where these "histories" do not happen in time, but rather, are the indeterminate flashings up of different times, rhythms, temporalities through one another.

Because *even* (but not only!) the smallest bits of matter, the electron, for example, engage in (all possible ways of) self-touching, the very nature of "self" is necessarily troubled. In the electron's touching itself, it intra-acts with the vacuum's indeterminate play of being-nonbeing and is in touch with all possible virtual-particles/fluctuations-of-nothingness in all possible ways.

> According to QFT, *even the smallest bits of matter are an enormous multitude!* Each "individual" is made up of all possible histories of virtual

intra-actions with all others; or rather, according to QFT, there is no such thing as a discrete individual with its own roster of properties. In fact, *the "other"—the constitutively excluded—is always already within (but not fully enclosed by the self): the very notion of the "self" is a troubling of the interior/exterior distinction.* Matter in the indeterminacy of its being un/does identity and unsettles the very foundations of non/being. Together with Derrida we might then say: "identity . . . can only affirm itself as identity to itself by opening itself to the hospitality of a difference from itself or of a difference with itself. Condition of the self, such a difference *from* and *with* itself would then be its very thing . . . the stranger at home." What is being called into question here is the very nature of the "self"; all "selves" are not themselves but rather all matter of time-beings. *The self is dispersed/diffracted through being and time.* In an undoing of the inside/outside distinction, it is undecidable whether there is an implosion of otherness, or a dispersion of self throughout spacetimemattering.[195]

These are admittedly very strange matters, especially in light of the predominant scientific and Western metaphysical assumption that matter is an eternal and fixed sameness that travels in the void. Clearly, the very nature of matter and nature are being radically reworked in quantum theory. And it is these very forms of self-touching (self-intra-actions) through which the infinite makes itself felt. *Ontological indeterminacy—the flashing up of the infinite in the making of the finite, a dynamism that constitutes the eternal transience of matter, the recurrent passing-away and opening-up of eternally reconfigured im/possibilities in the very undoing of linear causality and determinism—is at the core of mattering, in its very constitution.*

Hence, to say that the messianic is written into the very structure of matter-time-being is not to claim that matter fully includes, encloses, or encompasses the eternal. Rather, it is to say that the "flashing up"—the in/determinate play of non/being as it flashes-up/returns-to the vacuum—is integral to the very nature of matter. It is, for example, to say that the indeterminacy of time-being, the flashing up of other possible worldings—the messianic—is shot through every bit of matter, every moment. Moments are not infinitely thin-slices of the present, but rather "now-time" (*Jetztzeit*)—the infinitely thick-now, "blasted out of the continuum of history" and "shot through with chips of Messianic time"—the possibilities for reconfiguring spacetimemattering in an instant: "a leap in the open air of history."[196]

On the one hand, the fact that the eternal is not fully contained within matter is evidenced by the fact that the "flashing up" is itself a characteristic of the nothingness that "surrounds" as well as is threaded through every bit of matter. On the other hand or, rather, at the same time, the sense of an "inside" versus an "outside" to matter is compromised by the very fact that there is no absolute separation between "matter" and "void" or, for that matter, between "inside" and "outside," which is not to deny distinctions, but rather to come to terms with the radical openness that is matter in its eternally iterative rematerialization—its materializing "itself" otherwise, reconfiguring "itself" anew. *Matter is indeed agentive in its very constitution.*[197]

What about Butler's point that in Benjamin's "Fragment" the messianic is the name for the break in referentiality that occurs when something historical seeks to relate itself to the messianic and necessarily fails? Recall that the process of making finite what "starts out" as infinite is very involved and complicated, whether in QFT or in Lurianic Kabbalah. In QFT, the process of "renormalization"—the "handling" of infinities in QFT—is *not* a mechanism for placing the infinite or the eternal within the realm of representation, "in some final and readable form within worldly or worldly terms." On the contrary, the individual mathematical maneuvers are not translatable in worldly terms; "renormalization" is a patchwork set of mathematical acrobatics that lacks a sound mathematical foundation, barely managing a safety net, and the severe (but not wholly intractable) difficulties of "taming" the infinities marks a failure of the possibility of representing and explicating the nature of the infinity of infinities.

And yet through the renormalization procedure it is nonetheless possible to conclude that for all this self-touching-the-other, through the "bare" particle's inseparability from the vacuum, *the result of the eternal flashing up of the infinite, is that each bit of matter (no matter how large or small) far from being fixed and eternal ultimately winds up having only a finite lifetime.* Although the focus has been on the smallest bits of matter, in order to show that matter has this messianic structure written into its finitude, *no matter how small a piece*, this is surely true of all material beings, each of which *is an enormous entangled multitude*. This is a good reminder that we are not just talking about the "microworld" (as if one existed—indeed, any such conception is clearly being undone here!) nor "inanimate" matter (as if that category any longer had any meaning). In particular, the very materiality of our being, indeed *all beings*, participates in this rhythm of eternal transience.

Hence, on this account the messianic can be understood as a constitutive alterity, a flashing up of other time-beings with/in and through matter *in its very constitution*. Is there some way in which we might understand this operation as a flashing up that is *constitutive* of matter itself *without* thereby "collapsing the spheres of this-worldly and the otherworldly"?

This question goes to a broader set of questions about how distinctions and separations are figured within this relational ontology of agential realism, which is the basis for my reading of QFT (and which also constitutes a further elaboration of the theory). It is important to realize that the usual options—no separation/distinction and absolute separation/distinction—do not exhaust the possibilities. *Agential separability* provides one such alternative, and this goes to the core of the notion of intra-action, which troubles conventional conceptions of causality (and spacetimemattering).[198] In brief, *agential separability* is the agential realist conception of separability that results from *intra-actions* (not interactions): Differentiations do not precede intra-actions but rather result in and through them, such that the "differentiations" so constituted are entangled or ontologically inseparable (as part of one phenomenon). Agential separability is a differentiation within entanglement: a cutting together-apart, differentiating-entangling as one move.[199] "Differentiating is not about radical exteriority but rather agential separability. That is, differentiating is not about othering or separating [off,] but on the contrary about making connections and commitments."[200]

Significantly, my agential realist reading of QFT entails intra-actions, not interactions, and so it is possible to acknowledge that the flashing up of the messianic is written into the very structure of—is indeed constitutive of—matter-time-being itself, without thereby collapsing the spheres of this-worldly and otherworldly. In particular, the this-worldly and the otherworldly are neither entirely separate realms nor overlapping or equivalent realms; rather, they are agentially separable, which constitutes a radical reworking of the immanent/transcendent dichotomy. The infinite flashes up within and through the constitution of the finite, but this is not to say that it is possible to fully contain the infinite within the finite. And furthermore, to fully jettison the transcendent is an act of hubris, the kind of hubris that breeds the violence of state-sanctioned, state-funded technoscientific practices like the Manhattan project, which presumed that it was possible to get a handle on matter down to its "smallest parts" in order to harness the energy of "a thousand suns." However, to deny the immanent, to deny matter the fullness of its agency, is to cut short the practical sensuous material possibilities for change. It is to forget one of the central contributions of Lurianic Kabbalah, the fact that the messianic is not something that arrives (one day) from on high, but rather is here-now in the form of "messianic chips" whose energies are released through collective praxis—material activities of reconfiguring the conditions of im/possibilities of change/transformation/revolution, to enliven and activate the possibilities for justice that are here-now, in this very instant. Benjamin caught hold of Luria's notion of praxis and the revolutionary possibilities for redemption and read him through Marx, and vice versa. The repair of the world (*tikkun olam*) is an immanent praxis shot through with and made possible by the flashing up of the infinite from within the finite.

And this point brings us full circle, back to Benjamin's "Theological-Political Fragment," wherein "a worldly restitution leads to the eternity of downfall, and the rhythm of this eternally transient worldly existence, transient in its totality, in its spatial but also in its temporal totality, the rhythm of Messianic nature, is happiness. For nature is Messianic by reason of its eternal and total passing away." Butler's inspired interpretation of downfall and happiness brings this home. She writes:

> Can we understand happiness as the apprehension of the rhythm of transience, and even consisting in the abandonment of an anthropocentric relation to loss? . . . The embrace of transitoriness implies the loss of the very first-person perspective that would make the embrace; to hear or sense the rhythm of transience is precisely to allow one's own loss, even the loss of one's own finite personhood, to become at once small and iterable. . . . Happiness . . . is that rhythmic movement or sound by which each and every living process is washed away, and thus linked with one another in their vanishing. One might understand this as a radical break with narcissism, including most importantly, the negative narcissism of guilt. . . . The nonanthropocentric apprehension of transience is what releases the human subject from guilt, self-vilification, and cycles of retribution . . . which would include accusations of guilt and rationales for war. Further, from the perspective of eternal transience, my own life is equal to the life of every other, that is, equally subject to this eternally recurrent dying away. . . . The deconstitution of the human subject into its "living soul" is precisely the access to the nonanthropocentric basis of his or her happiness, a link with an eternal rhythm that washes away traces of guilt, and affirms transience as an eternal link among all living beings.[201]

And perhaps it is this eternal link among all living beings, all beings in their aliveness, this shared transience, and the possibilities for renewal that follow downfall, that is needed in confronting the rise of fascism in its connections with late capitalism, the normativity of state-sanctioned violence against the oppressed, and the ongoing devastation of the planet and all its inhabitants. Facing the im/possibilities of living on a damaged planet, where it is impossible to tease apart political, economic, racist, colonialist, and natural sources of homelessness (otherwise called "the problem of refugees"), will require multiple forms of collective praxis willing to risk interrupting the "flow of progress"—not by bombing the other but by blasting open the continuum of history. THE END

NOTES

Acknowledgments: In engaging with the methodology of diffraction/constellation, time and again I found myself overcome by the multitudinous patternings that came to the fore while I traced the flashes, lighting up intermittently here and there, watching the configurations crystallize and then rearrange. I found myself taken in by the configurations and felt amazement as each one expanded, shifted and reconstellated again; the project felt like it would burst under this pressure, as it kept expanding in more and more directions. Perhaps this was a taste of what Benjamin may have experienced in his unending travails on the *Passagen-Werk*. As a result, and also because I found myself face to face with the transience of the life of loved ones, I missed innumerable deadlines for this essay, and every one of them was greeted with understanding, encouragement, and grace by my supremely patient editors, Catherine Keller and Mary-Jane Rubenstein. I am enormously grateful to them for hanging in there with me despite all the missed deadlines and beyond all expectations, and for their generous feedback. I am also infinitely indebted to Fern Feldman for making it possible for me to continue living in deadline mode for months on end, for her careful reading of and feedback on the essay, for amazingly helpful and inspiring conversations, and for all her ideas that I can't even begin to separate out from mine in this essay and are wholly inadequately acknowledged here. I am deeply grateful to Victor Silverman and Judith Butler for their important comments on earlier drafts of the essay. Finally, this essay was written during a year in my life when I lost my father, Harold Barad, and my dear friend Professor Ibrahim Baba Farajajé. This essay is dedicated to them. May their memories be for a blessing.

1. "Thick-now" is an identifying phrase that I use to invoke the thick sense of multiple historicities and temporalities that exist in any given moment. More on this below. In particular, the "thick-now of the present," as I have used it, is resonant with Benjamin's now-time (*Jetztzeit*), as I understand it. See, for example, Karen Barad, "TransMaterialities: Trans*/Matter/Realities and Queer Political Imaginings," *GLQ* 21, no. 2–3 (2015): 387–422, and Karen Barad, "Troubling Time/s and Ecologies of Nothingness: On the Im/Possibilities of Living and Dying in the Void," in *Eco-Deconstruction: Derrida and Environmental Philosophy*, ed. Matthias Fritsch, Philippe Lynes, and David Wood (New York: Fordham University Press, forthcoming 2017).
2. Walter Benjamin, "Theses on the Philosophy of History," in *Illuminations*, trans. Harry Zohn (New York: Schocken, 1969), Thesis XVII, 263.
3. Judith Butler, *Parting Ways: Jewishness and the Critique of Zionism* (New York: Columbia University Press, 2012), 103. Butler's Benjamin quotations are taken from "Theses on the Philosophy of History," 261.
4. Benjamin, "Theses," Thesis XVI, 262; see also Thesis XVII, 263.
5. The quotation marks are Benjamin's: "'Construction' presupposes 'destruction,'" quoted in Susan Buck-Morss, *The Dialectics of Seeing: Walter Benjamin and the Arcades Project* (Cambridge, MA: MIT Press, 1989), 219.

6. Susan Buck-Morss, *Dialectics*, 250. Buck-Morss here refers to Benjamin, V, 578 (N3, 4).
7. Although Derrida distances deconstruction from Benjaminian destruction (Derrida, "Force of Law," trans. Mary Quantaince, *Cardozo Law Review* 11, nos. 5–6 [1990]: 919–1046), I hope it will become evident here that there may be creative materialist possibilities for reconsidering Derrida's distancing response.
8. Butler, *Parting Ways*, 92–98.
9. Benjamin, "Theses," Thesis IX, 258.
10. Ibid., Thesis IX, 257–58.
11. Ibid., Thesis XVI, 262.
12. This story appears in different versions in *Yerushalmi Brachot* (2:17a, b) and *Lamentations Rabbah* (16:51). From the latter: "A hard fate is in store for my child. 'Why' he asked; and she answered 'Because close on his coming the Temple was destroyed.' He said to her, 'We trust in the Lord of the Universe that as close on his coming it was destroyed so close to his coming it will be rebuilt.' . . . After some days the man said, 'I will go and see how the child is getting on.' He came to the woman and asked, 'How is the child?' She answered, 'Did I not tell you that a hard fate is in store for him? Misfortune has dogged him. From the time [you left] there have been strong winds and a whirlwind came and carried him off.' He said to her, 'Did I not tell you at his coming [the Temple] was destroyed and at his coming it will be rebuilt?' [footnote: The fact that the storm carried him away was evidence that the child was to be used for supernatural purposes]" (*Lamentations Rabbah* [London: Soncino, 1983], 137). The tradition was that the Messiah would be born on the day that the Temple was destroyed. I am grateful to Fern Feldman for bringing this to my attention.
13. Benjamin, "Theses," Thesis XVII, 263.
14. Ibid., Thesis II, 254.
15. Butler, *Parting Ways*, 94, my emphasis.
16. Ibid.
17. Benjamin, "Theses," Thesis VIII, 257.
18. Ibid.
19. Butler, *Parting Ways*, 94.
20. Ibid., 123.
21. Butler writes, "It is important to remember not only that divine power destroys mythical power, but that divine power *expiates*. This suggests that divine power acts upon guilt in an effort to undo its effects. Divine violence acts upon lawmaking and the entire realm of myth, seeking to expiate those marks of misdeeds in the name of a forgiveness that assumes no human expression. . . . For Benjamin the Jewish God does not induce guilt and so is not associated with the terrors of rebuke. Indeed, divine power is described as lethal without spilling blood. It strikes at the legal shackles by which the body is petrified and forced into endless sorrow, but it does not strike, in Benjamin's view, at the soul of the living. Indeed, in the name of the soul of the living, divine violence acts" (Butler, *Parting Ways*, 80).

22. Cited in Susan Buck-Morss, *Walter Benjamin and the Arcades Project* (Cambridge, MA: MIT Press, 1991), 250.
23. From Ernest Rutherford's Nobel Prize lecture: "A minute fraction of the radium atoms is supposed each second to become unstable, breaking up with explosive violence. A fragment of the atom— an α-particle—is ejected at a high speed, and the residue of the atom, which has a lighter weight than before, becomes an atom of a new substance, the radium emanation" (Ernest Rutherford, "The Chemical Nature of the α-Particles from Radioactive Substances," in *The Collected Papers of Lord Rutherford of Nelson*, vol. 2 [New York: Routledge, 2014], 139). The announcement of major new discoveries in nuclear physics by Rutherford at a scientific meeting in 1933 inspired the headlines and an article by Rutherford in *The Times of London*. A five-stacked headline from *Illustrated News* announced that nothing less than "a new world" resulted from this discovery (Richard Reeves, *A Force of Nature: The Frontier Genius of Ernest Rutherford* [New York: W. W. Norton and Sons, 2008], 149).
24. *Spacetimemattering* is an agential realist term that acknowledges the iterative materialization of space-time-matter. See Karen Barad, *Meeting the Universe Halfway: Quantum Physics and the Entanglement of Matter and Meaning* (Durham, N.C.: Duke University Press, 2007). More on this below. I ask for patience with the many citations to my own work in this essay. I am trying to condense much into this essay, and I am forgoing any systematic explication of agential realism here. The many citations are meant to direct the reader to more detailed discussions of points only hinted at here.
25. Benjamin, "Theses," Thesis A, 263.
26. As I tried to make clear in my earlier work, "posthumanist" (for lack of a better word when I first used it in 2002) thinking for me is *not* about moving beyond the human, but rather about understanding how the cut gets made that differentially marks the "human" in relation to its constitutive others. That is, the "posthumanist" point is that distinguishing cuts must enter into the frame of analysis rather than beginning the analysis after these divisions are made, thereby enabling an exploration of the limits that human exceptionalism in its differential constitution of the "human" poses to theory construction, political praxis, and scientific practices. See Barad, *Meeting the Universe Halfway*, esp. pp. 32 and 136, and chap. 4.
27. "At the lab bench" is a shorthand for radically reworking scientific practices from within so that questions of justice (including questions of what constitutes justice and for whom) are part and parcel of scientific practices, as opposed to the usual after-the-fact ethical analyses performed by professional ethicists where the "ethical implications" regarding possible (mis)uses of scientific findings are analyzed in a limiting way that leaves science itself in the rarefied realm of pure knowledge. The latter approaches, including ELSI (ethical, legal, and social implications) analyses that are an afterthought on grant applications, leave science untouched, miss crucial issues of justice, and have the wrong temporality: They are too little, too late. For more details see J. Reardon, J. Metcalf, M. Kenney, and K. Barad, "Science and Justice: The Trouble and the Promise," *Catalyst: Feminism, Theory, Technoscience* 1, no. 1 (2015): 1–36.

28. I only hint here at some of the important entanglements. See *Meeting the Universe Halfway*. More recently I have been publishing works on the entanglements between *quantum field theory* and the atomic bomb, which are of relevance here: See, for example, "Troubling Time/s," and "No Small Matter: Mushroom Clouds, Ecologies of Nothingness, and Topologies of Spacetimemattering," in *Arts of Living on a Damaged Planet*, ed. Anna Tsing, Heather Swanson, and Elaine Gan (University of Minnesota Press, forthcoming, 2017). My most recent writings on quantum field theory seek to shed light on the violence as well as the radical political potential written into quantum field theory. This essay, engaging with the work of Walter Benjamin, is drawn from a larger body of work in progress, tentatively titled *What Flashes Up*.
29. A critique of science that only attends to its overblown authority, without also taking account of its empirical reliability, and the fact that science and fact-making are under attack by right-wing forces has never been sufficient, and surely isn't now.
30. For example, the scientific notion of diffraction (already suggested by Benjamin's references to crystallization) might be useful—see below. A diffractive methodology (Barad, *Meeting the Universe Halfway*) would allow the following type of claim to be made without theocratic implications: The fact that Benjamin finds Kabbalah to be an important resource for his work is not in spite of, but because of, his political, indeed Marxist, commitments. In particular, despite the fact that several prominent scholars of Benjamin's work, including his friend Gershom Scholem, find an irreconcilable tension between his commitments to Marxism and Kabbalah, it is precisely those times when Benjamin refers to Marxism that he is drawing most directly from Kabbalah, and vice versa. Which is *not* to say that they are the same in his mind or that he collapses the two—but rather, that he diffractively works their insights through one another—understanding that each is in fact always already threaded through the other: so that, for example, it is not only the fact that the messianic takes the form of interrupting commodity time, but also that praxis takes the form of *tikkun olam* (repair of the world through material acts); neither is fully enclosed in the other but each is diffractively threaded through the other.
31. See Butler, *Parting Ways*.
32. On questions of water justice see, for example, Nicole Harari and Jesse Roseman, *Environmental Peacebuilding Theory and Practice: A Case Study of the Good Water Neighbors Project and In-Depth Analysis of the Wadi Fukin/Tzur Hadassah Communities* (Amman, Bethlehem, and Tel Aviv: EcoPeace/Friends of the Earth Middle East, 2008). Website http://foeme.org/uploads/publications_publ93_1.pdf.
33. Benjamin, "Theses," Thesis XIII, 261.
34. Isaac Newton, Scholium to the Definitions in *Philosophiae Naturalis Principia Mathematica*, bk. 1 (1689); trans. Andrew Motte (1729), rev. Florian Cajori (Berkeley: University of California Press, 1934), 6.
35. Benjamin, "Theses," Thesis XVII, 263. For more on the materialization of time see the agential realist notion of *spacetimemattering* in *Meeting the Universe Halfway*, and also as further elaborated in Barad, "Quantum Entanglements and Hauntological Relations of Inheritance."

36. Benjamin, "Theses," Thesis XVII in its entirety, 262–63, my emphasis.
37. Walter Benjamin, *Arcades Project*, [N10, 3], 475, my emphasis.
38. Ibid. [N2, 6].
39. Anne Sayre, *Rosalind Franklin and DNA* (New York: W. W. Norton, 1975).
40. Different kinds of beams are used to determine structure, depending on particular parameters of the crystalline or molecular structure and what level of detail is sought. X-rays (high energy light), electrons, and neutrons are often used. Sometimes an inverse process is used to determine the characteristics of the substance that makes up the beam rather than the crystal. Crystallography has been used to confirm that electrons (and other subatomic entities generally considered to be particles) can exhibit wave behavior: In particular, aiming a beam of electrons at a crystal produces a diffraction pattern. With regard to Benjamin's remark that the "total event" is contained in a small bit, two facts about diffraction are worth noting: (1) The diffraction pattern is a "fourier transform" of the spatial structure, which in essence means that it is taking account of all possible wave interferences with the structure, and (2) diffraction is the basis for holography—the making of holograms—where beams of light coming from lasers directed from different directions and bouncing off the object combine to make a diffraction pattern—that is, a hologram. Each fragment of a hologram contains the entire image of the object.
41. On a materialist reading of Derrida's notion of *hauntology*, especially as appropriate in understanding quantum physics, see Karen Barad, "Quantum Entanglements and Hauntological Relations of Inheritance: Dis/continuities, SpaceTime Enfoldings, and Justice-to-Come," *Derrida Today* 3, no. 2 (2010): 240–68. For more on temporal diffraction see Barad, "Troubling Time/s" and references therein.
42. See my analysis of the physics-politics of the "quantum eraser experiment," and see also the agential realist notion of *spacetimemattering* as a process of materialization, in Barad, *Meeting the Universe Halfway*, esp. chap. 7, and Barad, "Quantum Entanglements and Hauntological Relations of Inheritance."
43. Benjamin, *Arcades Project*, <M°, 14>, 854.
44. Albert Einstein, "Zur Elektrodynamik bewegter Körper," *Annalen der Physik* 17 (1905): 891.
45. Benjamin, *Arcades Project*, <H°, 16>, 845.
46. Ibid., [N2a, 3], 462, my emphasis.
47. Ibid., [N1, 1], 456.
48. Martin Uman, *Lightning: Physics and Effects* (Cambridge: Cambridge University Press, 1986).
49. Karen Barad, "TransMaterialities: Trans*/Matter/Realities and Queer Political Imaginings," in *GLQ* 21, no. 2–3 (2015): 398. See also Karen Barad, "Nature's Queer Performativity (the authorized version)," in *Kvinder, Køn & Forskning* (Women, Gender & Research), no. 1–2 (2012): 25–53, and Vicki Kirby, *Quantum Anthropologies: Life at Large* (Durham, N.C.: Duke University Press, 2011).
50. Discovery Channel television program "Discovery Wonders of Weather: Lightning Phenomena" (September 2007), http://science.howstuffworks.com/nature/natural-disasters/lightning.htm, quoted in "Nature's Queer Performativity," 35.

51. I make this argument in more detail in "Nature's Queer Performativity" and in "TransMaterialities." Related to lightning and also to Benjamin's "dialectical reversal," see the discussion of the quantum eraser experiment in Barad, *Meeting the Universe Halfway*, chap. 7, and Barad, "Quantum Entanglements and Hauntological Relations of Inheritance."
52. Benjamin, *Arcades Project*, [K1, 2], 389.
53. Rolf Tiedemann, "Dialectics at a Standstill: Approaches to the *Passagen-Werk*," in *The Arcades Project*, 942, my emphasis.
54. Benjamin, "Theses," Thesis XIV, 263 and Benjamin, *Arcades Project*, [NII, 3], respectively, as quoted in Tiedemann, "Dialectics at a Standstill," 942.
55. See below for more details on quantum field theory (QFT) and its notion of thick-now as a condensation of temporalities and possibilities.
56. Significantly, all three are also important figures in Kabbalah. See Barad *What Flashes Up* (in progress) for more details.
57. The quoted phrases are from Benjamin, "Theses," Thesis VII, 256.
58. Tiedemann, "Dialectics at a Standstill," 931.
59. Benjamin's montage methodology has many interesting resonances with my "diffractive methodology"—a specifically materialist methodology that is designed to be consistent with and is informed by agential realism's ontoepistemology (Barad, *Meeting the Universe Halfway*, chap. 2). That the physical phenomenon of crystallization already entails diffraction hints at the resonances here; for more details see Barad, *What Flashes Up*.
60. On "dis/continuity" and its relationship to writing histories that do not assume linear time see Barad, "Quantum Entanglements and Hauntological Relations of Inheritance."
61. See Barad, *Meeting the Universe Halfway* for more details.
62. Barad, *Meeting the Universe Halfway*. I have continued to elaborate both the ontoepistemological framework of agential realism and the agential realist interpretation of quantum physics following the publication of *Meeting the Universe Halfway*. Of particular relevance to this essay are my more recent publications on quantum field theory (QFT). These include Karen Barad, *What Is the Measure of Nothingness? Infinity, Virtuality, Justice / Was ist das Maß des Nichts? Unendlichkeit, Virtualität, Gerechtigkeit*, dOCUMENTA (13): 100 Notes—100 Thoughts / 100 Notizen—100 Gedanken | Book N°099 (English & German edition, 2012); Karen Barad, "On Touching—The Inhuman That Therefore I Am," in "Feminist Theory Out of Science," special issue, *differences: A Journal of Feminist Cultural Studies* 23, no. 3 (2012): 206–23; Barad, "TransMaterialities"; Barad, "Troubling Time/s and Ecologies of Nothingness"; and Barad, "No Small Matter."
63. Karen Barad, "Diffracting Diffractions: Cutting Together-Apart," in "Diffracting Worlds, Diffractive Readings—Onto-Epistemologies and the Critical Humanities," ed. Kathrin Thiele and Brigit Kaiser, special issue, *Parallax* 20, no. 3 (2014): 168–87.
64. Judith Butler, "One Time Traverses Another: Benjamin's 'Theological-Political Fragment,'" in *Walter Benjamin and Theology*, ed. Colby Dickinson and Stéphane Symons (New York: Fordham University Press, 2016), 272–85.

65. Ibid., 272, my emphasis.
66. Ibid., 274, my emphasis.
67. Ibid., my emphasis.
68. Ibid., 273.
69. Marc-Alain Ouaknin, *The Burnt Book: Reading the Talmud*, trans. Llewellyn Brown (Princeton, N.J.: Princeton University Press, 1995), 169. The citation for the quotations used by Ouaknin here are from Emmanuel Levinas's *Totality and Infinity* (Pittsburgh: Duquesne University Press, 1988), 283.
70. Ouaknin, *Burnt Book*, 171. The citation for the quotations used by Ouaknin here are from Maurice Blanchot's *L'entretien infini* (Paris: Gallimard, 1969), 260. I have changed the spelling of the Hebrew word *Hidush* to keep the transliteration conventions consistent in my essay.
71. Ibid., 94.
72. Ibid., 62; this is Ouaknin quoting Vladimir Jankélévitch.
73. Ibid., 89.
74. Genesis (*Bereishit*) 1:1
75. Rashi, *The Torah: With Rashi's Commentary*, translated, annotated, and elucidated by Rabbi Yisrael Isser Zvi Herczeg (New York: Mesorah Publications, 1995, 1999), *Bereshit/Genesis* 1:2.
76. See Barad, "Quantum Entanglements and Hauntological Relations of Inheritance," 244.
77. Benjamin, "Theses," Thesis XIV, 261.
78. Barad, "Nature's Queer Performativity," 40.
79. Jacques Derrida, *Specters of Marx: The State of the Debt, the Work of Mourning, and the New International* (New York: Routledge, 1994), 29.
80. Barad, "Quantum Entanglements and Hauntological Relations of Inheritance," 265.
81. On these points see Barad, *Meeting the Universe Halfway*; Barad, "Quantum Entanglements and Hauntological Relations of Inheritance"; Barad, "Nature's Queer Performativity," Barad, "No Small Matter."
82. On the *dillug* in the Passover story see R. Sholom Noach Berezovsky, *Sefer Netivot Shalom al Moadim u'Midot* (Jerusalem: Yeshivat Beit Avraham Slonim, 5761), 2:313.
83. Benjamin, "Theses," Thesis XV, 261.
84. Benjamin, "Theses," Theses XIV, XV, 261.
85. Rashi, *Shemot/Exodus*, 3:106.
86. The four new years are: The first of *Nisan*, the first of *Elul* (for the tithe of animals), *Rosh Hashanah* (literally Head of the Year, which is the first day of the seventh month, *Tishrei*; New Year for the years and for planting and for vegetables, commemorates creation of the world), and *Tu B'Shvat* (New Year of the Trees). *Mishnah Rosh Hashanah* 1:1.
87. Ouaknin, *Burnt Book*, 171.
88. R. Shalom Noach Berezovsky, *Netivot Shalom al haTorah*, 2:323, trans. Fern Feldman. My gratitude to Fern Feldman for this reference.
89. See *Sefer Yetzirah: The Book of Creation*, trans. and commentary by Aryeh Kaplan (Boston: Weiser Books, 1997), chap. 4.

90. See below on matters of time-being in quantum field theory. Benjamin mentions both *Jetztzeit* (now-time) and *Jetztseins* (now-being, waking-being), *Arcades Project*, <O°, 5>, 857; "waking-beings" is Tiedemann's translation on page 942.
91. Barad, "Quantum Entanglements and Hauntological Relations of Inheritance," 258.
92. Ibid., 260–61.
93. Benjamin, *Arcades Project*, [K1, 2], 389.
94. See below on matters of time-being in quantum field theory. For more details see Barad, "Troubling Time/s".
95. Benjamin, *Arcades Project*, [M1a, 1], 418.
96. Butler, *Parting Ways*, 102.
97. Barad, "Quantum Entanglements and Hauntological Relations of Inheritance," 261.
98. Benjamin, "Theses," Thesis B, 264.
99. Rabbi Akiva, as quoted in *Pirkei Avot* 3:19.
100. Every Passover Haggadah has this line: "In every generation a person is obligated to see themselves as if they had left Egypt [*Mitzrayim*: the narrow place]" (*Mishnah Pesachim* 116b).
101. Deuteronomy (*Devarim*) 15:15.
102. Barad, "Quantum Entanglements and Hauntological Relations of Inheritance," 265.
103. Ibid., 266.
104. Benjamin, *Arcades Project* [N11, 3].
105. Deuteronomy 16:20. See also Butler, *Parting Ways*.
106. Benjamin, "Theses," Thesis VI, 255.
107. *Pirkei Avot* is traditionally studied between Passover and Shavuot.
108. *Pirkei Avot*, 1:17.
109. Karl Marx, "Theses on Feuerbach," Thesis 11, in *The Portable Karl Marx* (New York: Penguin, 1983), thesis 11, 158.
110. *Midrash Beraishis Rabbah* 1:1.
111. *BT Pesachim* 7a.
112. Isaac Luria (1534–72). For more information about Luria see Larry Fine, *Physician of the Soul, Healer of the Cosmos: Isaac Luria and His Kabbalistic Fellowship* (Stanford: Stanford University Press, 2003). Luria wrote little in his short lifetime. His teachings were written down by his students. The most studied versions of his teachings, and there are multiple redactions, are by R. Chayyim Vital. Page numbers here refer to an English/Hebrew version of *The Tree of Life (Etz Chayyim)*, vol. 1, with an introduction by Donald Wilder Menzi and Zwe Padeh, 2nd ed. (New York: Arizal, 2008).
113. Hava Tirosh-Samuelson, "Philosophy and Kabbalah: 1200 to 1600," in *The Cambridge Companion to Medieval Jewish Philosophy*, ed. Daniel H. Frank and Oliver Leaman (Cambridge: Cambridge University Press, 2003), chap. 11.
114. *Tree of Life*, 13.
115. Ibid.

116. Ibid., 24.
117. Ibid., 13.
118. Lawrence Fine, "The Contemplative Practice of Yiḥudim in Lurianic Kabbalah," in *Jewish Spirituality*, vol. 2, ed. Arthur Green (Crossroad, 1997), 66.
119. On matter's self-touching see Barad, "On Touching."
120. Ouaknin, *Burnt Book*, 182.
121. Derrida, *On Touching—Jean-Luc Nancy*, trans. Christine Irizarry (Stanford University Press, 2005), 369.
122. On the notion of different sizes of infinities see the work of the mathematician Georg Cantor. Georg Cantor, *Contributions to the Founding of the Theory of Transfinite Numbers*, trans. Philip E. B. Jourdain (New York: Dover, 1915).
123. On the reshimu see Gershom Scholem, Kabbalah (New York: Meridian Books, 1974), 312, and Saul Baumann, Keys to the Wisdom of Truth, with an introduction and commentaries by Reb Zalman Schachter-Shalomi (ALEPH Canada: Alliance for Jewish Renewal [pamphlet]).
124. Barad, *What Is the Measure of Nothingness?* 8–9.
125. See Luria, *Tree of Life (Etz Chayyim)*. On co-creation see especially on *tikkun olam*, raising the sparks.
126. Barad, "TransMaterialities," 396; see also Barad, *What Is the Measure of Nothingness?*
127. Barad, "TransMaterialities," 395.
128. Ibid., 396.
129. Barad, "No Small Matter."
130. On *mati v'lo mati* see *Tree of Life*; on agential separability see Barad, *Meeting the Universe Halfway*.
131. Baumann, *Keys to the Wisdom of Truth*.
132. Barad, "No Small Matter." For more on time diffraction see Barad, "Troubling Time/s."
133. The physics term "vacuum fluctuation" is unfortunate, but it has stuck. The notion of "fluctuation" which suggests a process that happens *in* time is a remnant of a misleading narrative, a story in part for the public, that physicists began repeating at some point about the temporal indeterminacy at the core of things. An unfortunate consequence is that this has led to misunderstanding of the nature of virtuality; nonetheless, this misunderstanding has been and continues to be the basis for some philosophical accounts of the virtual. The tale speaks of the void as a kind of dishonest banker who plays fast and loose with the energy books (the energy banker can cheat as long it does it quickly enough that no will notice). But this way of putting it misses the radical features of temporal indeterminacy. See Barad, *What Is the Measure of Nothingness?* and "On Touching."
134. Barad, "Troubling Time/s"; see also Barad, *What Is the Measure of Nothingness?*
135. Butler, "One Time Traverses Another," 275.
136. Barad, *What Is the Measure of Nothingness?* 13.
137. Ibid.
138. Butler, "One Time Traverses Another," 274.

139. Like the word "fluctuation" (see n. 133 above), "transience" is also perhaps not the most apt term to use when it comes to the vacuum, since both suggest processes that happen *in* time over a finite interval, whereas temporal indeterminacy is a defining feature of the vacuum. However, I use this term here not only to point out resonances with Butler's use of the term, but also because it is precisely this virtual feature of the vacuum (named either fluctuation or transience) that is materialized as a property of matter once the intra-actions between the vacuum and matter are taken into account. In other words, the *virtual eternal transience of the vacuum* ultimately manifests as *the eternal transience of all matter*. (This point will be discussed forthwith.)
140. Butler, "One Time Traverses Another," 275.
141. Barad, "No Small Matter." This line from the Bhagavad Gita was famously quoted by physicist J. Robert Oppenheimer (his translation from the Sanskrit) in the wake of the first atomic bomb explosion. See also Barad, "Troubling Time/s" for more on the relationship between the bomb and quantum field theory.
142. Butler, "One Time Traverses Another," 274–75.
143. A. Zee, *Quantum Field Theory in a Nutshell*, 2nd ed. (Princeton, N.J.: Princeton University Press, 2010), 3.
144. Barad, *What Is the Measure of Nothingness?* 9.
145. Ibid., 14–15.
146. See Barad, "TransMaterialities," 410.
147. Scholem, *Kabbalah*, 136.
148. Ibid., 140.
149. Benjamin, "Theses," Thesis XVII, 262–63.
150. Benjamin, "Theses," Thesis XIV, 261; Thesis A, 263.
151. See Scholem, *Kabbalah*, and *Tree of Life*.
152. Barad, "No Small Matter."
153. Butler, "One Time Traverses Another," 273.
154. Benjamin, "Theses," Thesis V, 255.
155. Butler, "One Time Traverses Another," 273.
156. Ibid., 273–74.
157. Ibid., 277–78.
158. Barad, "On Touching," 214.
159. Karen Barad, "Deep Calls unto Deep: Queer Inhumanism and Matters of Justice-to-Come," Transdisciplinary Theological Colloquium, Drew University, Madison, N.J., March 29, 2014.
160. Barad, *What Is the Measure of Nothingness?* 16–17.
161. Benjamin quoted in Tiedemann, "Dialectics at a Standstill," in *Arcades Project*, 944 (orig. 1:1231).
162. Benjamin, "Theological-Political Fragment," 313.
163. This claim is indebted to and builds directly on Butler's insights regarding the nature of nature and the messianic in Benjamin. More on this below.
164. The Jewish tradition has many references to a hopeful vision of the downfall of empires. See, for example, Midrash Tanchuma-Yelammedenu, va-Yetze 2, the story

of Jacob's vision of the angel's ascending and descending a ladder, a rhythmic ascent and downfall. In this case, the specific empires mentioned are Babylon, Media, Green, and Edom (i.e., Rome). My gratitude to Fern Feldman for this reference.

165. Benjamin, "Theses," Thesis VI, 255.
166. Ibid., Thesis II, 254.
167. See my account of the "quantum eraser" experiment in *Meeting* and "Quantum Entanglements and Hauntological Relations."

    I am indebted to Judith Butler for her specific framing and articulation of the possibility of this linkage and an email discussion of this point (July 2016). Any mistakes or misunderstandings in the presentation here are mine alone.
168. Butler, "One Time Traverses Another," 278.
169. Ibid., 273.
170. Benjamin, "Theses," Thesis XVII, 262.
171. This also goes for materialisms as well. It would be wrong to think Marx should be turned aside and disposed of in some "turn" to "new materialisms." Feminist science studies and its new materialist innovations are deeply indebted to Marx, whose ghosts are very lively and most welcome. I, for one, understand feminist science studies to have firm roots in Marxist theory and feminist engagements with it. (This legacy is surely evident in classic works such as Donna Haraway's "Cyborg Manifesto" and Sandra Harding's work on standpoint epistemologies; see, for example, Harding, *Whose Science? Whose Knowledge?*) Feminist science studies and feminist studies more generally (if such generalizations are useful) did not and do not dispose of the economic in favor of less material matters, but rather have been at work theorizing how to understand the nature of the material in ways that include but are not restricted to economic forces and how they contribute to the materialization of bodies that matter—all bodies—in their differential constitution. The "new" ought not be understood indexing a successor account and replacement for the "old" according to the logics of the temporality of progress. How the relationship between the "new" and the "old" is understood is part of what is at issue in understanding the nature of matter, materiality, and materialization. See especially Barad, "Diffracting Diffractions."
172. Butler, "One Time Traverses Another," 275.
173. Ibid., my emphasis.
174. Ibid., 277.
175. Ibid., 273–74.
176. In "One Time Traverses Another," Butler is focused on decoding Benjamin's "Fragments," and she therefore stays close to this text. In particular, she doesn't assume there is any consistency or even compatibility across the key articles that frame Benjamin's writings—between the "Fragments" (1919) and the "Theses" (1940). My approach also doesn't assume that Benjamin is unfurling a single philosophy over his lifetime; at the same time, my approach is to work with the fragmentary nature of things, and all matter of quantum leaps, at least experimenting with them and trying them out. And if it is indeed the case that Benjamin's philosophy does not unfold in some continuous fashion, but rather leaps here and there, this does

not necessarily mean that all coherence and significant patterns of *différancing* are lost or rendered impossible. For example, in leaping with Butler across various Benjamin writings, there is nonetheless a consistency in the feature of breaking or disjuncture that runs through his various definitions of the messianic (e.g., break in referentiality ["Fragments"], break in state-sanctioned violence ["Critique of Violence"], break in the continuum of history ["Theses"]), and it is the very nature of this break that it is important to contemplate. More on this below.

177. Butler, *Parting Ways*, 70.
178. Ibid.
179. Butler, "One Time Traverses Another," 274, my emphasis.
180. Ibid., 273–74.
181. Benjamin, "Theological-Political Fragment," 313.
182. Butler, "One Time Traverses Another," 278; second emphasis is mine.
183. Butler, *Bodies that Matter: On the Discursive Construction of "Sex"* (New York: Routledge, 1993).
184. Butler, "One Time Traverses Another," 274–75.
185. Ibid., 275.
186. Ibid.
187. Ibid.
188. Ibid., 276, my emphases.
189. Ibid., my emphasis of this last sentence.
190. Ibid.
191. Butler's goal in "One Time Traverses Another" is to explicate Benjamin's understanding of the nature of nature and the messianic, rather than using Benjamin's philosophy to further elaborate her own philosophy. Given Butler's prior work on the nature of nature, one might expect her to note any uneasiness with Benjamin's conception of nature if he were thinking of the eternal as marking passive nature and imprinting it with eternal transience. Indeed, Butler argues that Benjamin's conception of *nature is constituted in and through its relationship with the eternal* such that there is no nature that endures while awaiting the mark of the eternal. The question remains, nonetheless, how to understand where this leaves nature with regard to the question of its own agency, especially given the plethora of active verbs attributed to the eternal in its acting in and through nature: such as "traverse," "enter into," "inform," "interrupt," "breaking in/out," "flashing up." That is, when it comes to how the eternal makes itself felt as the transient, it is not clear whether nature also has an active/agentive nature or whether it is the action of the eternal that produces a nature that is necessarily transient but is not itself agentive.
192. Butler, "One Time Traverses Another," 277.
193. See n. 24 above on *spacetimemattering*.
194. Barad, "On Touching," 215.
195. Barad, "Troubling Time/s," and "On Touching." This is true of moments of time as well as bits of matter (being), each of which is indeterminately infinitely large and infinitesimally small, where each bit is specifically constituted through an infinity of intra-actions with all others.

196. Quotes from Benjamin's "Theses," Theses XIV, XV, A, 261, 263.
197. The unending dynamism that is ontological indeterminacy (in action) results in matter's ongoing touching and reconfiguring of itself. This is surely not to say that humans have no role to play in revolutionary politics, as if the world will take care of itself, for there is no world separate from the "human" in its differential constitution and its ongoing rematerializations—which is not to say that humans cannot become extinct, or that the existence of humans is a precondition for the world worlding itself, but on the contrary, to acknowledge the human as being *of* the world as currently configured. (Humans do not live in an environment, but rather humans in their differential constitution are specific material configurings *of* the world.) And at the same time, agentive participants such as floods, mosquitoes, melting ice caps, drought, together with water availability, market forces, colonialism, militarism, and so on, in their inseparability, are also part of the field of revolutionary possibilities in their ongoing reconfiguring. (For more details see Barad, *Meeting the Universe Halfway*.) And while Benjamin in the "Theses" seems to suggest that the messianic flashes up during times of urgency (e.g., "to seize hold of a memory as it flashes up in times of danger" [Benjamin, "Theses," Thesis VI]), there is an undoing or dispersion of the notion of event in the agential realist formulation offered here, so that it is possible to hold both points at once: that is, that matter is constituted through the ongoing flashing-up of other times, other histories, other possibilities, and this eternal dynamism offers possibilities for seizing hold of other times as they flash up (and are perhaps most needed in times of danger), as noted in the beginning of this section.
198. For more details on "agential separability" see Barad, *Meeting the Universe Halfway*.
199. For more details see *Meeting the Universe Halfway*. On the ethical dimension see esp. 392–93. See also Barad, "Diffracting Diffractions," and Trinh T. Minh-ha, "Not You/Like You: Post-Colonial Women and the Interlocking Question of Identity and Difference," *Inscriptions*, special issues 3–4, "Feminism and the Critique of Colonial Discourse," (1988), http://culturalstudies.ucsc.edu/PUBS/Inscriptions/vol_3-4/minh-ha.html. [26/02/2014] (my emphasis).
200. Barad, *Meeting the Universe Halfway*, 392.
201. Butler, "One Time Traverses Another," 279–80, 283, 284.

❧ Vegetal Life and Onto-Sympathy

JANE BENNETT

In "I Eat an Apple," Annemarie Mol wonders what would happen to accounts of human agency if they were to begin with the figure of the human-that-eats, rather than, say, thinks or wills or makes:

> *I eat an apple.* Is the agency in the *I* or in the *apple*? . . . Here is the apple, there am I. But a little later (bite, chew, swallow) I have become (made out of) apple; while the apple is (a part of) me. *Transubstantiation.* What about that for a model . . . ?[1]

The eating self, it seems, would both highlight the I's entanglements with *outside* bodies (foodstuffs, farm workers, agribusiness machines, etc.) and also underscore the I's *internal* heterogeneity, including a vegetality.[2] The eating self would present human agency as a multimedia effort distributed across a variety of material elements and shapes.[3] Or, in Mol's words, it would lead us away from a "muscular" model of agency and toward something like the gut's "transubstantiation" model[4]—wherein there is an activeness that is neither quite "free" nor fully "determined": "Thus we leave the model of the muscular body and its particular centred masculinity behind—as well as its 'other': the actor controlled from elsewhere (be it a boss, a ruling class, a structure). The eating 'I' is no longer eager to stay a free man lest he becomes a slave, but an altogether different kind of being."[5]

But can we say more about what this "different kind of being" is and does? How to describe this being and his or her or its powers of action? How, in other words, to paint a fuller picture of the life of this nonautonomous but underdetermined creature? In this essay I follow Mol's lead and pursue the trail of the plant within us. I also enlist the help of three naturalists—Henry David Thoreau, Henri Bergson, and Charles Darwin—each of whom, like Mol, believed that human agency was entwined with botanical life. Thoreau presents himself as made up of the (mostly vegetarian) things he eats, claiming to be a worshipper of Hebe, who was herself the "daughter of Juno and wild lettuce."[6] Bergson contributes the notion that the plant and the animal are distinguishable only by different accents within a shared set of tendencies. And Darwin makes the vegetality of the human more palatable by highlighting the ways in which plants themselves have an impressive range of capacities, such as a creative (rather than predetermined) responsiveness and coordinated powers of movement.

Of special interest to me is the suggestion, made most strongly by Thoreau, that the vegetality of humans and the vegetality of plants can somehow *communicate*. Why should I not, asks Thoreau, "have intelligence with the earth? Am I not partly leaves and vegetable mould myself?"[7] (Walt Whitman, whose *Leaves of Grass* is rich in homo-botanical eros, concurs: "I find I incorporate gneiss, coal, long-threaded moss, fruits, grains, esculent roots.")[8] I take Thoreau (and Whitman) to be pointing to a certain overlap (a partial coincidence, not full identity) between the materialities of humans and that of plants. There is, in other words, a kind of reaching out from each body toward what it senses to be its likeness "out there." Leo Bersani has described similar such leanings as "homoness," which he defines as a "communication of forms" or a "correspondence" between shapes, "a kind of universal solidarity not of identities but of positionings . . . in space, a solidarity that ignores even the apparently most intractable identity-difference: between the human and the nonhuman."[9]

Where Bersani speaks of "homoness" and "love," I experiment with the term "sympathy":[10] sympathy as a resonance that does not collapse into a simple sameness.[11] What is sought is a way to name a cross-species rapport or exchange, wherein things engage on the basis of affinities of shapes that are "inaccurate replications"[12] of each other. (More on how shape matters later.) Sympathy was an important term for various nineteenth-century attempts to disclose a mode of encounter between naturally social bodies, as my examples from Darwin, Thoreau, and Bergson will show. Since then, the figure of sympathy has fallen onto hard times, especially within political theory, where critiques of organicism and trends toward defining the political exclusively as disruption or as the struggle for power have rendered it a morally suspect term. Neither has sympathy fared well within an Arendtian framework, where it appears as a subjective sentiment ("the passion of compassion")[13] at odds with the reasoned work of political freedom: "Because compassion abolishes the distance, the worldly space between men where political matters . . . are located, it remains, politically speaking, irrelevant and without consequence."[14] The larger project of which this essay is a part is to see whether it is possible and desirable today to rehabilitate the figure of sympathy as a politically relevant term. One first step, pursued in this essay, is to consider a notion *of* sympathy that is not in the first instance a subjective emotion or sentiment but an impersonal mesh of affiliations between natural bodies. To do this, I turn to my nineteenth-century poet-naturalist-philosophers, Thoreau, Bergson, and Darwin, who were intrigued by the strange attachments and dispatches between animal and vegetable bodies—intrigued by a mode of communication that I call *onto-sympathy*.

The mode of inquiry was that of an engaged amateur. They combined the technique of close observation—as when Thoreau writes, "O the evening robin, at the end of a New England summer day! If I could ever find the twig he sits upon! I mean *he*; I mean *the twig!*"[15]—with speculative or poetic narration. They did so with the aim of heightening attentiveness to subaltern and incipient goings-on, to processes and events eclipsed within default regimes of perception, feeling, thinking.[16] More recently, the artist Claire Pentecost has advocated a similar stance, of "the public amateur" who enters into specialized discourses of science and technology in order to reveal how those discourses function politically, and to encourage and empower citizens to think carefully for themselves about how science and technology function in their daily lives.[17] Carla Hustak and Natasha Myer pursue a similar goal in their account of the "chemical ecologies" of plant encounters: They depict/induce a "world [that] is full of 'propositions' waiting to be registered by *interested* bodies."[18] I revisit Thoreau, Bergson, and Darwin and their plants in order to give witness today—to become more *interested* in—the vegetal life in which we participate.

LEAVES AND NEEDLES

In the "Solitude" chapter of *Walden*, Thoreau details his rich social life with animals, plants, water, and stones. These friends gather at the pond or outside Thoreau's cabin; they run into each other on walks; and the man's experience is solitary only if sociality is arbitrarily restricted to interhuman relations: "I am no more lonely than a single mullein or dandelion in a pasture."[19] For Bersani, the humans who are in the best position to discern the kind of cross-species communication he called homoness are "failed subjects," a descriptor that Thoreau, I believe, would have embraced for himself.[20] Failed subjects both fail to reach and fail to strive for conventional selfhood, and their inward sense of being alive develops instead through a process of being inaccurately replicated in the sounds, shapes, and scents of the outside bodies they encounter—an "outside that is 'mine' without belonging to me."[21]

One such fabulous encounter happens on a "delicious evening" as Thoreau "walks along the stony shore of the pond": He finds that "sympathy with the fluttering alder and poplar leaves almost takes away my breath; yet, like the lake, my serenity is rippled but not ruffled."[22] (We can imagine that the rhythm of Thoreau's walking has started to sync with that of the fluttering leaves.) Later on, Thoreau again detects a force of sympathy; but this time its initial trajectory is the reverse, coming toward rather than from him: "Every little pine needle expanded and swelled with sympathy and befriended me. I was so distinctly made aware of the presence of something kindred to me, . . . and also that the nearest of blood to me . . . was not a person nor a villager."[23]

We have here two occasions of cross-body filiation. Though my focus in this essay is on human-plant sympathies, it is worth noting that for Thoreau sympathy is not exclusively an interorganism affair. Stones, like Thoreau's body, also have the capacity to affect and be affected: "If I were to discover that a certain kind of stone by the pond-shore was affected, say partially disintegrated, by a particular natural sound, as [from] a bird or insect, I see that one could not be completely described without describing the other. I am that rock by the pond-side."[24] Thoreau also acknowledges the resonance between a vital force in people and the flame of a candle: "All material things are in some sense man's kindred, and subject to the same laws with him. Even a taper is his relative—and burns not eternally . . . but only a certain number of his hours."[25] What is more, there are occasions of sympathy to which people are not a party at all, except after the fact, as in Thoreau's example of an exchange between the geological bodies of pond and weather:

> The pond does not thunder every evening, and I cannot tell surely when to expect its thundering; but though I may perceive no difference in the weather, it does. Who would have suspected so large and cold and thick-skinned a thing to be so sensitive? Yet it has its law to which it thunders obedience when it should as surely as the buds expand in the spring. *The earth is all alive and covered with papillae.*[26]

As this last example indicates, there is a certain vitalism alongside Thoreau's transcendentalism, according to which "there is nothing inorganic."[27]

But let us return to the encounter between Thoreau, leaves, and needles. What, we might ask, is the mechanism, the *how*, of these exchanges? Can we, heuristically, parse the operational process of sympathy into phases? Let us call them gravitation, corporation, and annunciation. *Gravitation*: Thoreau's body would have leaned toward edible plants, drawn by their color or scent or shape and by the way these material aspects have been inflected by rituals, memories, and words. This could lead to the taking in of the outside materials. In eating the leaves, berries, or seeds of plants, a new *corporation* is formed, in which veggies and flesh mutually transform. It is worth noting that eating is for Thoreau only one of the ways in which human and nonhuman bodies mingle forces, sometimes to find a modus vivendi and sometimes to clash violently. He also acknowledges scent-infusion as a mode of exchange: In "Walking" he notes that "the skin of the Eland, as well as that of most other antelopes just killed, emits the most delicious perfume of trees and grass," just as "the trapper's coat emits the odor of musquash . . . ; . . . a sweeter scent to me than that which commonly exhales from the merchant's or the scholar's garments."[28] Bodies take on the scents they meet, just as you are what you eat. And elsewhere Thoreau focuses on how sounds too are vehicles of transmission (a kind of speech exchanged) between bodies.[29]

In another round of gravitation, the interiorized (eaten or smelled or heard) vegetality would lean into its material counterparts (distant kin) in the plants on the outside, and vice versa. Examples of this are when pine needles swell out toward Thoreau or when Thoreau is drawn along the wooded paths of his walks. We might think of such activities as a kind of recognition. In Hegel's famous account of "Lordship and Bondage," recognition refers to an intersubjective process whereby one person acknowledges another as like oneself in its capacity for freedom and subjectivity: "Self-consciousness exists in itself and for itself, in that, and by the fact that it exists for another self-consciousness; that is to say, it is only by being acknowledged or 'recognized.'"[30] But Thoreau's fleshly nod to the (alder or pine) other is not human-centered and is more materialized than Hegel's. It gives witness not to a shared personhood but to the existence of partial, protean, nonsubjective affinities, affinities between the materialities of dividuated bodies, affinities that make possible the emergence of a conjoint rhythm or resonance.[31] We have here something like a "being-with outside ourselves [that] is prior to any particular emotional experience of sympathy."[32]

In another phase or movement of sympathy, this recognition, which had hitherto been at work below the level of conscious human detection, announces itself a bit more clearly, entering more explicitly into Thoreau's awareness. *Annunciation*: the presence of a current of connection is, for example, reported to Thoreau via a sense of being favorably disposed toward the plants. "I was so distinctly *made aware* of the presence of something kindred to me." An uncanny sense of "kinship" would be the way that complex bodies acknowledge a degree of overlap in the way each expresses its vitality. "Kinship" is Thoreau's term, but it may accent too strongly the *sameness* of human and leaf, whereas the communicative relays at work in *Walden* are not relationships of strict identity. I would like to inflect what Thoreau calls a kinship (between him and the needles and leaves) toward Bersani's notion of inaccurate replications or Deleuze and Guattari's refrain of rhythms and motifs.[33]

Even so, less enchanted readers might wonder whether the sympathy Thoreau reports remains but an intrahuman event, a product of Thoreau's and our poetic imaginations. Why speak of sympathy rather than, say, simile, wherein pine needle and naturalist are *positioned by human thought* as like one another? A Thoreauian might reply: Because there would still have to be something *in or of* the pine needles themselves that provoked people to register an encounter as a resemblance: There would have to have been *some* overlap (albeit an "inaccurate" repetition). What is more, the experimental scientist in Thoreau insists that the possibility that the needles do not merely acquiesce to an experience of sympathy but actively *call* for it should not be ruled out in advance. In a journal entry of 1850, Thoreau seems to locate the ultimate source of this call in a divine He: "*I* did not *make* this demand for a more thorough sympathy. This is not my idiosyncrasy or disease. He that made the demand will answer the demand."[34] But we might leave open the theological question—taking Whitman's advice to "argue not concerning God"[35]—and focus instead on the extent to which the force of things is sensuously materialized, is, that is, a function of shape, smell, sound, and so on. One way to denote, to not forget the presence of this kind of force or agency would be to give it a name, say, onto-sympathy. Under such a banner, the subjective sentiment of kinship and the literary technique of simile would be called forth by a larger (I would say "impersonal") cosmic tendency for bodies to follow attractions, to (mis)recognize, and to form affiliates.

In the following quotation we see Thoreau alternating between simile ("as if") and an affirmation of an actual capacity of bodies (human ears, bells, pine needles, woods, echoes) to sound out and communicate their presence to each other:

> Sometimes, on Sundays, I heard the bells. . . . At a sufficient distance over the woods this sound acquires a certain vibratory hum, *as if* the pine needles in the horizon were the strings of a harp which it swept. . . . There came to me in this case a melody which the air had strained, and which had conversed with every leaf and needle of the wood, that portion of the sound which the elements had taken up and modulated and echoed. . . . The echo . . . *is* not merely a repetition of what was worth repeating in the bell, but partly the voice of the wood; the same trivial words and notes sung by a wood-nymph.[36]

The notion of onto-sympathy invokes the presence of a tendency coursing through/in bodies of all sorts; it has the advantage of allowing us to dwell a bit longer with the thought of nonhuman agencies. When Thoreau was "distinctly made aware of a presence kindred to me" at the pond, we are able to think the thought that the pine needle too had some agency in producing that effect, beyond the fairly passive role of datum for perception or raw material for simile.[37]

PLANT AND ANIMAL TENDENCIES

I've been suggesting that some of the encounters at Walden Pond model a cross-species affiliation or onto-sympathy, wherein material elements within one complex body enter into a rhythm of coordination with a somewhat amenable set of elements within another complex body that is not of its "kind." Sympathy, then, as a function of overlapping, coincidental, and cooperating materials partially shared. A sympathy that came to presence, for example, when the vegetal elements in Thoreau resonated with the vegetal elements in the leaves. (It is proper to speak of vegetal *elements* in the plants, for just as Thoreau is not all-man or even all-animal, plants are not *all* vegetal.) The strand of Thoreau that I have emphasized is one concerned less with unequivocal natural "laws" than with a pulsating order of gravitating, leaning, resonating bodies. This is a cosmos governed not by a fixed logic but by the eccentric systematicity of what Henri Bergson called "tendencies." A "logic" has a precision and an internal consistency that a tendency does not. A logic has a telos; a tendency has a trajectory. A logic is singular; a tendency must be one of many. A logic can be surveyed from the outside; a tendency must be inhabited to be understood. A logic is pure; a tendency is messy.

For Bergson, there is "no definite characteristic [that] distinguishes the plant from the animal.... There is no manifestation of life which does not contain, in a rudimentary state—either latent or potential,—the essential characters of most other manifestations. The differences are in the proportions."[38] For Bergson, plants and animals alike possess capacities for "mobility," "feeling," and "consciousness," and both harbor tendencies toward "torpor," "insensitivity," and "unconsciousness"—but in different "proportions." In the plant the powers of motility, affect, and awareness are mostly "asleep," although "not so sound asleep that they cannot rouse themselves when circumstances permit or demand it." Likewise, "the evolution of the animal kingdom has always been ... dragged back, by the tendency it has kept toward the vegetative life. However overflowing the activity of an animal species may appear, torpor and unconsciousness are always lying in wait for it."[39]

Bergson's assertion of the shared (but not operationally equal) capacities of animal and vegetable is grounded in his belief in élan vital, "a primitive, original impetus, anterior to the separation of the two kingdoms," in which every plant and animal participates.[40] Élan vital is defined primarily as an "effort," albeit one that is "rudimentary," "vague," and "diffused."[41] Not despite but because of this protean quality, élan vital is an effort of far greater "depth" and resilience than the endeavoring of any formed individual.[42] Bergson also turns to the language of physics and biology to describe élan vital: It is a "continuity of genetic energy ..., [a] current passing from germ to germ through the medium of a developed organism. It is as if the organism itself were only an excrescence" of it.[43] Because of their participation in this current, plants and animals are *creative*, which is to say that the course of their lives is not the unfolding of a preexisting regimen but a genuine and free responsiveness to ever-changing milieus.[44]

Bergson was responding to Darwin, whose theory of evolution Bergson thought needed a more robust acknowledgment of creativity (as the spontaneous generation of novelty). When Bergson says that "vegetable and animal are descended from a common ancestor which united the tendencies of both in a rudimentary state,"[45] his reference to a "common ancestor" is a nod to Darwin on natural selection. "All living species," writes Darwin

> have been connected with the parent-species of each genus . . . and these parent-species, now generally extinct, have in their turn been similarly connected with more ancient species; and so on backwards, always converging to the common ancestor of each great class.[46]

We can take Bergson's notion of élan vital to be a specification of the power of action of these "ancestors": Élan vital highlights what it is that Darwin's "common ancestors" *do* within the process of selection. What they do is start it off; they initiate or spark evolutionary changes in the physical forms of living things. This impetus is directional in the sense of being an effort, but creative rather than being the realization of a preexisting plan.[47] One could say that with Bergson, Darwin's notion of "common ancestor" itself "converges" to the even more primordial élan vital.

In presenting plants and animals as outgrowths ("excrescences") of a protean creative force called élan vital, Bergson repeats Thoreau's intuition that there is some force linking dividuals, something that periodically bubbles up as a sense of kinship. For Bergson as for Thoreau, "The earth I tread on is not a dead inert mass."[48] But whereas Bergson specifies the *action-style* of the common ancestor, Thoreau offers a specification of what it might have *looked like*, what its physical form, profile, or contour was. For part of what made it possible for Thoreau to notice that "something kindred" in the plants was an imperfect similarity or homoness between the shapes of his body and the shapes of vegetables. I like the idea that there might have been an attraction between Thoreau's angular phenotype and the pine needle's pointy-ness.[49] But Thoreau himself focused on another shape, that of the drop. I turn now to his fascination with that shape and its role in the operation of sympathy.

THE SHAPE OF THE DROP

After sympathy with the fluttering alder and poplar leaves almost takes away his breath, but before he detects the pine needles' sympathy for him, Thoreau reports being "conscious of a slight insanity in my mood."⁵⁰ It is raining, and he becomes "suddenly sensible" not only of his kinship with plants but of an "unaccountable friendliness" between himself and the "pattering of the drops." Drops reappear in the "Spring" chapter in the famous and drippingly sensuous scene at the railroad embankment.

One early spring day, the snowy banks of the ditch give way to ejaculates of sand, which "flow down the slopes like lava.... Innumerable little streams overlap and interlace . . . , exhibiting a sort of hybrid product, which obeys half way the law of currents, and half way that of vegetation." These drops or globules of water-sand eventually yield to "the forms of sappy leaves or vines, making heaps of pulpy sprays a foot or more in depth."⁵¹ Both the water-sand and the "sand foliage" had erupted suddenly, as if, says Thoreau, he had witnessed the bursting of the "vitals" of the earth. Here we again detect elements of a Thoreauian vitalism: On the railroad bank, Bergson might say, vital nature splays itself out, self-diversifying as it yields the repeating shapes of flesh and cellulose and much more.

Thoreau then makes the psychedelic claim that each of the different materials—clay, water, sand, vines, human flesh—are variations on the theme of the droplet or "moist thick *lobe*":

> What is man but a mass of thawing clay? The ball of the human finger is but a drop congealed. The fingers and toes flow to their extent from the thawing mass of the body. . . . The nose is a manifest congealed drop

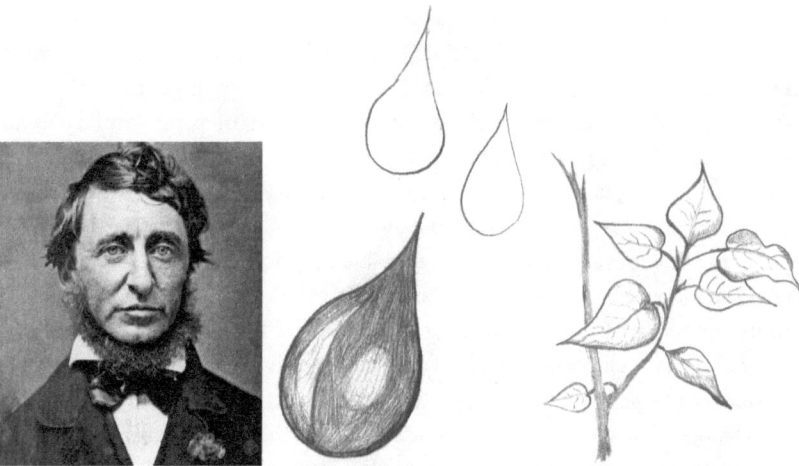

Figures 1–3

of stalactite. The chin is a still larger drop, the confluent dripping of the face. The cheeks are a slide from the brows into the valley of the face. . . . Each rounded lobe of the vegetable leaf, too, is a thick and now loitering drop . . . ; the lobes are the fingers of the leaf. (203)

What at first appears as a discrete entity—a nose, a chin, a leaf—reveals itself to Thoreau to be a congealed drop of the flow of nature.[52] (See Figures 1–3.) The drop is, one could say, the last (or first) moment in a cosmic process of individuation-deformation. The drop is a "fall" back into a well of indetermination, but it is also a fall out of the protean sky into a world of actual things.[53] In the "drop" passage we see Thoreau using shape as a figure of connectivity: Everything is a riff on the phenotype of a drop falling out of a protean, viscous substance. Or, as Bersani proposes, "All being moves toward, corresponds with itself outside of itself," in a "replication" that is both universal in scope and inaccurate in its iterability: Each thing "reoccurs [slightly] differently everywhere."[54]

DARWIN AND CLIMBING PLANTS

"Conscious of a slight insanity" in my own mood, I'd like to push this theme of a homoness disclosed by *shapes* a bit more, this time by exploring the role played by the botanical shapes of the drawings in Darwin's *On the Movements and Habits of Climbing Plants* (1875). Bergson had cited that text in *Creative Evolution*, as support for his claim that "though the plant is distinguished from the animal by fixity and insensibility, movement and consciousness sleep in it as recollections which may waken."[55] And indeed, Bergson's eschewal of a sharp, ontological division between plant and animals for a continuum of "tendencies" accords with Darwin's understanding of nature as organized according to the principle of what *The Descent of Man* (1871) called "numberless gradations."[56] But Bergson drifts from Darwin when he names "torpor" as the predominant mode of vegetal life and finds instances of plant motility to be "exceptional."[57] Bergson emphasizes the *latency* of the power of movement and sensitivity in plants, whereas Darwin emphasizes its *ubiquity*: "The capacity of acquiring the revolving-power on which most climbers depend is inherent . . . in almost every plant in the vegetable kingdom."[58]

On the Movements and Habits distinguishes three types of climbing plants: spiral-twiners (with "the subordinate divisions of leaf-climbers and tendril-bearers"), "hook-climbers, which . . . can climb only in the midst of an entangled mass of vegetation," and "root-climbers, which are excellently adapted to ascend naked faces of rock."[59] Embedded in that schema is a claim about plant purposiveness, that is, the plant's quest for "free air" pursued at a minimal of "expenditure of organized matter" and effort.[60] Darwin also distinguishes between different styles of vegetal motility. There is *"getting into position,"* as when "tendrils place themselves in the proper position for action, standing, for instance in the *Cobaea*, vertically upwards, with their branches divergent and their hooks turned outward, and with the young terminal shoot thrown on one side."[61] There is the movement of *response*, which is not an automatic reaction to light or gravity or touch, but a nondeterministic endeavoring in accord with the specificity of the encounter.[62] Here it seems that Darwin comes quite close to Bergson's *creative* evolution. Another movement named by Darwin is a botanical equivalent of the irrepressible energy of young children who *have to run around*: This is a "spontaneous revolving . . . which depends on no outward stimulus, but is contingent on the youth of the part and on its vigorous health." There are also "movements—often rapid movements—from contact with any body," as when plants, "after bending from a touch, become straight again, and again bend when touched." Finally, there is the coiling or tightening the grip of a plant's "prehensile organs," as when "tendrils, soon after clasping a support, but not after a mere temporary curvature, contract spirally."[63]

On the Movements and Habits works on the reader, then, by means of both close description of specific plants and a synthesizing taxonomy.[64] These can be understood as two "techniques of self" (to use the Foucauldian notion), each of which helps sharpen and refine Darwin's capacity to read nonhuman modes of expression and to enter into sympathy with them. These techniques have also drawn generations of plant-loving readers into the satisfying project of identifying and sorting differences and similarities. But there is one additional technique employed by the text, the botanical drawings made by Darwin's son George (see Figures 4–7).

Included also is a sketch whose authorship is shared a bit more by a plant, drawn by "the movement of the upper internode of the common Pea, traced on a hemispherical glass, and transferred to paper; reduced one-half in size" (see Figure 8).

I think that Darwin inserted these drawings in his quest to promote a stronger sympathy for—a more acute discernment of—the life of plants and the entwined lives of plants and animals (including humans). The visual presence of the spirals of tendrils or the lobes of leaves can help induce in the reader a stronger sense of the way he or she is already attached to plants. One could say that Darwin stages scenes of a materialist recognition of kinship between shapes. Look how the tendril of a young *Bignonia unguis* has the shape of a claw! It "curiously resembles the leg and foot of a small bird with the hind toe cut off" (see Figure 9).

If I add my own drawing to those of George and Pea, we can also see how the branches of the *Ampelopsis hederacea* repeat the spirals of a person's hair (see Figures 10, 11).

Figures 4–7

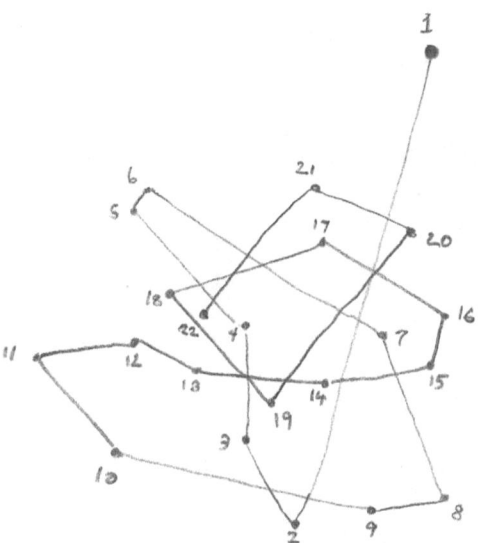

Figure 8

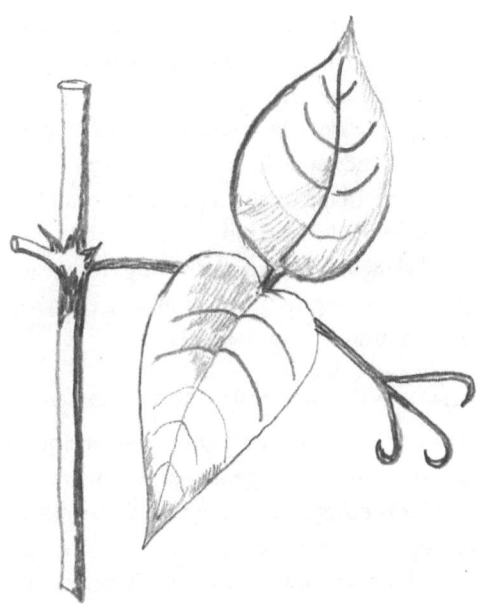

Figure 9

Figures 10 and 11

Perceived resemblances in shape can trigger the thought that there also exist some shared *capacities* across the differences between plant and animal, as when *Bignonia* brings itself "into contact with some twig, and then . . . seizes fast hold of the twig, exactly like a bird when perched."[65]

Darwin returns us to Thoreau's idea of a mode of communication that proceeds via shape. To the shape of the drop, Darwin adds these shapes: the pointy leaf, the serpentine tendril, the spiral twist, the star-flower, the root tangle, the three-pronged claw. The botanical drawings foreground resemblances in form, and this foregrounding reminds us of our everyday (subterranean) acts of affiliation and communication with plants.

ECOLOGICAL SENSITIVITY AND WINGED THOUGHTS

"It is remarkable," writes Thoreau, "how closely the history of the apple-tree is connected with that of man. The geologist tells us that the order of the *Rosaceæ*, which includes the Apple, also the true Grasses, and the *Labiatæ*, or Mints, were introduced only a short time previous to the appearance of man on the globe."[66] What Thoreau, along with Mol, Bergson, and Darwin, call us to do is to think subjectivity in conjunction with vegetality. This is both Herculean and a quite ordinary task. It is Herculean because it clashes with a powerful strain of anthropocentrism in our thinking and feeling and because we remain haunted by the image of a great, vertical chain of being, and perhaps because vegetality resists linguistic representation even more than the other aspects of our selves. It is a task of everyday life because instances of human-plant affinity abound once you become alert to them, and then there's a slight but real kaleidoscopic shift in everything you see, hear, smell, touch, taste, and think. What becomes a little more sensible, conceivable, and plausible is the existence of a web of cross-body, shapely communications that subvents or is only vaguely hooked into more word-reliant networks.[67] One may start to experience oneself less as an intersubjective being and more as an *inter- and intra-twined* shape. And that *experience* is needed if the more philosophical project is to get off the ground.

The claims about sympathy and vegetal life pursued in this essay might qualify as what Thoreau called "winged thoughts," which "like birds" do not tolerate too much handling.[68] Neither ought it be implied that sympathy can bear the entire weight of an ethos, ecological or otherwise. I think of it as one element within a larger constellation of ethical ideas and political practices. The sympathy that exists must struggle against, assert and reassert itself in response to, or wane in the face of other forces at-large, other blocs of affects and percepts, including those we might term antipathy, aggression, agon. These circulate widely, but sympathy flows too, like a scent passed from one body to another, as when "the trapper's coat emits the odor of musquash."[69]

NOTES

1. Annemarie Mol, "I Eat an Apple: On Theorizing Subjectivities," *Subjectivity* 22, no. 1 (2008): 28–37. For their comments and criticisms of this essay, I am grateful to Lisa Baraitser, Michael O'Rourke, Eileen Joy, Nigel Clark, João Florêncio, Milla Tiainen, and the other participants of the Birkbeck workshop "A Feeling for Things"; to Jeff Dolven and Brooke Holmes and the other participants in "The Secret Lives of Plants" symposium at Princeton University's Interdisciplinary Doctoral Program in the Humanities; to the participants of my summer 2013 seminar at the Cornell School for Criticism and Theory; to Rochelle Johnson, Kristen Case, William Rossi, Wai Chee Dimock, Jennifer Gurley, and the participants on the Thoreau panels at the 2014 American Literature Association meetings; and to William Connolly, Jennifer Culbert, Catherine Keller, Brianne Donaldson, Nathan Gies, Anatoli Ignatov, Katrin Pahl, Joshua Ramey, and Jenna Supp-Montgomerie.
2. This self would be, in Louise Amoore's words, a "dividuated subject . . . composed of an amalgam of elements from other subjects and objects, held together by associations." Amoore highlights how dividuation is subjectivating (in a Foucauldian sense), enmeshing human bodies in a dense web of surveillance and control

apparatuses (*The Politics of Possibility: Risk and Security beyond Probability* [Durham, N.C.: Duke University Press, 2013], 131). The eating self could also lend itself to the idea that individuality is a dividuation of an ongoing ontological or cosmological process, as in Whitehead's *Process and Reality* (ed. David Ray Griffin [New York: Free Press, 1978]) or Catherine Keller's notion of "a manifold that is not a mere many, not a bunch of single ones, not a set of single identities externally related . . . , [but a] process [that] is open-ended, *infini*, unfinished" (Catherine Keller, "And Truth—So Manifold!" *Feminist Theology* 22, no. 1 [2013]: 77–87).

3. I offer an account of distributed agency in *Vibrant Matter: A Political Ecology of Things* (Durham, N.C.: Duke University Press, 2010).

4. In using the term "transubstantiation," Mol seems to invoke a kind of alchemy whereby one substance becomes another. In *Meeting the University Halfway* (Durham, N.C.: Duke University Press, 2007), Karen Barad takes issue with the interactionism of such a model, with its image of preexisting substances that subsequently engage each other. Barad offers instead a universe of intra-action, a process more akin to the way waves engage each other and produce swells than the way lead turns into gold. For Barad, entanglements *diffract* more than *transubstantiate*. I think that both models—Mol's and Barad's—are useful for rethinking what counts as "agency": Mol's helps us to home in on instances of domination and suffering of bodies, and Barad's helps us to better attend to the ongoingness of any shape and to the new political openings accompanying a new diffraction.

5. Mol, "I Eat an Apple," 30.

6. Henry David Thoreau, "Spring," in *Walden*, in *Walden and Civil Disobedience*, ed. Owen Thomas, Norton Critical Edition (New York: W. W. Norton, 1966), 93.

7. Ibid.

8. Walt Whitman, "Song of Myself," sec. 31, in *Leaves of Grass*, ed. Michael Moon (New York: W. W. Norton, 2002).

9. Leo Bersani, "Gay Betrayals," in *Is the Rectum a Grave? and Other Essays* (Chicago: University of Chicago Press, 2010), 43–44. I am grateful to Nathan Gies for alerting me to these passages from Bersani.

10. Bersani also proposes the "ontological reality" of a world where "all being moves toward, corresponds with itself outside of itself" ("Sociality and Sexuality," in *Is the Rectum a Grave?* 118).

11. My thanks to Jenna Supp-Montgomerie for this formulation, and for underscoring this point.

12. The phrase is Bersani's in "Gay Betrayals," 43.

13. Hannah Arendt, *On Revolution* (London: Penguin, 1990), 84.

14. Ibid., 87. In "Thinking with Things: Hannah Woolley to Hannah Arendt" (*Postmedieval: A Journal of Medieval Cultural Studies* 3, no. 1 [2012]: 63–79) Julia Reinhard Lupton provides a fascinating reading of Arendt on "judgment" that may complicate Arendt's disparagement of sympathy. Lupton invokes Arendt's claim (in *Between Past and Future* [New York: Penguin Classics, 2006]) that "insight" is a key part of judgment: insight as the power by which "our strictly private . . . senses . . . can adjust themselves to a nonsubjective . . . world which we have in common and

share with others" (73). Insight here seems similar to the not-quite-willed capacity to prehend other bodies; Lupton likens Arendtian judgment to Woolley's cooking practice, which proceeds via "sympathy" and "the anticipatory alliance of hand, eye, mind, tools and materials achieved in the exercise of craft" (69).

15. Thoreau, "Spring," 206.
16. "In his biography of 1873 William Ellery Channing used the phrase 'poet-naturalist' to resolve the dichotomy in Thoreau's work between subjective interpretation and objective reporting" (Nina Baym, "Thoreau's View of Science," *Journal of the History of Ideas* vol. 26, no. 2 [1965]: 221–34).
17. Claire Pentecost, "When Art Becomes Life: Artist-Researcher and Biotechnology," *eipcp* (European Institute for Progressive Cultural Policies) 01/2007, accessed at eipcp.net/transversal/0507/pentecost/en/base_edit. See also Thom Donovan, "5 Questions for Contemporary Practice with Claire Pentecost," *Art 21 Magazine*, January 31, 2012, accessed at http://blog.art21.org/2012/01/31/5-questions-for-contemporary-practice-with-claire-pentecost/#.V-fnEUsUViF.
18. Carla Hustak and Natasha Myers, "Involutionary Momentum: Affective Ecologies and the Sciences of Plant/Insect Encounters," *differences: A Journal of Feminist Cultural Studies* 23, no. 5 (2012): 105. Hustak and Myers describe how Darwin's bodily practice of careful observation generated a stronger sympathy for plants and animals, a sympathy that in turn enhanced his ability to detect aspects of their lifeworlds (90–92). See also Michael Pollan's "The Intelligent Plant," *New Yorker*, December 23 and 30, 2013. accessed at http://www.newyorker.com/magazine/2013/12/23/the-intelligent-plant.
19. Thoreau, *Walden*, ed. Thomas, 92. Or, as Channing put it, Thoreau "admired plants and trees: truly, he loved them" (William Ellery Channing, *Thoreau: Poet-Naturalist* [Boston: Roberts Brothers, 1873], 202).
20. "At his or her best, the homosexual is a failed subject, one that needs its identity to be cloned, or inaccurately replicated, outside of it. This is the strength, not the weakness, of homosexuality, for the fiction of an inviolable and unified subject has been an important source of human violence" (Bersani, "Gay Betrayals," 43). Thus, failure is not a virtue in itself but a first step in reconstituting atomistic notions of self and agency.
21. Bersani, "Sociality and Sexuality," 118.
22. Thoreau, *Walden*, ed. Thomas, 87.
23. Ibid., 89.
24. Henry Thoreau, *Journals*, February 20, 1857, accessed at https://www.walden.org/wp-content/uploads/2016/02/Journal-9-Chapter-7.pdf. In this same journal entry, Thoreau invokes the notion of all things "being made" for each other: "What is the relationship between a bird and the ear that appreciates its melody, to whom, perchance, it is more charming and significant than to any else? Certainly they are intimately related, and one was made for the other. It is a natural fact." Is this an appeal to a divine Creator or to something more like the natural process of evolution? On the topic of Thoreau's relationship to "developmental" theories of science of his day, see William Rossi, "Transcendentalism and Evolutionary Theory," in *The*

Oxford Handbook to Transcendentalism, ed. Joel Myerson, Sandra Petrulionis, and Laura Dassow Walls (New York: Oxford University Press, 2010), 583–96.

25. Thoreau, *Journal*, Fall 1846, at Walden Pond, in *Material Faith: Henry David Thoreau on Science*, ed. Laura Dassow Walls (New York: Mariner Books, 1999, 9). My thanks to Rochelle Johnson for alerting me to this passage.
26. Thoreau *Walden*, ed. Thomas, 199, 204, emphasis added.
27. Thoreau, "Spring," in *Walden*, ed. Thomas, 207. On this point, Rochelle Johnson cites James R. Guthrie's discussion of Thoreau's fascination with the phenomenon of corn growing in the night, which was for him emblematic of the "intense unconscious" activity of nature. (See Rochelle Johnson, "'This enchantment is no delusion': Henry David Thoreau, the New Materialisms, and Ineffable Materiality," *ISLE: Interdisciplinary Studies in Literature and Environment* 21, no. 3 (2014): 606–35.
28. Henry David Thoreau, "Walking," para. 21, at http://thoreau.eserver.org/walking2.html.
29. Johnson notes how Thoreau "indicates his own *un*sound state when he fails to heed sound: that is, his own soundness, like that of nature, depends on sound" ("This enchantment is no delusion"), 620.
30. G. W. F. Hegel, *Phenomenology of Spirit*, para. 178. Given the ontology of *Geist*, the intersubjective process of recognition is also simultaneously psychological, historical, and cosmic. Theorists "who approach the concept of recognition as part of a philosophical treatment of intersubjectivity . . . often regard recognition as a ubiquitous mechanism by which meaningful social relations are constituted, deliberately or otherwise" (Patchen Markell, "Recognition and Redistribution," in *Oxford Handbook of Political Thought*, ed. John Dryzek, Bonnie Honig, and Anne Phillips (Oxford: Oxford University Press, 2006), 454.
31. In *A Thousand Plateaus*, trans. Brian Massumi (Minneapolis: University of Minnesota Press, 1986), Gilles Deleuze and Félix Guattari use "rhythm" to suggest a model of engagement that is not based on sameness or the condition of coming from the same root: "Rhythm is the Unequal or the Incommensurable that is always undergoing transcoding. . . . It does not operate in a homogeneous space-time, but by heterogeneous blocks." For them, each living thing has several "milieus," and "the milieus pass into one another, they are essentially communicating." "Rhythm" is that "coordination between heterogeneous space-times." (126)
32. This is Linda Ross Meyer's Heideggerian formulation in *The Justice of Mercy* (Ann Arbor: University of Michigan Press, 2010), 33.
33. On the latter terms, see Deleuze and Guattari in "The Refrain," in *A Thousand Plateaus*, 322.
34. Henry David Thoreau, July 19, 1850, Journal, in *Walden, Civil Disobedience, and Other Writings*, ed. Rossi, 356.
35. Walt Whitman, "Preface 1855—Leaves of Grass, First Edition," in *Leaves of Grass and Other Writings*, ed. Michael Moon, Norton Critical Edition (New York: W. W. Norton, 2002), 622.
36. Thoreau, "Sounds," in *Walden*, ed. Thomas, 83, emphases added.

37. In a journal entry of December 31, 1851, Thoreau rejects the notion that the earth is "a dead inert mass": "There is motion in the earth as well as on the surface; it lives and grows. It is warmed and influenced by the sun—just as my blood by my thoughts.... The earth I tread on is not a dead inert mass. It is a body—has a spirit—is organic—and fluid to the influence of its spirit—and to whatever particle of that spirit is in me.... Even the solid globe is permeated by the living law. It is the most living of creatures" (cited in David Skrbina, *Panpsychism in the West* [Cambridge, Mass.: MIT Press, 2005], 224).
38. Henri Bergson, *Creative Evolution*, trans. Arthur Mitchell (Mineola, N.Y.: Dover, 1998), 106.
39. Ibid., 113.
40. Ibid., 119.
41. Ibid., 109–10.
42. Ibid., 87.
43. Ibid., 27.
44. "It would be futile to try to assign to life an end.... To speak of an end is to think of a pre-existing model which has only to be realized. It is to suppose, therefore, that all is given, and that the future can be read in the present.... Life, on the contrary ... is undoubtedly creative, *i.e.* productive of effects in which it expands and transcends its own being. These effects were therefore not given in it in advance, and so it could not take them for ends" (ibid., 51–52).
45. Ibid., 113. The topic of Bergson's notion of "sympathy" (in relation to "intuition") deserves further attention here. See Brian Massumi's excellent account in *What Animals Can Teach Us about Politics* (Durham, N.C.: Duke University Press, 2014).
46. Charles Darwin, *On the Origin of Species*, 1st ed., chapter 9.
47. The late nineteenth-century movement toward a "critical vitalism," to which Bergson, Hans Driesch, Walt Whitman, William James, and others contributed, is laudable for the way it punched holes in the anthropocentrism of the dominant way of conceiving the relationship between plants, animals, and humans. It shifted the focus from the question "What is unique about humanity?" to "What is the larger Life in which humanity partakes?" But although the vitalists rightly insisted on the presence of some kind of energetic, free agency—an element of spontaneity that could not be captured by a mechanistic model of nature—élan vital is for me but a holding action. It maintained an open space that more immanent and even more naturalistic philosophies—of complexity, of process, of new materialism—could then seek to fill. For the vitalists, matter required a not-quite-material supplement, a divinely infused élan vital, to become animate and mobile; but today it is possible to invoke a materiality that needs no animating accessory but expresses itself in multiple types and modes of actants.
48. Whitman, "Preface 1855—*Leaves of Grass*, First Edition," in *Leaves of Grass and Other Writings*, 622.
49. My silly idea sends tendrils into the more respectable direction of comparative plant morphology: the comparative study of the physical shape or external structures

of plants, and "an attempt to interpret these on the basis of similarity of plan and origin" (Harold C. Bold, C. J. Alexopoulos, and T. Delevoryas, *Morphology of Plants and Fungi*, 5th ed. [New York: Harper Collins, 1987], 3).

50. Thoreau, *Walden*, ed. Thomas, 88.
51. Ibid., 201.
52. As with an LSD trip, in *Walden* discrete objects usually defined by their ordinary use or function now appear, in Dietrich Diederichsen's account of psychedelic thought, "as suddenly severed from all context" and as simultaneously "sublime/ridiculous." But while the "context" that is interhuman culture is indeed severed, an *other* context takes its place for Thoreau, that of an ecology of natural bodies and forces (Dietrich Diederichsen, "Animation, Dereification, and the New Charm of the Inanimate," *e-flux* journal, no. 36 [2012]).
53. Thanks to Jeff Dolven for this point.
54. Bersani, "Sociability and Cruising," in *Is the Rectum a Grave?* 55.
55. Bergson, *Creative Evolution*, 119.
56. According to Darwin, "If no organic being excepting man had possessed any mental power, or if his powers had been of a wholly different nature from those of the lower animals, then we should never have been able to convince ourselves that our high faculties had been gradually developed. But it can be shewn that there is no fundamental difference of this kind. We must also admit that there is a much wider interval in mental power between one of the lowest fishes, as a lamprey or lancelet, and one of the higher apes, than between an ape and man; yet this interval is filled up by *numberless gradations*" (*Descent of Man*, 66–67, emphasis added). Later in the text, Darwin repeats the point that the discerning naturalist will find that body types which "at present appear to be specifically distinct" often turn out instead to "graduate into each other by close steps" (182). A focus on similarities, gradations, and tendencies did not prevent Thoreau, Darwin, or Bergson from acknowledging the value of drawing distinctions. The naturalist-philosopher both detects resemblances/resonances and constructs categories to mark differences in function and form.
57. According to Bergson, "In the exceptional cases in which a vague spontaneity appears in vegetables, it is as if we beheld the accidental awakening of an activity normally asleep" (*Creative Evolution*, 109).
58. Charles Darwin, *On the Movements and Habits of Climbing Plants* (Cambridge: Cambridge University Press, 2009), 117. Gustav Fechner, a contemporary of Darwin, and author of *The Soul Life of Plants*, goes so far as to say that this power of movement suggests the presence of some analog of "mind" within the plant: "I cannot conceive how running and crying have a peculiar right, as against blooming and the emission of fragrance, to be regarded as indications of psychic activity and sensibility . . ." (Gustav Fechner, *The Soul Life of Plants*, in *Religion of a Scientists: Selections from Gustav Th. Fechner*, ed. and trans. Walter Lowrie, Pantheon Books, 1946, 169). Lowrie describes Fechner as a "pronounced materialist; but his materialism did not exclude spirit, did not concede even that matter and soul could be separated . . .

the one was the way things must be conceived from without, the other . . . from within" (46).
59. Darwin, *On the Movements*, 108. Associated with each class is a distinctive movement style: Spiral twiners have the simplest choreography and the fewest moving parts, with tendril bearers being the most baroque.
60. Ibid., 107–8.
61. "or, as in *clematis*, the young leaves temporarily curve themselves downwards, so as to serve as grapnels" (ibid., 115).
62. In contrast to many of his peers who understand phototropism in purely mechanistic terms, as if the movement of a plant toward the light was the same "as the elongation of a bar of iron from an increase in its temperature," Darwin argues that "seeing that tendrils are either attracted to or repelled by the light, it is more probably that their movements are *only guided and stimulated* by its action" (ibid., 116, emphasis added).
63. Ibid., 111, 117.
64. The exquisitely detailed accounts in Darwin's texts are based on hours and hours of observation. Take, for example, this scene of Darwin carefully measuring out a teeny-tiny loop of thread and tickling the plant with it: "*Clematis montana.*—The long and thin petioles of the leaves, whilst young, are sensitive, and when lightly rubbed bend to the rubbed side, subsequently becoming straight. . . . A loop of thread weighing a quarter of a grain caused them to bend; a loop weighing only one-eighth of a grain sometimes acted and sometimes did not act. The sensitiveness extends to the angle between the stem and leaf-stalk. . . . A young inclined shoot . . . made a large circle opposed to the course of the sun in 4 h. 20 m., but the next day, being very cold, the time was 5 h. 20m" (ibid., 27).
65. Ibid., 50.
66. Henry David Thoreau, "Wild Apples," *Atlantic Magazine*, November 1862. http://www.theatlantic.com/past/docs/issues/1862nov/186211thoreau.htm.
67. Perhaps this is what Achilles Mbembe calls a neo-animism. See also the "Animism" issue of the journal *e-flux*, ed. Anselm Frank (2012).
68. Henry David Thoreau, *A Week on the Concord and Merrimack Rivers* (ed. C. F. Novde [Princeton, N.J.: Princeton University Press, 1980], 339). In "Walking," Thoreau also refers to "winged thoughts": "We are accustomed to say in New England that few and fewer pigeons visit us every year. Our forests furnish no mast for them. So, it would seem, few and fewer thoughts visit each growing man from year to year, for the grove in our minds is laid waste . . . and there is scarcely a twig left for them to perch on . . . . Our winged thoughts are turned to poultry." Thoreau's concern not to seek "too great fullness & detail" in expression again appears in an August 22, 1851 Journal entry: "It is the fault of some excellent writers—De Quincy's first impression on seeing London suggest it to me—that they express themselves with too great fullness & detail. They . . . lack moderation and tendentiousness—they do not affect us by an ineffectual earnesst and a reserve of meaning—like a stutterer—they say all they mean" (in *Walden*, ed. Rossi, 357).

69. Henry David Thoreau, "Walking," para. 21, at http://thoreau.eserver.org/walking2.html. *Walden* may not be the best text for an exploration of Antipathy, for there Thoreau tends to keep even his acts of violence toward nature within the frame of friendship. In the "Economy" chapter, for example, Thoreau says, "Before I had done I was more the friend than the foe of the pine tree, though I had cut down some of them, having become better acquainted with it" (ed. Thomas, 28). My thanks to Anatoli Ignatov for this point.

◆ Tingles of Matter, Tangles of Theology

CATHERINE KELLER

Intra-acting responsibly as part of the world means taking account of the entangled phenomena that are intrinsic to the world's vitality and being responsive to the possibilities that might help us and it flourish.
—KAREN BARAD, Meeting the Universe Halfway

I believe in one matter-energy, the maker of things seen and unseen.
—JANE BENNETT, Vibrant Matter

In the context of this transdisciplinarily theological conversation, I thought one of us at least should ask: just what kind of theology might be materializing in this engagement of the so-called new material-isms? If theology as such is to survive the intra-action, will it have articulated a materialism of its own? Is theological materialism, at least of the Christian ilk, a mere contradiction in terms? What theology actually identifies itself as a Christian materialism? Despite a half-century of dialogue with Marxism, liberation theology, for example, does not wave such a banner, though it was of course dismissed from the start by its conservative opponents precisely *as* materialism. Similarly, a range of ecological and cosmological theologies have for decades insisted upon nondualist affirmations of materiality, but not on their own materialism. And although the Christian left decries the consumer capitalism of the prosperity gospel (according to which health and wealth are a blessing of the Lord), the latter's evangelists hardly trumpet their and Jesus' "materialism" as such.

The biblical heritage never after all escapes its dense and complex relation to a matter always deemed God's creation and "good." If all that stuff was later thrust down as the lower partner of a Platonized Christian dualism, still it never ceased to haunt every supernaturalism with the cloudy crowd of its materialities. No Christianity could quite throw off the narrative weight of fleshly incarnation, material justice, bodily resurrection. But what theology, I wondered, would actually insist on the term "Christian materialism"?

Imagine my surprise when this answer revealed itself: the Spanish Josémaría Escrivá, canonized in 2002 as a saint, writes that we can "rightfully speak of a Christian materialism, which is boldly opposed to those materialisms which are blind to the spirit." For "authentic Christianity, which professes the resurrection of all flesh, has always quite logically opposed 'dis-incarnation.'"[1] What progressive Christian would not assent to this sentiment? Escrivá, however, happens to be the founder of Opus Dei. While not as colorfully sinister as the Da Vinci Code version, that organization has from the start been criticized by democratically inclined Catholics for its alliance with authoritarian regimes, its misogyny, its secretive access to power. As to the movement's widespread and emphatically "materialist" practices of mortification of the flesh, particularly with the use of the cilice, the thigh clamp, or the corded whip (however creative a BDSM counterreading a queer theorist could offer) they aim at sensations quite different from the tingle referred to in my title.

Nonetheless, this mortifying Christian materialism participates in the immense biblical force field of the mattering of the world and indeed in the rich manifold of the material practices of the world's spiritual traditions. If this whole problematically complex history, with its extreme contradictions and persistent evolutions, is to come into theological self-explication, what could be more important than fresh insight into matter itself? And where better to find such rethinking than to subject theological assumptions to the challenge of the resignification of matter currently under way?

In order not to get lost in the vast history of those assumptions, I feature a version of this resignification that produced its own theological movement, that of process theology. To this end I want to pull especially Karen Barad's philosophical physics into conversation with the physical cosmology of Alfred North Whitehead. His early twentieth-century response to the (then very) new physics produced its own theological movement. It is a movement whose current liveliness is linked to its irreducibly ecological commitments. So the comparison of Whitehead's early unfolding of ontological relationalism with Barad's fresh version will, I hope, allow us to cut to the theological quick. The quickened notion of materiality, if not of materialism, that emerges from such an exchange, may help reorient our endlessly entangled bodily lives and their spirited practices. I suggest that any materialism—however new—that honors its matter will recognize itself as inextricably, if imperceptibly, linked to the whole contradictory history of religious bodies. Those contradictions congeal particularly around figures of "the body of Christ." So the recognition of the entangled history of the bodies will not diminish the differences, ethical or ontological: Rather it situates them within a "spacetime mattering" (Barad) that will carry the theological charge of "the Eros of the universe" (Whitehead) into a "polyamory of place" (Bauman). In the attempt to energize eco-ethical resonance between the impossibly diverse bodies of the world, I am hoping against hope—against hope for salvation *from* the world—for help from the deep bottom up: even from the vibrant "energy-matter" of the quantum minimum.

SUPERSTITIONS AND SUPERPOSITIONS

In meditating on mattering bodies, Christian theology confronts first of all the dense global mass of "the body of Christ." An amalgam of metaphoric and material entanglements, the history of the body of Christ—as the always redoubled figure of the church itself and of the church's founding figure—includes quite a lively mass of bodies in addition to that ubiquitous, tortured one. The formidable biblical apparatus of the creation and all its good and sinning bodies, its codes for material practice, its prophetic insistence on material—not ethereal—justice for vulnerable bodies, beyond the first and second testament spectrum of always bodily resurrections, endlessly recirculates all that mattering matter.[2] Then there come all the later practices of "material Christianity."[3] Think, for instance, of the relics, little bits of the flesh, perhaps of bones, tooth, or foreskin, of shroud or cross, of some spiritual ancestor. As the rules for moving, "translating," and then subdividing corpses relaxed, the Second Nicene Council (787 CE) prescribed at least one saintly relic inside every altar. (Someone will no doubt soon interpret the affective efficacy, the requisite physicality, of the reliquary as quantum entanglement channeled for popular piety.) Or consider the contemplation of icons, understood to be materializations, never mere representations, of the holy ones depicted: "The flesh transfigured by the icon transfigures the gaze turned upon it."[4] Contemplation of the body of Christ became participation in his incarnation, transfiguration, resurrection. The depictions of resurrection and its bodiliness shift theologically, hallucinatorily. In the twelfth century the resurrection of the dead means "the regurgitation or reassemblage of exactly the body we possess on earth." So in one depiction, for instance, "those who rise include skeletons still lying prone in their coffins, bodies emerging from sarcophagi entangled in shrouds, and body parts vomited up by birds, bees, and fish."[5]

Of course, these iconographies and these practices arise amid the incense and music, the sacramental rituals performed in multisensory drama, or the homey habits that compose so much of every religion.[6] The particularly Christian bodies multiply them with rosary beads, the founding of hospices and charities, peasant revolts, apocalyptic communes, reactionary counter-revolutions. On and on, Catherine of Siena and Angela of Foligno kiss the lepers' wounds, and Saint Francis preaches to the birds. The endless pilgrimage, the milagros, healing or weeping Guadalupes, the layings-on of hands—these hardly end with the Middle Ages.[7]

Superstitions or superpositions? At any rate these holy matters are nothing if not vital. Yet the matter itself—tongue, wound, incense, icon—was not normally understood by theologians to be inherently alive or meaningful, so much as enlivened by invisible forces. If explicated, the metaphysics would be more dualistic supernaturalism than, say, Jane Bennett's naturalist version of vitalism as "heterogeneous monism."[8] The enchantment of life is conferred from above. But first of all these were practices, not theories. And there would be dense parallels in every spiritual tradition, including all those the church suppressed as idols.

Within Christianity all these numinous materializations thinned out rather dramatically with the Reformation, which marked this density of material practice as works righteousness, papist superstition, and Catholic idolatry. Of course, Protestantism did not expel the goodness from the creation, salvation from incarnation, or sacramentality from the edible body. But the way was being vigorously and inadvertently cleared for the (then) new materialism of natural science, and therefore for the whole modern history of the bifurcation of science and religion, of matter and meaning. Religion appeared to back slowly and spasmodically into the private sphere, its social, political, and sacramental bodies ever more immaterial; while science left "meaning" to religion for hoi polloi, to philosophers for the few. It focused instead on the facts of the matter at hand. And as mere matter of fact, matter ceases to matter spiritually or ethically.

The repressed returns, as it often does, in conservative gestures. But the new-old orthodoxy of Christian materialism does not only take the reactionary form of Opus Dei or of the pro-capitalist prosperity gospel or of the natalist neovitalism of the "culture of life" that Bennett tracks.[9] It has also taken, for example, the canny form of radical orthodoxy, as in John Milbank's argument that "materialist materialism is simply not as materialist as theological materialism." That former tautology would refer to secularized matter. The "alternative Catholic metanarrative" he champions features the human as "eroto-linguistic animal." It seductively celebrates the whole medieval aesthetic sensorium washed out by Protestantism. And it derives its materialism (unseductively for any feminist theology) from the "absolute truth" of the Incarnation of God the Son, which mediates the Father's transcendence and so "enchants" material reality through the "Trinitarian interpenetrations." Milbank's debate in this instance is with the Hegelianism of Žižek, which he finds "a sad, resigned materialism which appears to suppose that matter is quite as boring as the most extreme of idealists might suppose." Žižek, the self-designated "Christian atheist," loves this debate. He responds, "Yes, if by 'materialism' one understands the assertion of material reality as fully ontologically constituted, 'really existing out there,' which I emphatically do not: the basic axiom of today's materialism is for me the ontological incompleteness of reality."[10]

As much as I do *not* love this conversation, Žižek's specific retort is right on target for our present discussion. He celebrates a materialism that undermines the mere (indeed boring) matter of fact "from within, just as quantum physics, for example, undermines our common notion of external reality: beneath the world of simply existing material objects we discover a different reality of virtual particles, of quantum oscillations, of time-space paradoxes, etc., etc.—a wonderful world which, while remaining thoroughly materialist, is . . . breathtakingly surprising and paradoxical."[11] Žižek's version of the new materialism falls from this angle of his parallax closer, in other words, to the present transdisciplinarily theological conversation than to that of the orthodox Christian materialism he dialectically engages. But here I want to borrow from him only the quantum allusion, as it springs into play against the claim of a Christian materialist theological orthodoxy. The little quantum bristles in Žižek's countering version of Christian materialist discourse also with resistance to a capitalist economy that would master all the matter of the globe, down to the genetic codes. Such a capitalist materialism (identifiable neither with Milbank, Žižek, or "new" materialism) indeed reproduces matter as preexistently "out there," fully determinate or determinable. Of course, Marxist materialism helped crack open history, to confer subjectivity and vitality on the material bodies of workers, and to collapse the superstructure of supernaturalism; but it did little to challenge the closed determinism of Newtonian matter. And those who can be called new materialists do not align themselves with the matter of any totalizing orthodoxy. They are not therefore apolitical or indeed opposed to historical materialism, but entangle economics in a political multiplicity—ecological, sexual, racial—and so embody resistance to the newer capitalism of neoliberal economics.

In a universe of incompletion, quantum indeterminacy—if we may build already on Karen Barad's reading of Niels Bohr—destabilizes the entire apparatus of substantialist matter and its metaphysics. That metaphysics of substance dominated medieval scholasticism, however complicated or decorated by the myriad material practices of the religious. And it learned to do without any God-essence in modernity. The independent substance of a mastermind, observer, and finally owner takes center stage, with the substantial properties that are properly attributed to persons of substance. But the transcendently objective posture continues to mimic the Unmoved Mover of all matter, selfsame and impassive, at His masculine and ghostly remove, even when He comes as Logos clothed in the flesh of a single and exclusive materialization.

Perhaps then the appearance of the innocent little quantum in big debates about science and theology is neither new nor innocent; and among numerous publics, some present readers included, it still tingles with interest.[12] Will it point us—beyond the relics of antique material practices, beyond the orthodox reactions—to the really new, the *right*, materialism? Or rather to an uncertainty, indeed an indeterminacy of discourse, where the very name "matter" seems to mirror and mock the name "God"? In the meantime the phenomenal quantum arrives interlaced with a theological tradition uninhibited by orthodoxy, one that loops back—*reculer pour mieux sauter*—through work contemporary with and not unrelated to that of Bohr.

THE MATTER OF SIMPLE LOCATION

The mysterious quanta of energy have made their appearance.

ALFRED NORTH WHITEHEAD,
Process and Reality

It was the enigmatic attraction of the barely beginning quantum mechanics, together with the newly established principles of relativity, that provoked the mathematician Alfred North Whitehead to shift course in the mid 1920s. The math of the quantum delivered him such a dramatic challenge to commonsense substantialism that he began to write a speculative cosmology. He read substance metaphysics as the tacit presupposition of science from Descartes and Newton on. He articulates the challenge at the quantum level, where "mere endurance," which characterizes the "undifferentiated sameness" of substance, collapses into "the vibratory streaming of energy."[13] With this "dissolution of the quanta into vibrations" he zooms in on the common sense that formed the unconscious foundation of modern science, facilitating in the two centuries following Newton its triumphant freedom from questions of value or meaning.[14] To the ancient question "What is nature made of?" Whitehead writes, the modern "answer is couched in terms of stuff, or matter, or material—the particular name chosen is indifferent—which has the property of simple location in space and time. . . .What I mean by matter or material, is anything which has this property of *simple location*." Matter is present in a simple sense that "does not require for its explanation any reference to other regions of space-time."[15] "Simple location" characterizes for him the Newtonian atom, the massy, impenetrable, separate stuff that is what and where it is independently of its relationships. Such simplification is now read as an act of unrecognized abstraction by a dissociated observer. It is a symptom of the larger presumption that he famously names "the fallacy of misplaced concreteness."[16]

Whitehead could very well have then written the following: "We are not outside observers of the world. Neither are we simply located at particular places in the world." But in fact that happens to be Karen Barad, announcing that "we are part of the world in its ongoing intra-activity."[17] However, she writes in the name of matter itself, and shines as a star in the new galaxy of new materialists. And he reads "matter" *as* the content of the fallacy, and materialism as its perspective.

In other words, I've got a problem, semantic at the very least. I do not come to the present conversation apart from the Whiteheadian cosmology in its impact on both a margin of physics and a wider swath of theology, particularly feminist, ecological, economic. And from his point of view the fantasy of the "quiet extensive stone with its relationships of position and its quality of color" has yielded the metaphysical concept that has "wrecked the various systems of pluralistic realism." So materialism here—in this early Jamesian voice for a cosmic and not only social pluralism—appears as an unconsciously metaphysical antipluralism. In the light of quantum physics, he writes, "This materialistic concept has proved to be as mistaken for the atom as it was for the stone." As "these quanta seem to dissolve into the vibrations of light," materialism itself begins to dissipate.[18]

In the interest of a realistic pluralism today, *theologically* if indirectly kin to the "agential realism" unfolding in Karen Barad's mattering matter, I suggest it will be worth bearing with the apparent contradiction. The discursive tension only mounts in the enchanted buzz—so resonant with Whitehead's vibratory universe, of Jane Bennett's "vibrant matter." It forms the other stellar manifestation of the new materialism. Yet Whitehead wrote an entire alternative to materialism and its matter—notions that in his view ninety years ago seemed fundamentally frozen into both the dualisms and the monisms of substance. The alternative for him was neither idealism nor theology. He called it the philosophy of organism, in which every creature, or actual entity, including most expressly all those traditionally called inorganic, is an "event" of interrelation. A *live* event: It *is* "an actual occasion of experience"; it "feels" or "prehends," usually with no consciousness, the world from which it comes, the world of which it becomes. Every quantum of energy is thus metaphorically describable as a "throb of emotion" or a "subject of experience"—a "vibratory organism." Such a subject is not describable as an enduring entity unfolding in time, "having" experiences and attributes, but an occasion of becoming in relation:

> Every actual occasion exhibits itself as a process: it is a becomingness. In so disclosing itself, it places itself as one among a multiplicity of other occasions, without which it could not be itself.[19]

Whitehead's actual occasion thus transmutes the enduring substance—*res cogitans* or *extensa*—of separable individuals into the relationally constituted moments of becoming. This "becomingness" instigates a radical shift of ontology. Relation is no longer external. It is a matter, indeed a materialization, of "mutual immanence" or of "internal relations" constituting emergent subjects (superjects) rather than of attributes possessed by substances. For Whitehead, every subject—quantum or queen—experiences, feels, and responds spontaneously to its world. Each process of becoming counts as a responsive materialization of its world. In other words, relation no longer signifies an interaction between beings that exist before the interaction itself.

The resonance with Karen Barad's innovative language of "intra-active becomings" is dramatic. She has also, and no less radically, composed—drawing on a different, later and largely Continental philosophy—a full-fledged "relational ontology" as the basis for her "posthumanist account of material bodies." These bodies do not appear as classical agents, merely *inter*acting with their objects from the outside. "Rather, phenomena are the ontological inseparability/entanglement of intra-acting 'agencies.'" For her the "ontological inseparability" of the agencies of a becoming intra-action is often pictured as the scientist, the apparatus, and the quantum. (She stays much more consistently than Whitehead within the scene of physics.) The point is that the phenomena are "relations without preexisting relata."[20] I find Barad's work to be a great uplift and update of an ontological relationalism that comes already entangled in the phenomena of process, ecological, and feminist theology.

If Whitehead did not anticipate the neologism of intra-action, interactivity for him signifies always the mutual immanence of becoming occasions, which are not what they are in abstraction from one another—except in the mirage of "misplaced concreteness."

Far from simply located, every event in its unique perspective here and now is involved one way or another, significantly or trivially, positively or negatively prehended, in every other. The singular perspective in space-time is a momentary decision—it does not establish any boundary of mere nonrelation to anything. Meditating on the mysterious quanta, we find that the undoing of simple location registers its spookiest logic. "In a certain sense, everything is everywhere at all times. For every location involves an aspect of itself in every other location."[21] In a "certain sense"—as mainly enfolded in potentialities, most of which will remain forever irrelevant, "negatively prehended." But nonetheless the relativity of every location for the appearance (the phenomenon) of each actualization can fix no final boundary.

No wonder physicists such as David Bohm, Henry Stapp, and Shimon Malin, seeking support for a paradigm to help make intuitive sense of quantum entanglement, found in this early proposal philosophical aid and succor. Whitehead would in *Process and Reality* formulate this radically nonsimple location as his "ontological principle": "Everything is positively somewhere in actuality and in potency everywhere."[22] This potentiality, suggestive of the quantum potential or vacuum, also anticipates the French quantum physicist Bernard d'Espagnat's snapshot of a stone: "Its 'quantum state' is 'entangled' (this is the technical word) with the state of the whole Universe."[23] This is far from the stone we see as so solidly and stolidly—there. Such simplification is an abstraction, not necessarily wrong at all—until we mistake the abstraction for the concrete.

Whitehead does not apply to that stone the language of "quantum entanglement," the phrase which Schroedinger coined only a few years later. But for the purposes of the resonance machine building up in the name of the new materiality, it is not insignificant that William Connolly's book *The Fragility of Things* has a stirring chapter called "Process Philosophy and Planetary Politics." He draws therein from Whitehead a specifically *quantum* account of our vulnerable worldwide entanglements. "If Whitehead were writing today he would doubtless say that the fallacy refers in the first instance to those who still ignore that mysterious process by which two 'particles' separated after having been adjacent, now shift together simultaneously, even when at a great distance from one another."[24]

We witness here the entanglement of quantum theory, by way of Whitehead (with a little help from Nietzsche), in the energized current of a politics intensively sensitized to the matter-energy flows of the planet. There emerges the possibility of a cosmopolitics with *cosmos*.[25] Connolly's illustrations are germane:

> For Whitehead, misplaced concreteness means more broadly the tendency to overlook entanglements between energized, real entities that exceed any atomistic reduction of them, as when a climate pattern and ocean current system intersect and enter into a new spiral of mutual amplification, or when a cultural disposition to spiritual life befuddles the academic separation between an economic system and religion by flowing into the very fiber of work motivation, consumption profiles, investment priorities, and electoral politics.[26]

That "cultural disposition to spiritual life" might signify what he earlier and famously calls the "evangelical-capitalist resonance machine."[27] That would characterize the prosperity gospel and its political collusion with its ecumenically right-wing moral agenda. Or to the contrary, it may engage a theologically friendlier spirit of "existential faith"—such as the sort highly developed along the left flanks of theology, where process and liberation theologies have long labored—conducive to the cosmopolitics of a positive resonance machine.

It is in the interest of that planetary resonance—its "machine" read now as an intra-active Baradian agency, of course—that transdisciplinary theological engagements of the kind that this very volume represents proceed. The very notion of a bounded academic discipline is always already deconstructed by such an agential manifold. It is also for the sake of gathering such an ethico-political, ecotheological materialization—call it *ecosmopolitics* for short—that I foreground here the affinity between projects of becoming relationalism, and particularly between Whitehead and Barad. That Barad does not refer to Whitehead is just as well. The process vocabulary may exact too high a hermeneutical price. And the differences remain significant between these relational ontologies. For instance, although intra-active agency supersedes any individual agents in Barad, Whitehead's actual occasions count as individuals, albeit only as momentary events of relation. These are different strategies, equally committed to undoing individualism at every level.

And such a comparison only highlights the power of Barad's creation of a new interpretive apparatus, with current physics and with a timely range of poststructuralist and feminist theory otherwise lacking any explicate ontology. She notes that even Bohr, her primary source, "never spells out his ontological commitments."[28] This she does magnificently and in a way that attributes responsive value to every creature, including a particular starfish-like echinoderm. The brittlestar, a brainless creature with astonishing optic capacities, "may not get full credit for its superior ingenuity, which exceeds the current technological ingenuity of humans."[29] Her point is that in its enfolding parts of its environment and expelling parts of itself to the environment, its apparatus is an active part of the space-time manifold. It does not "have" but "is" an eye.[30] "Brittlestars are living testimony to the inseparability of knowing, being, and doing."[31] Her radical redistribution of mindfulness (like Whitehead's "mental pole" at work, usually unconsciously, in every actual occasion) cannot be written off as a romantic vitalism. Not that Bennett's work will any longer permit the dismissal of the ancestral observers of "vital things." This emergent sense of a knowingness entangled in the least conscious of creatures will take ever-new forms. The differential inseparability of knowing, being, and doing yields a rigorous rereading of matter itself.

What I have called elsewhere "entangled difference" demands an irreducible pluralism: Only so does the space of intra-action resist a monist reading. Indeed, difference here produces a bottomless ethic. For the responsiveness of things cannot be abstracted from its most extensive ethical "diffraction": "We are not merely differently situated in the world; 'each of us' is part of the intra-active ongoing articulation of the world in its differential mattering."[32] Does such matter correspond to that which stands accused of the fallacy of misplaced concreteness? Au contraire, does it not displace the dysrelational abstraction that poses as the concrete matter of (a not new) materialism? At any rate the mattering of matter is dismissed—as a pun or a projection—at the peril of everything that really does matter. For humans or brittlestars.

But for God? Developing the emergent conversation on materiality by means of such a Whiteheadian intra-action with Barad might be left to an arcane dissertation, were it not for two developing scholarly publics, one Deleuzian and the other theological. The first consists of the expanding circle of a Deleuzian-cosmological constructivism, exposing the extent of Whitehead's influence on Deleuze, and being now drawn into the philosophy of science by Bruno Latour and Isabelle Stengers.[33] Stengers especially funnels an unforeseen cone of Continental philosophical respect—after the sheer acosmology, the anti-metaphysics, of poststructuralism—toward a process-relational ontology. Of course, this all depends on the atheist street cred of these interpreters, who can thus dispel Whitehead's theological aura. Also the metamorphosis within poststructuralism toward "divinanimality"—attention to the nonhuman in its animal registers and its theological figurations—has begun, along with developments in phenomenology, to dislodge Continental anthropocentrism. As to the theological public: Especially in its Protestant forms it had remained word-centered, lacking all interest in material cosmology. In this it paralleled the philosophical preoccupation with language and culture. The marginal but tenacious web of Whiteheadian theologians has in the meantime been weaving a startlingly planetary public of process thinkers. Perhaps because of its ecological prescience, it can no longer be successfully silenced as heterodox or unhip in the theological mainstream.

It was in its pluralist redistribution of incarnation that process theology had first captured my interest.[34] Amid its "democracy of fellow creatures" (Whitehead) any event of materialization may carry sacramental significance. At the same time, its constituent relationalism was key, as I found early on, to articulating a feminist alternative to the separative ego. Of course, I admit that the metaphor of deity as "Eros of the universe," its persuasive lure displacing omnipotence, was the first seduction. It continues to destabilize the truth-power authorized by every religious absolute, including the materialism of radical orthodoxy. Process theology now collaborates across these registers with an expansively embodied relational pluralism.[35] In its avowal of the intra-activity of theology with the multiplicities of discourse, religious and irreligious, that have come entangled in theology itself all along, process theology joins a destabilizing "polydoxy."[36] In this, process theology has steadily advanced a pluralist practice of the nonseparability of religious from other, and not least of all from scientific, practices.[37]

Whitehead considered it the primary task of Western philosophy to heal the modern split of science and religion. He considered civilization at peril in its ongoing bifurcation of fact from value, matter from meaning. In other words, recognition of an ongoing entanglement of modern science in premodern theology—not any return to a presecularized worldview—may be called for. Non-European civilizations and indeed the Greeks demonstrated mathematical and cosmological genius; but why then, asked Whitehead, did experimental science develop only in the modern West? Empiricism itself betrays "the inexpugnable belief that every detailed occurrence can be correlated with its antecedents in a perfectly definite manner, exemplifying general principles. Without this belief the incredible labours of scientists would be without hope." Their labor, in other words, is motivated by the conviction "that there is a secret, a secret which can be unveiled." Materiality is not just illusion or distraction. This conviction stems, he argues, from "the medieval insistence on the rationality of God" as a kind of hybrid of the "personal energy of Jehovah" with "the rationality of a Greek philosopher." His point was that for good and for ill, the scrutable world of modern science "is an unconscious derivative from medieval theology."[38]

We might add that this trust in the unveiling of the secret derives from the original narrative revelation—"it is good"—with respect to each space and species of the whole intra-active creation. So then an ethico-aesthetic motivation infuses the epistemic trust: Matter is good, worthy, engaging. The scrutable thus implies the scrupulous. Its knowability does not entail its transparency so much as its relationality: as "Adam knew Eve." The metaphor of incarnation for a time intensified this trust. Yet the ontological dualism that soon captures Christianity so orders the God-world relation as to keep incarnation the exception that proves the rule—of external interaction. Indeed, it proves the dominance of an eternally preexisting being over a world of substances preceding their relations to each other. That supernatural dualism underlies the Cartesian substances on which, through Newton's atomic substances, the subjects and objects of pre-quantum science subsist.

In the liminal transition between the medieval and the modern another kind of science, inseparable from an alternative theology, was at least possible. I consider elsewhere the road not taken, as exemplified in the cosmological implications of Nicholas of Cusa's *docta ignorantia*, the "learned ignorance." Here such secrets as the infinite universe, with an earth neither at its center nor fixed, indeed with no fixed center, are revealed a century before Copernicus.[39] God the *complicatio/explicatio* enfolds as infinity the material manifold, which unfolds in free, indeed agential, inseparability: "All is in all and each in each."[40] Whitehead himself makes no reference to Cusa. I am here only suggesting that a dense force field of materiality pushes at once through medieval popular practice and cosmological speculation, not successfully contained by the disciplining dualism. It makes possible the modern science that the church also comes to oppose, in both the Reformation and Counter-Reformation repression of cosmological speculation (witnessed in the martyrdom of Cusa's follower Bruno).

If in response to the "secrets which can be unveiled" through relativity and quantum mechanics Whitehead takes up the thread of cosmology, the problem remains: He is arguing that the substantive "matter" signifies the fallaciously simple location. Matter suggests a solid, preexistent stuff. Actuality is for him actualization: it is embodiment; it is actualized as and in spatiotemporal bodies. Matter remained for him too static and stuffy a notion to capture the vibratory relationalism of any photon, atom, or scientist engaging them. Perhaps, however, nearly a century later, the quanta have given the noun "matter" a new chance.

Be that as it may, we might agree that Whitehead rejects materialism for approximately the same reason that the new materialists espouse it: to enact mindfulness of the live processes of *materialization*, minuscule and far-flung, of which we are a part, and for which we are always partially accountable. Process thought shares with the new materialists the sensibility of a vibrant and agential realism, unfolded as a full ontological relationalism. We cannot account for *any* reality apart from giving a scrutable and scrupulous account of ourselves—inclusive of the contextualities, opacities, and indeterminacies that shadow our becoming.[41] Whether you want to tag every vibrantly, queerly, or mindfully embodied relationalism as materialism will depend on your own shifting context of discursive accountability—on what language may enliven the ecosmopolitics of your context. Of your matter-energy.

APOPHATIC MATTER

For my part, I might experiment with the possibility of an *apophatic materialism*. The ancient mystical apophasis, the "unsaying" that characterizes negative theology, has recently come back into play, as theology recognizes the apparatus of its own inescapable contextualism. But it is never just "God" that comes unsaid. For there seems to be something hard to say, something that constantly comes unsaid, about matter itself—something revealed by the quantum phenomenon. Newton's atom was created in the faith "that God in the Beginning form'd Matter in solid, massy, hard, impenetrable, moveable particles."[42] If it has since morphed into the nuanced materiality of the quantum, matter itself turns woozy, wavy, wayward. Energy remains nonidentifiable with matter: Matter and energy translate in and out of each other; they form matter-energy; they are not the same "thing." And now dark matter is incomprehensibly different and more (almost eight times more) than baryonic matter—what had been understood to be matter—and dark energy differs from both as the sheerly unknown. A new common sense emerges: that "we live not in a cosmic clockwork, but in a cosmic network, a network of forces and fields, of nonlocal quantum connections and nonlinear, creative matter."[43] Scientific challenges to the received meaning of matter—and so to what materialism has meant—cross the spectrum of physics, chaos, complexity, biology, neuroscience.

If I can still rarely hear "matter" the noun, the substantive, without a clunk of reduction, a dissociation from the verbs, from ethical or existential mattering, from the engaging process of materialization, this hardness may be softening. Any supposedly new materialism will have deconstructed the fallacious concreteness of that lifeless, separative stuff. The naming now comes accompanied by fresh and vibrant language—and so always by an *unnaming* of what usually is named "matter." Such materialist apophasis doesn't make matter any less real (no more than theological apophasis makes God unreal). On the contrary it is precisely the felt reality, the affect of *mattering*, that causes language to shudder. The very intensity of our *mindful* relation to matter renders received language for it inadequate. And this happens with particular vividness among those hovering close to the phenomenon. In science the quantum phenomenon has provoked great bursts of apophatic incomprehension. And none knew this unknowability more than Bohr, laboring over *"this terrible riddle* which the quantum theory is."[44] Or as Richard Feynman put it: "There was a time when the newspapers said that only twelve men understood the theory of relativity. I do not believe there ever was such a time. . . . On the other hand, I think I can safely say that *nobody understands quantum mechanics.*"[45]

So you see how the problem of naming "matter" has come to rather hilariously mirror the problem of naming "God." *Si comprehendis non est deus* (Augustine). And from the seventh-century Pseudo-Dionysius on, the apophatic margin undid the language of God as any kind of "being." Eckhart captures it famously: "If you love God as he is God, as he is spirit, as he is person and as he is image—all this must go!"[46] But as with the quantum science, the knowledge of this unknowability, the knowing ignorance, does not repress, inhibit, or lose its subject "matter"—it produces new openings, new kinds of knowing.

Apophatic matter has been considered under the sign of "apophatic bodies."[47] It enacts one register of the transdisciplinary theological perspective of apophatic entanglement. Drawing on the ancient antecedents of the *docta ignorantia*, a theology that unsays any metaphor or idea of God, the apophatic entanglement unsays any certainty as to the *creatures* as well. For *in their material entanglements*, in their intra-active complexity, they "decay with imprecision, will not stay in place, / Will not stay put."[48] They refuse to remain the knowable, predictable, or controllable subjects of each other. Let alone of a knowably controlling Creator who always already knows–it-all before it comes to pass. The language of spirit, being, divinity, God, infinity, *posse ipsum*, possibility itself can, possibly—from this theological perspective—be otherwise and meaningfully activated, even after the death of God, the deaths of so many gods. But when the icons of an unknowable infinity cease to undergo their own iconoclasm—they freeze to idol.[49] And mask the entanglements of their own becoming.

Not just a mirror-game between the apophatic God and the apophatic matter is in play, but a chiasmus: The infinite folds in and out of the spontaneously materializing intra-actions. God as *complicatio/explicatio* nicknames that very enfolding and unfolding, is embodied, broken up, multiplied by it. There appears a crossover between negative theology and affirmative materialization in and as a world, which crosses back as the apophasis of matter at its mysterious limits, to all the affirmative bodily God-names of every tradition (mothers, fathers, lovers, daughters, sons, doves, snakes, fire, cloud, planets . . .). The interrelation of God and world in the ancient model swerves in Cusa's early modern experiment into an almost unprecedented formulation of material intra-activity: "Since the universe is contracted in each actually existing thing, it is obvious that God, who is in the universe, is in each thing and each actually existing thing is immediately in God, as is the universe."[50] So then an apophatically entangled theology may be said to diffract the entanglements of matter itself: for example, between earth and its "heaven" of subtle atmosphere of shifting parts per million of CO_2, between Christianity and its Jewish and pagan ancestors, between the apparatuses of science and those of religion, between the apophatic infinity of Cusa's cosmos and the incomprehensible infinity of infinities of the current multiverse.[51]

In such a God-mattering or matter-divining chiasmus the experimental metaphors of process theology may now benefit from such an apophatic gesture, preventing them from congealing into some ontotheological superentity. (We don't want God to misplace God's concreteness.) At the same time the ancient Neoplatonic lineage of apophasis needs the Whiteheadian figure of divinity as an open process of becoming, not simply as a free agent but rather as indeed intra-active energy entangled in, not determining in advance, all the creaturely becomings of the universe. Such a *theos* would be, inasmuch as it would be, always *materializing*. It was Whitehead's student Hartshorne who offered the trope of "the universe as the body of God."[52] The "most moved mover,"[53] affected by every ripple of matter, here displaces the immaterial changelessness of the Unmoved Mover.

"One matter-energy, maker of all things visible and invisible": Bennett's ironic mimicry of the Nicene Creed lands far from Nicaea but rather close—apophatically speaking—to such a chiasmus as this: "It is as true to say that God creates the world as that the world creates God." In other words, in its transdisciplinary apparatus the *theos* becomes determinate only outside of our constructive relation to it—not unlike the way the phenomenon remains inseparable from Barad's "agencies of observation." Of course, here one is not observing through measuring devices but through material practices by which one becomes an *observant* Hindu or Catholic, equipped with whichever apparatus of participation enhances that observance. And one then comes also equipped, in the more honest cases, to take responsibility for the mattering effects of one's observant agency (in the sun dance, the exodus, the mandala, Torah, gospels, the circumcision, incense, stained glass, the celice, the relics, the cantatas, the yoga mat, the march on Washington, the LGBTIQ wedding, the Black Lives Matter church marchers, etc., ad infinitum).

But in theologies supplemented by new materializations we discern animation in the materials themselves, even in the invisibility of the photons coming through the live molecules of wine and cup into the colorful display of your sophisticated retina—in the bread, flesh, sex, paper, tongues, community, in the vibrancy released or repressed of the participants. If some ultimate *complicatio* or *anima mundi* comes entangled in all of it, it would not be to confer life on a world external to it, let alone to create it ex nihilo, but rather to partake of the unpredictabilities and creativities and precarities of that shared life—as its own.

POLYAMOROUS PANENTHEISM

Okay, so I have talked about God, or whatever God nicknames. Otherwise, I would feel transdisciplinarily disingenuous. But let me make something else clear: The theological metaphors of Whitehead's thought—as, for instance, his "poet of the world"—do not turn up in the work of the quantum physicists he has influenced. Physics retains its distinct disciplinary register, which we may observantly *entangle in* theology but which does not thereby spookily at a distance, by superposition of the Holy Ghost, materialize the deity.

As the physicist Henry Stapp says, "Even if we discount the gods of various religions, it seems difficult to imagine how idea-like realities could emerge from a world completely devoid *of any such aspects*, and how physical laws could come to be fixed by a purely physical mindless universe."[54] He says "gods" only in this unsaying way, which does at the same time question any unquestionable atheism, along with the "purely physical mindless universe" that is pretty much the bottom line for any *old* materialism. He is thinking about the observer problem from a fresh point of view. He elsewhere notes that the way all actual occasions in Whitehead might be said to "observe" (nonconsciously but feelingly, and in Stapp's language "mindfully") obviates the absurd solipsism, which would have the moon exist only, as Einstein feared of quantum mechanics, when a human looks at it.[55] Whether in collapsing the wave function into the particle, or perceiving the wave as particle, each particular creature appears only in its observance of the other observant creatures. Stapp infers speculatively a whole "psychosomatic universe."

There is in this line of anachronistic thinking some echo of old vitalist or panpsychist countertraditions. Yet Whitehead's vocabulary is a bit different, as he reserved "psyche" as well as consciousness or mind for highly complex animals and considered vitalism still a dualistic compromise. But as noted above, he found not simply life, but a nonconscious "mentality" or awareness that prehends or feels in every quantum of matter. His "subjective aim" bears comparison with the "entelechy" that Bennett has found in the vitalism of Hans Driesch.[56] The process theologian David Griffin, who has written on the reenchantment of nature and of science, suggests "pan-experientialism." If some vitalist philosophies more closely approximate a pure immanence, or pantheism, than the pan*en*theism by which process theology sometimes labels itself, these are not mutually exclusive intuitions.[57]

For a theology of apophatic entanglement, God and matter name, indeed materialize, *different* but not *separable* becomings. Difference signifies not separation but relation. So theology cannot properly divide its deity from the matter of the world, or, for precisely that matter, panentheism from pantheism, without violating the entanglement that lends both "pan" and "theos" their flares of meaning. And by the same token, transdisciplinary explorations of natural science, or of "vital materialism," may fruitfully avoid all theistic avowal without thereby severing relations of historic, poetic, and ethical affinity to certain theologies.

This is not just a matter of polite exchange between disciplines, but of the creatively inseparable difference that what, for instance, Bennett develops as the political ecology of vital matter will require of us all. The apophatic element I hope contributes at least a theological hospitality to the uncertain reaches of pluralist engagement. It keeps scrutability scrupulous rather than smug. It may also serve, in the face of a still hopefully indeterminate planetary future, the responsible determinations needed for ecosmopolitan imperative. Thus Whitney Bauman concludes *Religion and Ecology* on this profoundly and becomingly apophatic note:

> Built into both science and religion is a sense of the unknown and ever changing process of becoming life: the truth regimes of science and religion are always changing and changing the worlds in which we live. When these truth regimes are taken to be reality *en toto* or as closer to nature *en esse*, then violence is perpetuated on the becoming planetary community.[58]

That essentialism, which he deconstructs with the help at once of polydoxical religion and queer theory, perpetrates the fallacy of misplaced concreteness across all the straitlaced regimes of disciplinary power. Those regimes impose the impermeable boundaries of the natural sciences, of Christian orthodoxies, of identitarian politics. Bauman destabilizes the ground they stand on by "queering nature"—in the full ecology of its cosmos. Then, with his sense of the unknown, which apophatically energizes the theological practice of his planetary ethic, Bauman is able to propose, far from the stasis of classical theism or materialism, a "nomadic polyamory of place."[59] That traces precisely the places of an apophatic materialism.

In other words an observant panentheism cannot do its work without periodic revitalization of its apophatic matter. That work now insists on the planetary ethic that locates all of our perspectives at the precise intersection of the microcosm and the macrocosm (where we meet the universe halfway). The quantum infinitesimals and the cosmic infinities entangle each other as the "great work" (Thomas Berry) of our planetary cosmopolis: Opus Earth. And that opus involves sometimes setting forth recognizably Christian alternatives to the unquestionable truth-regimes of the religious right (such as Opus Dei, Pat Robertson, Michele Bachmann, and Senator Inhofe). This cannot work as a simple opposition. For the certitudes of conservative Christian materialisms seem to arise in compensation-reaction to a certain disembodied liberalism, to the vacant immateriality and fading vitality characteristic of so much mainstream religious and secularized faith. The reactive formations may be perilously misguided, but they do react to a real problem. Yet, they do so within the terms of an ancient supernatural dualism, in which the divine participation in matter matters only by way of the one exceptional incarnation. Which then simply proves the rule of a dualistic transcendence. And in that gap between God and world there grows the standard exclusivism (*nulla salus extra ecclesiam*, Jesus saves, etc.). Whether in the fading mainstream or the empowered right, the christomonism feeds on the sense of the supernatural as what *makes* the natural and alone makes it worthwhile, and which alone makes humans, and humans alone, as Jane Bennett notes, "eligible for eternal salvation"—that is, rescue *from* the material world. Their supernaturalism can claim the mantle of Christian materialism precisely by intensifying its own binary oppositions—as, quite congruently, the scientific materialism of the West builds on the Cartesian dualism.

Modeling an entangled relationality, the panentheist alternative withdraws from its *theos* the status of an independent entity or substance that enters and enlivens from the outside, or that is contained within an inside. It can name the spirited oscillation of a seabird "vibrating" in the waves of the *mayim* (Gn 1:2).[60] And instead of a transcendent being "coming down" and donning external human flesh, there appears the figure of a becoming-body, in an incarnational poetics, or a phenomenology of the flesh of the world.[61] The singular incarnation undergoes radical redistribution. Already experimentation in multiple Christs—Black, Latino, Queer, or the female Sophia—has over a couple of generations effected what we might call a *superpositional christology*, in which communities in their material-ritual apparatuses diffract Christ differently. Yet in each case there can be no separation between the observant community and the bodily marking of their Christ. This is the one who in Matthew's account cannot be disentangled—to the consternation of the "Lord Lord" sayers—from the hungry, the imprisoned, the "least of these my siblings" (Mt 25:40). Marcella Althaus-Reid argues that inasmuch as in such liberation theologies "God has been the object of theological de-essentialization processes," fuller and queerer implications have yet to be drawn: There appears the "omnisexual God" of a "body-grounded theology," where God's omnitude "might be able to return the lost presence of the polyamorous body to its theological discourse."[62] That lost presence signifies no metaphysical substance to return to but a multicontextual ecology to create. There the love-body of Christ morphs into something very like Bauman's polyamory of place, the "love of many places as part of a larger, planetary community."[63] Such experiments in bodying God—decolonial, queer, ecological—dissolve the exclusionary body of Christ. In each case a concrete materialization is at stake. And taken together (from the viewpoint perhaps of a queerly planetary body), the intersections and coalitions may apophatically unsay any Christ-talk in order to form a superpositional cloud that—in the relational pluralism of religious and irreligious discourses—makes honest christologies possible.

But if such incarnational intra-activity breaks into an interreligious and transreligious diversity, it does not forfeit theological intensity. One may note indeed some unintended resonance between the theological polyamory and Whitehead's "Eros of the universe" luring all bodied occasions toward greater intensity, togetherness, and satisfaction.[64] For itself, too.

Such theological experiments have practical effects. They let us access the spiritually saturated materiality persisting in the diverse global materializations of spiritual practice. They make sure we do not abandon "Christian materialism" to any religiopolitical right. In this we do not merely disenchant supernaturalism but offer enticement to ecosmopolitical solidarities.

Not that any queer, black, female, or green-bodied divinity (let alone Whitehead's "primordial and consequent natures of God") will win over religious reactionaries. But versatile versions of relational theology that work sympathetically with *spirited* matter or holy flesh, across the ecumenical manifold, will remain indispensable in the attempt to energize diverse local solidarities for planetary revitalization. These apophatically entangled theologies may begin to release, within the cloud of impossible difference, the widening crowd of relations that may sustain a convivial planet. If such will now collude in fresh materializations of race, culture, gender, sex, and species—no simple transcendence of the old and evolving spiritual ways will be possible. Discernment between them will be.

And really, is it foolish to imagine, even across the surface of dispirited, reactionary, or dying religions, even here and there tangled in tawdry superstitions, the breaking of wave on wave of dimly felt superpositions? Felt unexpectedly, healingly, prophetically, amorously. And always with a tingle of mystery. The nonlocal—planetary, cosmic—materializes; it incarnates, locally. Sacramentally.

TINGLING TISSUE

As to the sacramental—Protestants did not stop performing the ritual of communion. But we certainly abandoned the notion of transubstantiation, with its embarrassingly direct physicality of eating and drinking a body mysteriously incarnate through the Eucharist. We might have cut off some access to the materialization of shared flesh, to the future unfolding of the enfolded past in the ritual entanglement of an elemental re-collection. "'Transubstantiation' is a religious term, and yet one that could just as well be applied to quantum phenomenon." So writes Vicki Kirby in *Quantum Anthropologies* (channeling Barad) of the "superposition of differences" of culture and nature, of mind and body.[65] Or as I must ceremonially add, of God and matter. The separated substances decompose even as, in the indeterminate interval, "naturecultures" (Haraway) and "matter-energy" (Bennett) and "ethico-onto-epistem-ology"(Barad) are composing and recomposing themselves unpredictably.

Each new insight into quantum entanglement has, I admit, caused my flesh to prickle in a way that the Eucharist rarely has. Not because God or anything else is getting proved, but because something about what we all, human and nonhuman, together *are*, what it is we are *part* of, is coming through just when we desperately, even politically, need it. And it comes with the dark luminosity of apophatic mysticism through the antimystical apparatus of a science that hadn't been looking for it. Maybe that is why it conveys the tingle of novelty, of discovery, so often stifled in traditions that already own the truth. The meaning of quantum entanglement—inseparability of once-linked bodies at unfathomable space-time distances—is spreading with the excitement of cutting-edge science.

So the tingle I have in mind is not just my own. It is the tingling of matter itself, in its newly revealed apophatic indeterminacy. A tingle—a prickle, sting, buzz, or quiver of excitement—captures something about the particle, the particular phenomenon, and so of any moment of matter. The wave of potentiality materializes qua particle (whether or not there is a "collapse of the wave function") only as a phenomenon of relation, subtle, sentient, vibrant. Each particle appears now less like a hard bit of stuff and more like an event of materialization: a tingle in the flesh of the world. What Mary-Jane Rubenstein calls a "strange wonder" insists itself upon us transdisciplinarily. Its recognition may be its own reward, but it is requiring more of us.

Here is a passage that fosters that "more," without in any way theologizing it, a text that tingles every time I read it:

If we hold on to the belief that the world is made of individual entities, it is hard to see how even our best, most well-intentioned calculations for right action can avoid tearing holes in the delicate tissue structure of entanglements that the lifeblood of the world runs through.[66]

The holes are growing. They take the shape of melting glaciers, spreading droughts, dying oceans, hungry millions. Our civilization's violation of its own materiality is building to its climax. It seems we now ignore the message of that minimum matter at peril of what matters maximally. The point, however, is not to cry doom but to stir an alternative. Prickles of fear are not far from tingles of attraction. If a "delicate tissue of ethicality runs through the marrow of being," no one needs to "believe" in God or materialism. Within the field of an apophatically entangled energy-matter—the transubstantiation of the blood of the world will happen willy-nilly. We can continue to spill it wantonly. No supernature will rescue us from us.

Or we can attend to the delicate tissue of our entanglements. Such responsiveness might make us more ethically affective and effective, capable of stronger coalitions, riskier conversations, collective actualizations, terrestrial communions. If so, we have to do not with imposition of any beliefs, but perhaps with the superposition of possibility: the chance of more convivial, maybe even ecosmopolitically sacramental, materialization.

NOTES

Thanks to Mary-Jane Rubenstein and Robert Kyle Warren for invaluable comments and crucial edits.

1. "The Christian Materialism of Blessed José Escrivá," in *Opus Dei: Tutta la Verita*, http://www.escriva.it/Ing/19970301.htm. Josémaría Escrivá de Balaguer, *Conversations with Mgr Escrivá de Balaguer* (Dublin: Scepter Books, 1968), 115.
2. Biblical materiality thus springs into play in the opening of Genesis, quickly yielding the bodily Edenic expulsion, the necessity of the law in all the Torah codes of

the Pentateuch, then the prophets with a new vocalization of material threat (the collective injustice) and promise (the new heaven and earth); in the later texts (esp. Ezekiel) the resurrection motif appears as the renewal of the people, gradually compressible into individual bodies.

3. Colleen McDannell, *Material Christianity: Religion and Popular Culture in America* (New Haven, Conn.: Yale University Press, 1995). See also Robert A. Orsi, *The Madonna of 115th Street: Faith and Community in Italian Harlem, 1880–1950* (New Haven, Conn.: Yale University Press, 1985).
4. Marie-José Mondzain, *Image, Icon, Economy: The Byzantine Origins of the Contemporary Imaginary* (Palo Alto, Calif.: Stanford University Press, 2005), 90.
5. Caroline Walker Bynum, *The Resurrection of the Body in Western Christianity, 200–1336* (New York: Columbia University Press, 1995), 186–87.
6. Dick Houtman and Birgit Meyer, eds., *Things: Religion and the Question of Materiality* (New York: Fordham University Press, 2012).
7. For further discussions of the complex materialities of Christian practices, see Caroline Walker Bynum, *Christian Materiality: An Essay on Religion in Late Medieval Europe* (New York: Zone Books, 2011).
8. Jane Bennett, *Vibrant Matter: A Political Ecology of Things* (Durham, N.C.: Duke University Press, 2010), 121.
9. Ibid., chap. 6, "Stem Cells and the Culture of Life."
10. Slavoj Žižek and John Milbank, *The Monstrosity of Christ: Paradox or Dialectic?*, edited by Creston Davis (Cambridge, Mass: MIT Press, 2009), 206 (materialist materialism), 131 (alternative Catholic metanarrative), 125 (eroto-linguistic animal).
11. Slavoj Žižek, "Dialectical Clarity versus the Misty Conceit of Paradox," in *Monstrosity of Christ*, 240.
12. Quantum entanglement is the subject of the chapter "Spooky Entanglements: The Physics of Nonseparability," in Catherine Keller, *Cloud of the Impossible: Negative Theology and Planetary Entanglement* (New York: Columbia University Press, 2014). A shorter version has been published online: "The Entangled Cosmos: An Experiment in Physical Theopoetics," *Journal of Cosmology* (September 2012), http://journalofcosmology.com/JOC20/Keller_rev1.pdf.
13. Alfred North Whitehead, *Science and the Modern World: Lowell Lectures, 1925* (New York: Free Press, 1967), 35.
14. Alfred North Whitehead, *Process and Reality: An Essay in Cosmology*, corrected edition, edited by David Ray Griffin, and Donald W. Sherburne (New York: Free Press, 1978), 94.
15. Whitehead, *Science and the Modern World*, 49.
16. Ibid., 51.
17. Karen Barad, *Meeting the Universe Halfway: Quantum Physics and the Entanglement of Matter and Meaning* (Durham, N.C.: Duke University Press, 2007), 184.
18. Whitehead, *Process and Reality*, 78–79.
19. Whitehead, *Science and the Modern World*, 175–76.
20. Barad, *Meeting the Universe Halfway*, 139.
21. Whitehead, *Process and Reality*, 91.
22. Ibid., 40.

23. Bernard d'Espagnat, *On Physics and Philosophy* (Princeton, N.J.: Princeton University Press, 2006), 19.
24. William E. Connolly, *The Fragility of Things: Self-Organizing Processes, Neoliberal Fantasies, and Democratic Activism* (Durham, N.C.: Duke University Press, 2013), 154. Connolly presented some of this work at Drew Theological School's 2013 Transdisciplinary Theological Colloquium.
25. Keller, *Cloud of the Impossible*, 263.
26. Connolly, *Fragility of Things*, 154.
27. William Connolly, *Capitalism and Christianity, American Style* (Durham, N.C.: Duke University Press, 2009), 39ff.
28. Barad, *Meeting the Universe Halfway*, 122.
29. Ibid., 372.
30. Ibid., 375.
31. Ibid., 380.
32. Ibid., 381.
33. Isabelle Stengers, *Thinking with Whitehead: A Free and Wild Creation of Concepts*, trans. Michael Chase, foreword by Bruno Latour (Cambridge, Mass.: Harvard University Press, 2011).
34. John B. Cobb, *Christ in a Pluralistic Age* (Philadelphia: Westminster Press, 1975).
35. Roland Faber and Catherine Keller, "A Taste for Multiplicity: The Skillful Means of Religious Pluralism," in *Religions in the Making: Whitehead and the Wisdom Traditions of the World*, ed. John Cobb (Eugene, Ore.: Cascade Books, 2012).
36. See Catherine Keller and Laurel C. Schneider, introduction to *Polydoxy: Theology of Multiplicity and Relation* (New York: Routledge, 2011). Also see Mary-Jane Rubenstein and Kathryn Tanner, guest eds., *Polydox Reflections* (Malden, Mass.: Wiley Blackwell, 2014), *Modern Theology* 30, no. 3 (2014).
37. A stream of thought such as that so importantly advanced by Philip Clayton unfolds the dialogue of science and religion in neighborly proximity to, but without direct dependence on, Whiteheadian thought.
38. Whitehead, *Science and the Modern World*, 15–16.
39. See Karsten Harries, *Infinity and Perspective* (Cambridge, Mass.: MIT Press, 2001). Also see my chapter "Enfolding and Unfolding God: Cusanic Complicatio," in *Cloud of the Impossible* and Mary-Jane Rubenstein's *Worlds without End: The Many Lives of the Multiverse* (New York: Columbia University Press, 2014), especially the section entitled "End without End: Nicholas of Cusa" in chapter three of Rubenstein's book.
40. Nicholas of Cusa, *De docta ignorantia*, in *Nicholas of Cusa: Selected Spiritual Writings*, trans. H. Lawrence Bond (New York: Paulist, 1997), 140.
41. Speaking of precarious opacities, fortunately even Judith Butler now also uses a language of ontological relationalism and has taken steps to correct her own anthropocentrism in the context of a reading of Whitehead. Roland Faber, Michael Halewood, and Deena Lin, eds., *Butler on Whitehead: On the Occasion* (Lanham, Md.: Lexington Books, 2012). See also my chapter on Butler, "Unsaying and Undoing: Judith Butler and the Ethics of Relational Ontology" in *Cloud of the Impossible*.
42. Sir Issaac Newton, *Opticks; or, A Treatise of the Reflections, Refractions, Inflections and Colours of Light* (London: William Innys, 1730), 376.

43. Paul C. Davies and John Gribbin, *The Matter Myth: Dramatic Discoveries that Challenge Our Understanding of Physical Reality* (New York: Simon & Schuster/Touchstone, 1992), 17.
44. Louisa Gilder, *The Age of Entanglement: When Quantum Physics Was Reborn* (New York: Knopf, 2008), xv.
45. Richard Feynman, BBC Publications, 1965, 129.
46. *Meister Eckhart: The Essential Sermons, Commentaries, Treatises, and Defense*, trans. Edmund College and Bernard McGinn (Mahwah, N.J.: Paulist, 1981), 53:208.
47. Chris Boesel and Catherine Keller, eds., introduction to *Apophatic Bodies: Negative Theology, Incarnation, and Relationality* (New York: Fordham University Press, 2010) and my essay "The Cloud of the Impossible: Embodiment and Apophasis," in *Apophatic Bodies*.
48. T. S. Eliot, "Burnt Norton." In *Four Quartets* (New York: Houghton Mifflin Harcourt, 1943).
49. Nicholas of Cusa, *De docta ignorantia*, 126. "Therefore the theology of negation is so necessary to the theology of affirmation that without it God would not be worshiped as the infinite God but as creature, and such worship is idolatry, for it gives to an image that which belongs only to truth itself."
50. Ibid., 140. I have written about Cusa in *Apophatic Bodies*. His apophatic relationalism with its explication in cosmology and religious diversity is key to my book *Cloud of the Impossible*.
51. For multiversal meditations along these very lines, please see "Ends without End: Nicholas of Cusa" in chapter three of Mary-Jane Rubenstein's *Worlds without End*.
52. Charles Hartshorne, *Omnipotence and Other Theological Mistakes* (Albany: State University of New York Press, 1984). His briefly proposed metaphor of the "universe as the body of God" is then expanded by Sallie McFague as part of the ecofeminist archive. Sallie McFague, *The Body of God: An Ecological Theology* (Minneapolis: Fortress Press, 1993).
53. Hartshorne's answer to the classical Unmoved Mover of orthodoxy, a development of Whitehead's "consequent nature of God," the aspect of the divine that becomes, and so has actuality, through the materializations of actual occasions.
54. Henry P. Stapp, *Mindful Universe: Quantum Mechanics and the Participating Observer* (Berlin: Springer, 2007), 332. Emphasis added.
55. N. David Mermin, "Is the Moon There When Nobody Looks?: Reality and the Quantum Theory," *Physics Today* 38, no. 4 (1985): 38–47.
56. See Jane Bennett, "A Vitalist Stopover on the Way to a New Materialism," in *New Materialisms: Ontology, Agency, and Politics*, ed. Diana H. Cool and Samantha Frost (Durham, N.C.: Duke University Press, 2010); also see Bennett, *Vibrant Matter*, chap. 5, "Neither Vitalism nor Mechanism."
57. I have written about two thinkers considered vitalists or panpsychists: Anne Conway (about whom I learned from Carol Wayne White's *The Legacy of Anne Conway* [Albany: State University of New York Press, 2009]), "Be a Multiplicity: Ancestral Anticipations," in *Polydoxy: Theology of Multiplicity and Relation*, ed. Laurel C. Schneider and Catherine Keller (New York: Routledge, 2010). And Gustav Fechner,

"The Luxuriating Lily: Fechner's Cosmos in Mahler's World," in *Mahler im Kontext/ Contextualixzing Mahler*, ed. Erich Wolfgang Partsch and Morten Solvik (Vienna: Boehlau Verlag, 2011).
58. Whitney A. Bauman, *Religion and Ecology: Developing a Planetary Ethic* (New York: Columbia University Press, 2014), 172.
59. Ibid., esp. chap. 6, "Developing Planetary Environmental Ethics: A Nomadic Polyamory of Place."
60. Literally vibrating in Genesis 1:2, *meherephet*. See my *Face of the Deep: A Theology of Becoming* (London: Routledge, 2003).
61. See Mayra Rivera Rivera on Merleau-Ponty, in *Poetics of the Flesh* (Durham, N.C.: Duke University Press, 2015).
62. Marcella Althaus-Reid, *Indecent Theology: Theological Perversions in Sex, Gender and Politics* (London: Routledge, 2000), 53–54.
63. Bauman, *Religion and Ecology*, 128.
64. Alfred North Whitehead, *Adventures of Ideas* (New York: Macmillan, 1933).
65. Vicki Kirby, *Quantum Anthropologies: Life at Large* (Durham, N.C.: Duke University Press, 2011).
66. Barad, *Meeting the Universe Halfway*, 396.

Agents Matter and Matter Agents: Interpretation and Value from Cells to Gaia

PHILIP CLAYTON AND ELIZABETH SINGLETON

There is not so much Life as talk of Life, as a general thing. Had we the first intimation of the Definition of Life, the calmest of us would be Lunatics!
—EMILY DICKINSON TO ELIZABETH HOLLAND, 1877, Letter 492

For centuries, treating simple organisms as agents was viewed as sheer human projection. After all, Descartes assured us, animals are mere machines. In these pages we wish to defend the opposite affirmation: *Every participant in the dynamics of Darwinian evolution is an agent with "interests," acting with and through its environment.* That means literally every participant, from the simplest cell to Gaia, the earth's ecosystem as a whole.

What is changed when we consider every living being—from unicellular organisms to humans, and beyond—as perceiving, experiencing, valuing, and valuable agents? We suggest that the argument works best when one develops it simultaneously from both ends of the spectrum. What results is a sort of symmetry, like the famous Whiteheadian symmetries, but in this case a symmetry of part and the whole, of the very small and the very large. Symmetries invite a double movement. One follows them downward or inward, tracing agency all the way back through evolution to the smallest units on which natural selection operates; and one follows them upward or outward, all the way to the biosphere as a whole.

This pervasively agential approach to biology—running from the smallest, simplest unicellular organism to Gaia as a whole, and simultaneously in the other direction as well—spawns a compelling ethic of embodied responsibility. We hope also to show that it transforms standard views of mattering and minding and, as a result, challenges long-held assumptions about science, theology, and their (co-)evolving intertwinings.

AGENCY

Immanuel Kant's commitment to Newtonian principles and to a mechanistic universe ran deep. Thus it is doubly a reflection of his genius that he intuited an agent-based framework for comprehending the nature of life. As Kant wrote in the *Critique of Judgment*,

> An organized being is then not a mere machine, for that has merely moving power, but it possesses in itself formative power of a self-propagating kind which it communicates to its materials though they have it not of themselves; it organizes them, in fact, and this cannot be explained by the mere mechanical faculty of motion.[1]

Consider the humble eukaryotic cell.[2] Cells do self-propagating work. Work, in the most basic sense, means constraining and employing energy for a purpose. A cell captures energy inside its cell wall and employs that energy to carry out processes that sustain the structure and the functioning of the cell—processes, in short, that keep it alive. As cells carry out this complex web of work, constraining and employing energy and doing things with it (such as DNA replication and enzyme synthesis), the astonishing fact is that they attain a kind of closure: Each successful cell eventually builds a viable replica of itself.

Formally speaking, this process of self-closure corresponds closely to the self-propagating organization that Kant identified. Note that the self-propagating organization one finds in cells does not involve matter alone, energy alone, information alone, or entropy alone. It is a process that involves all these factors—and something more, as well. It appears that this self-propagating organization, "communicate[d] to its materials though they have it not of themselves," is a new form of energy-matter organization in the world; it is *living matter*, ontologically emergent. The structural and functional features of the eukaryotic cell meet Kant's requirement for ontological emergence: The whole has causal powers not possessed by the parts. Because the "whole" of the cell is capable of building copies of itself, it is capable of evolution by natural selection. All that's necessary is that mutant variants of the minimal autonomous agent, or real cells such as bacteria, can themselves have heritable variants that are selected for or against in their particular environment.

Volumes have been written on human agency. The majority of the theological tradition shares a crucial assumption in common with the humanist tradition, which it spawned: the assumption that humans stand at the pinnacle of creation. The agency-centered ontology that we offer here turns that assumption on its head. No more can we use human traits as the standard for moral consideration, measuring the value of all bodies based on their likeness to human bodies or selves. We may well *start* with our own embodied existence; Donna Haraway's recognition that all knowledge is "situated knowledge" leaves us no alternative.[3] All who reflect have a mother language, a first experience of embodied thought. But finding analogous agencies both as we move downward to smaller parts and as we move outward to broader systems quickly decenters our starting point, and thus ourselves. We are, it turns out, neither center nor pinnacle. Would that we could find "new theologians," like Nietzsche's "new philosophers" in *Beyond Good and Evil*,[4] who could think with us part and whole, pan-sacred and pan-profane, in a pan-agential world.

What happens, then, when we begin to understand agency at the ends of the biological spectrum? Ascriptions of agency and purpose represent a standard form of explanation in accounts of human behavior. In this chapter we use those well-established results as a touchstone—and, in part, as a point of contrast—for developing a theory of biological agency. After all, humans regularly offer teleological or "means-ends" explanations in which reasons appear as causes of behavior: She proclaimed the defendant innocent because she was not convinced by the prosecutor's arguments; he repeated his point because he did not believe they heard him the first time.[5] If human action, reasons, motives, intents, and purposes do not offer an adequate vocabulary and framework for the attribution of teleological explanations, nothing does. Whatever theoretical stance one takes regarding full-blown human action, it is a stunning fact that the universe has given rise to entities that do, daily, modify the universe to their own ends. Philosophers call this capacity *agency*.

THE FIRST AGENTS

With this touchstone in place, we turn to a new question: What is the minimal natural system to which one might attribute teleological or purposive explanations?

The first systems to which one can meaningfully attribute agency and (some measure of) autonomy are, we suggest, the simplest biochemical molecular systems on which natural selection can operate. Such systems must be able to reproduce themselves. In standard works on the origin of life, this entails (among other things) having a particular kind of membrane that separates "inner" and "outer." The "inner" includes information that codes for building a successor organism. There must also be some perception of the environment, presumably via osmosis through the membrane, so that the organism "knows" when the surrounding chemical solution contains a sufficient quantity of the building blocks needed for its reproduction. When the organism's receptors signal the appropriate conditions, its membrane needs to open or dissolve, so that chemicals can bond with receptor sites (proto-RNA) that were formerly inside its membrane, in order to make a duplicate of the original informational system. New membranes must then grow around both the original and the copy, so that the original organism now exists as two.[6] (The full story is far more complex and far more fascinating than any brief summary can capture. In particular, there are multiple interesting proposals concerning how a system can evolve these reproductive capacities before it is able actually to reproduce.)

Darwinian natural selection can operate on this simple biochemical system, this proto-organism, only if two conditions are met. First, the copying process can't be perfect;[7] some variance needs to enter into the process of reproduction. Second, the environment needs to select for some copies and against others. The natural result of differential selection is that some variants will increasingly come to dominate a particular environmental niche, whereas others will go extinct. Once this Darwinian process begins, the rest is history—or so biologists believe. The patterns and outcomes of emergent complexity, once launched in unicellular organisms, continue to do their work across the eons of evolution. One need not believe that the process is necessary or pre-designed in order to wonder at the profusion of agents, adaptations, and functions: intercellular communication, organ systems, symbiosis across species, social learning, mental representations—all linked within a single global system, Gaia.[8]

What plausible conjectures can we make about biological agents across evolution? Can we say that they are, in some sense, "agents like us" without an illicit projection of distinctively human attributes onto them?

> As noted, even simple agents have an "inner" and an "outer" world. It follows that they are connected to an environment in a way that no mere biochemical solution is. (We return to the implications of this amazing emergent phenomenon in the next section.)
> The dynamic within which biological agents exist cannot be explained in purely chemical terms. As living organisms, they exist within webs of competition and cooperation. Only the tools and concepts of biology suffice to account for what it is for them to exist. Pre-biological theories cannot fully explain the natural dynamic that defines the unique type of entities that organisms are.
> Put differently, that organisms exist within an all-consuming Darwinian dynamic means that they have interests as no mere chemical compound does. If their membrane opens in an environment containing toxins, the individual dies. Dying is not one of the thousand natural shocks that chemicals are heir to; it's not a property that one can ascribe to any chemical compound as such. By contrast, if one species

releases large amounts of toxins into its environment (or changes the global climate, as one species has now managed to do), the entire system of related organisms can become extinct.

Having interests means that, for organisms, not every outcome is equal. Being bathed in nutrients is good, and being bathed in toxins is bad. Of course, the simpler agents lack the kind of internal feedback system (nerve cells, central nervous system) that would be necessary for them to become consciously aware of, or to consciously feel anything about, these results. "Good" and "bad" are products of the task bequeathed to them by that Darwinian dynamic that defines their existence: to survive and reproduce.

We thus need to use some values language to define their existence. Of course, since simple organisms could not use this language themselves, speaking in this way is in some sense a projection: We are doing the valuing here. But in another sense it's not pure projection. Uranium-238 radiates alpha-particles and decays into uranium-234. This may be good or bad for us, but it doesn't matter to the uranium. By contrast, obtaining food sources and reproducing are part of succeeding for an organism. They matter to it, even when it has no thoughts or feelings about the matter.

Indeed, a second dimension of value emerges with organisms. As complexity increases over evolutionary time, it becomes less and less likely that starting the process over again would yield exactly the same organism. Greater organismic complexity, and increasing dependence on particular ecosystems, produces an ever-greater probability that a second history of random variation and selective retention would not produce the same species over again.

The extinction of a species is (with a high level of probability) the definitive end of that particular lineage and its members. Whatever structures, functions, and experience may have characterized these organisms will be permanently lost. They are, in this precise evolutionary sense, unique. Over time, then, the biosphere begins to include a new form of contingency—an emergent form of "possibly not being."

Biodiversity is an instrumental good for humans, since it increases our own odds of survival. But arguably it is also an intrinsic good that the world would contain a rich range of types of beings and experience—at least theologians have traditionally argued in this way. If so, then the permanent loss of unique life-forms and their experience brings with it a loss of value.

LIFE AS SEMIOTIC AGENCY

Biosemiotics offers one of the most powerful tools in theoretical biology for conceiving the meaning of the emergence of life. Jesper Hoffmeyer from the University of Copenhagen provides a cogent presentation of the biosemiotic standpoint:

> Some billions of years ago it happened that a number of daughter cells from a single monocellular organism developed a series of symbiotic relationships with one another and, in time, even entered into a process of shared ontogenetic differentiation—so that there emerged a small multicellular organism, consisting of cells with connected life histories but with differentiated roles in their newly collaborative mode of being. These cells having reached this state, it would no longer be sufficient for us as scientists to describe the activity of each single constituent cell in isolation. Rather, one would hereafter have to consider the presence of a new holistically autonomous actor, an interpreter—or a system of interpretants as Stanley Salthe (1993) has formulated it—that is able to organize the semiotic life processes of negotiating an external environment for the benefit of the collective, and at the expense of the interests of single cells. Thus, in a certain sense, the appearance of a multicellular organism might be seen as the appearance of a new kind of causality in the natural world, i.e., a formal causality, as suggested in the Aristotelian scheme.[9]

We might want to quibble that what Hoffmeyer describes is not only formal but also final causality in Aristotle's sense. Still, the concept of biosemiotics is compelling, and the rapid growth of this field has in our view been fruitful for biology. For biosemioticians, an organism is an entity with interests, a "holistically autonomous actor." It interprets its environment in light of those interests, and as it does so events in its environment become signs for it. The classic features of semiotics emerge: interpreters, interpretations, and interpreted phenomena (signs). Indeed, if one accepts that semiotic relations are the basis of knowledge, it follows that all organisms know (in some extended sense of "know").

What is powerful about the biosemiotic approach is that it introduces a robust notion of (teleological) agency *without needing to appeal to an antinaturalistic or dualistic metaphysics*. A new dynamic is manifest in the natural world, one that evidences final causation. By contrast, it's not clear that introducing mental substances—a move completely anathema to modern science—is philosophically sufficient to explain the qualities of agency.

For the biosemiotic thinkers, a unicellular organism is already a unit of meaning, interpretation, and purpose.[10] In Darwinian terms, its purpose is to survive and reproduce, and actions that help bring about this purpose can be called purposive actions. This starting point, biosemioticians argue, is sufficient to provide all living organisms with a "for-the-sake-of" structure. Stuart Kauffman describes organisms' interests in humorous terms: Every organism is "out to make a living" in its world.[11] With interests and goals come values for the organism itself, long before the emergence of consciousness.

When a single-celled organism spins its flagellum in order to move up a glucose gradient and obtain more nourishment, it interprets the higher glucose concentration as a good and acts in order to ingest more of it. Of course, the interpreting here proceeds solely at the chemical level, through osmosis and chemical bonding to the external cell membrane; we have no evidence of thoughts, mental attitudes, feelings, or other subjective experience. Still, the three basic elements of interpretation ("semiosis") are already present: There is an interpreter; there is a state of affairs being interpreted as a sign; and there is an interpretation (in this case, the movement of the organism toward nourishment). If semiosis is knowledge, then there is knowledge. And there are also consequences: If the organism misinterprets a toxin as nourishment and moves toward it, it will die.

In a sense, it doesn't really matter whether one calls these actions "real interpretations" or merely "proto-interpretations"; the broader point still remains. A continuum, a similarity-in-difference, runs from the simple interpretations that unicellular organisms make when they respond to their environment, all the way through to the multidimensional interpretations that primates and other complex animals form as they creatively vary their behavior in response to new stimuli. Step by step, the interpretations become more complex, more multifaceted, and more comprehensive as they respond to, and thereby help create, ever-more-complex eco-contexts for action.

According to the emerging systems perspective that we defend, the program of biology is to reconstruct the ever-growing complexity of organisms, structures, and behaviors as one moves along the continuum of complexity from the simplest organisms to the most complex ones. One will expect to find analogs to many human functions in more simple organisms. Simple organisms perceive their environment, though without eyes; they know or are aware of features of their surroundings, though without conscious awareness. They act to fulfill goals, though without forming conscious intentions. In short, they are interpretive agents, agents of interpretation—which makes life a sort of hermeneutical agency. In some contexts biologists emphasize the continuities across multiple regions of the spectrum of complexity, and at other times they emphasize the discontinuities.

INDIVIDUAL AND COMMUNITY, PART AND WHOLE

The biosphere was packed with living, interpreting systems well before human beings came onto the stage. These agents possess many of the properties that one sees manifested today in "higher" organisms, albeit in less developed forms.

Multicellular organisms are communities *cum* individuals—a community that has become an agent, a center and locus of action. Biosemioticians believe (rightly, in our view) that explaining the dynamics of individual organisms requires a threefold framework of sign, interpreter ("interpretant"), and interpretation. This framework, which semiotics draws from Charles Sanders Peirce, introduces the concepts of meaning and reference, and by implication value, all the way across the biosphere. Without these three concepts, we have argued, one cannot understand the evolution of life.

Note that the semiotic framework comes in more and less constricting forms. A deflationary approach, for example, is taken by Terrence Deacon in *Incomplete Nature*. Here semiotics functions to *limit* the kind of ontologies that may be introduced. Deacon insists that there are only three qualitatively distinct kinds of dynamics in the natural world: homeodynamics, morphodynamics, and teleodynamics.[12] Once goal-directed dynamics arise (which for Deacon occurs even before the first self-reproducing cell), all further developments of agents are only variations on this one theme. By contrast, we see no reason to exclude further qualitative developments. Social learning, animal cultures, mental representations, consciousness—all of these denote new emergent patterns, new forms of agency. Although we are pleased that Deacon has found a home in Buddhist metaphysics, we fear it may also prejudice his theory of agency. Where we perceive an agent-centered ontology—an ever-growing profusion of agencies throughout the biosphere—Deacon sees goal-directed patterns but the absence of agents as such.

Biosemiotic approaches do however share a more integrated understanding of the relationship between parts and wholes. Jesper Hoffmeyer formulates it as well as any other author:

> In most biological models, as well as in everyday folk psychology, the prototype *organism* remains essentially a vertebrate, like ourselves. Vertebrates are always well-integrated, coherent organisms with well-defined forms. They consist of genetically uniform cells and have well-defined life cycles, starting with a single cell and ending in reproduction via the transmission of germ cells. However, by far most organisms of this world are *not* vertebrates—and most of them do not obey the aforementioned criteria very well.[13]

For most life on this planet, the relationship between an organism, other members of its species, and its ecosystem are far more fluid than is the case for vertebrates with their more bony stabilities. As Elisabet Sahtouris writes, "No being in nature, outside our own species, is ever confronted with" the choice between individual and whole. She adds, "If we consult nature, the reason is obvious. The choice makes no sense, for neither alternative can work. No being in nature can ever be completely independent, although independence calls to every living being, whether it is a cell, a creature, a society, a species, or a whole ecosystem."[14]

In short, an adequate biology views living things as systems, as interconnected processes. This viewpoint clashes deeply with the anthropocentric, and clearly androcentric, view that to be an individual is to be separate from others. It's time to leave that myth behind. New resonances arise; one begins to think instead of Donna Haraway's companion species, Lynn Margulis's host of biotic others, Jane Bennett's animal-vegetable-mineral sonority complex, Karen Barad's agential realism, and other allies.[15] As soon as one begins to hear these harmonies, and the overtones that they produce, the symmetry between the very small and the very large comes into full view, bringing with it some rather revolutionary implications.

GAIA: THE BIOSPHERE-AS-A-WHOLE AS AGENT

As the biosemioticians courageously (and justifiably) extend semiotic agency and interpretation even to the simplest organisms, some theorists are (with equal courage) extending agency all the way outward to the Earth's biosphere as a whole. The "Gaia hypothesis," proposed by James Lovelock and further developed by Lynn Margulis, has exactly this function. On the Lovelock/Margulis hypothesis, the biosphere as a whole is a living entity that regulates itself in specific ways in order to support the continuation of life. If there are regulatory patterns that occur only at the level of the biosphere as a whole and not merely as an aggregate of its parts, as appears to be the case, then Gaia also has features of agency. Analogous to the simple microorganisms that make up most ecosystems, it (she?) too is a whole that is greater than the sum of its parts. Also like other organisms, it makes adjustments and adaptations so that the living parts of which it consists can continue to emerge and flourish.

It is difficult to convey how controversial this suggestion has been. Bruno Latour, for example, describes the responses of friends and colleagues as he was preparing his Gifford Lectures on the Gaia theory.[16] No real biology could make such anthropocentric, eccentric claims, he was told, and no scientist in his or her right mind would be associated with such views. Lynn Margulis, who wrote extensively on the Gaia hypothesis and who was the keynote speaker at the fourth international conference on Gaia theory in 2006, came under equally sharp attack for her advocacy.

What does it mean to attribute agency to the biosphere as a whole? The Gaia hypothesis leads one to look for features in the biosphere that are analogous to the qualities of other organisms. Each organism is a system of living systems; so too is the biosphere itself. One thus finds emergent properties in the web of life that are beyond human control, just as one finds value in this self-regulating living system of living systems that is not merely derived from its parts. Lovelock explains Gaia with scientifically infused wonderment:

> The entire range of living matter on Earth, from whales to viruses, and from oaks to algae, could be regarded as constituting a single living entity, capable of manipulating the Earth's atmosphere to suit its overall needs and endowed with faculties and powers far beyond those of its constituent parts.[17]

Of the various features of Gaia theory that have caused Lovelock and Margulis to be shunned by much of the scientific community, none has been more derided than the teleological suggestions implicit in the theory. Affirming that the earth-system has agency requires a broader definition of agency than most scientists have so far accepted, one that includes responsiveness of a system to changes within itself and its environment. On this definition, a living organism is a system of systems that function together, creating a new emergent system that manifests agency. Emergent agents are engaged with, even parts of, other systems. But their causal role in such interactions is as more than an aggregate of their parts or subsystems; they become actors in their own right.

Gaia theories range from cautious to bold. The most cautious introduce Gaia as pure metaphor; they speak of the biosphere *as if* it were the anthropomorphic Greek goddess in order to encourage more responsible environmental policies and lifestyles. Lovelock did affirm that homeostasis occurs at the level of the biosphere as a whole (the *biota*). On our view, this criterion is enough to constitute Gaia as agent and (if all agents are inherently valuable) as a locus of intrinsic value.

Lovelock became (regressively) less bold over time, however. In the end, he pulled back from his early claim that Gaia acts intentionally or with purpose, denying for example that "planetary self-regulation is purposeful, or involves foresight or planning by the biota."[18] Lovelock's retreat was not necessary, however. Homeostasis is goal-directed, whether or not it is consciously planned and maintained. As we have maximized value by extending it to the smallest biological agents, value should be maximized as one moves outward to the largest system that manifests agential action. Gaia's homeostasis plays a crucial role in maintaining an earth environment that is favorable for life.

If agency does indeed extend outward to the biosphere as a whole, then we are also responsible to the embodied whole of life. To allow this responsibility to resonate more fully with human intuitions about moral obligation, we prefer to speak of the earth's biosphere as a whole not as "it" but as "she": Gaia.[19] We are interlocked and interdependent within her single web of life. Although we have the ability to do unspeakable harm to the systems of which she is composed, we are also able to contribute in more positive ways, helping the systems function smoothly and organically.

To consider the biosphere as an interpreting agent is to view ourselves and all other living agents within the biosphere as, together, parts of a larger living entity. As we know, human agents have the ability to cause malfunctions within this system. In their much-cited 1982 paper in *Science*, Jack Sepkoski and David Raup identified five mass extinctions,[20] and scientists are increasingly concluding that we are now in the midst of the sixth extinction, this one however, unlike the previous ones, caused by human activities.[21]

The most obvious implications of the Gaia hypothesis are that we are not responsible only to isolated individuals (or to those species we find either useful or cute), but also to that communion of bodies that together make up the system of all living things, and that we must find a way to live among and within this living system of life. Elisabet Sahtouris, in her book *EarthDance: Living Systems in Evolution*, writes that "it is one thing to be careful with our environment so it will last and remain benign; it is quite another to know deeply that our environment, like ourselves, is part of a living planet."[22] If the earth's biosphere itself is a living system with agency, our moral responsibility is not to the earth as lifeless matter, but to Gaia herself and to all the specific living systems and environments that constitute and sustain her.

On the one hand, agency that extends to living systems as a whole means that we should be fearful because we cannot control or predict Gaia's response to the mess we have made.[23] The earth possesses emergent properties that are unpredictable. On the other hand, it also means that we should be hopeful *because* we cannot control or predict the response of the earth to the mess we have made. There is hope beyond human hands and possibilities beyond human imagination as Gaia interprets and responds to the changes inside of her. After every disaster life has flourished in a new way, growing differently and more complex—creating systems within systems that have agency and express a will to life by responding to their contexts in ways that express the creativity of living agents.

When combined with the affirmation of the agency of all living things, this approach means that agents are not valued only in and of themselves.

Agents are also valued as participants in the entire system of life, which is held together by networks of bodies and objects. As Sahtouris writes, "No being in nature can ever be completely independent, although independence calls to every living being, whether it is a cell, a creature, a society, a species, or a whole ecosystem."[24] Gaia thrives when conscious, powerful agents like us limit our consumption for the sake of the whole (kenosis) and when we foster the unique contributions of those agents whose existence is more precarious than our own.

To recognize that Gaia exercises her own agency requires a balance between valuing individuals and valuing collectives. All of us living things depend on this balance; every organism depends on its ecosystem, and every ecosystem on its organisms. Also, every organism is itself a network, dependent on the proper functioning of its parts (as cancer so painfully reminds us). This interdependence is even more dramatically true of Gaia. Drought in Africa is not drought for Asian farmers, and the cancer in your friend's body is not cancer in yours. But all droughts and all cancers are internal to Gaia. Her network encompasses them all; her body includes them all; her fate hangs in the balance in each case. Once we begin ascribing agency to networks, we recognize the drastic implications that follow for the way we conceive of morality and relationships, as well as for the way we understand the living agents among whom we live.

To consider the network of life as a whole as perceiving, aware, valuing, and valuable suggests that there is a larger teleology for life that extends beyond, but may be inclusive of, human flourishing. As Lynn Margulis and Dorian Sagan write, "A biological system acting cybernetically gives the impression of teleology. If only the results and not the feedback process were stated, it would look as if the organisms had conspired to ensure their own survival."[25] We argue that they not only give "the impression of" teleology; they *really are* teleological. Each of us is an agent among agents, in interfolding networks that join together and manifest Gaia, the largest living agent, in whom we live and move and have our being.

To endorse the Gaia hypothesis is to step clearly beyond inherited notions of agency. Elisabet Sahtouris formulates these implications most clearly:

> For now what matters is to understand this new way of seeing that all evolution—of the great cosmos and of our own planet within it—is an endless dance of wholes that separate themselves into parts and parts that join into mutually consistent new wholes. We can see it as a repeating, sequentially spiraling pattern: unity → individuation → tension/conflict → negotiation → resolution → cooperation → new levels of unity, and so on.[26]

Sahtouris argues that we continue to miss the clues about the nature of biological organisms because "we have not understood ourselves as living beings within a larger being, in the same sense that our cells are part of each of us."[27] The Cartesian myth, beholden to its strict Cartesian coordinates and to the Western legacy of dualism, has no place for the universal arising of all things from each other (*Pratītya-samutpāda*) of Buddhist thought, no place for the interdependent existence of part and whole, certainly no place for the embedded/embodied spiritualities of indigenous traditions. Only when one abandons the idea of unique soul-substances, each a "thinking thing" (*res cogitans*) with "his" own independent identity—as well as the God created in "his" image—can one begin to take on board the lessons that contemporary biology is teaching.

IMPLICATIONS OF THE PAN-AGENCY VIEW

This chapter plays but a small role in a larger (eco-)collection. The volume of which it is a part, taken as a whole, offers multiple ways of conceiving the natural world beyond the dualisms of matter versus mind, value-laden versus value-free, independent part versus all-encompassing whole. One can only hope that each succeeding generation will find these ideas easier to grasp, more congenial—more, well, natural.

Because what must occur is nothing less than a shift of worldview, and because science will need to play a major role in bringing about this shift, we have considered the entire spectrum of life from the very small to the very large. In a more lengthy presentation, one could explore specific features of interpreting, valuing agency at more of these individual levels—from the earliest unicellular organisms, the first participants in the Darwinian dynamics of natural selection, to microbiomes, to broad ecosystems, and finally to the entire system of life as we know it, the Earth's biosphere.

We have paid special attention to the origins of life in this chapter because the realm of the very small has so often been used as the reason for reducing life to nonlife. Atoms, molecules, the equations of biochemistry, genetics, proteomics—these building blocks are often presented as the antithesis to organisms, agents, and a life-centered perspective. We have suggested that the opposite conclusion follows. As Stephan Harding writes,

> Thus the great archetypes of Gaia and *anima mundi* that figure so importantly in the human soul could well be prefigured in some mysterious way not in some abstract realm far from this world, but in the very molecules and atoms that constitute our palpable, sensing bodies.[28]

In this age of systems biology and ecosystems theory, microbiology need no longer stand in opposition to Gaia. Biologists are only beginning to formulate the common principles that hold for all living, interdependent systems. We believe that the data and the experiments will eventually settle the matter on the side of agents. Or, to use the verbal form, what biology teaches us is that matter "agents." Conversely, since agents respond in material ways, we can again use the verbal construction: Agents "matter." In the spirit of that core principle of constructive science-religion dialogue, "creative mutual interaction," these two short phrases together offer a sketch of what we believe will eventually be the complete picture of the Earth's biosphere and the organisms that dwell therein.

Nor is this a matter of abstract academic dialogue and specialized disputation. This biology-inspired worldview supports an ecological way of being in the world. In *The Universe Story*, Brian Swimme and Thomas Berry make it the rallying cry of what they call the emerging Ecozoic Age.[29] Should humanity fail to live its way into an organic worldview, it is in our view unlikely that humans will find sufficient motivation to limit their consumption and change their lifestyles (and political systems) soon enough, and extensively enough, to avoid catastrophic environmental—and thus also social and economic—consequences. In the end, Gaia really does have the last word.

Western thinkers are only in the early stages of learning to think in the context of pervasive interdwelling systems, processes, and values. Non-Western cultures are millennia ahead of us in this regard: indigenous cultures and lifeways, India's *advaita Vedanta*, the Jain call to do no harm (*ahimsa*) to any living being, and the wealth of cultures nourished on the Buddhist teachings of Dependent Arising and interbeing. Indeed, the whole dispute between Gaia's advocates and their opponents could be reread as a battle between the worldview or "metaphysic" of interbeing and that of independent or substantial being.

As Gaia brings discomfort to a traditional model of biology, she brings equal discomfort to traditional theologies. Gaia's lifeworld entails a thought-world long forgotten, even repressed, in the intellectual categories that dominated paradigmatic Western philosophy and theology. Dead matter, isolated atoms, freestanding substances, and controlling empires offer little help in comprehending the interdependent world of Gaia and her indwelling systems. Any theology that would render support and nurture to a Gaian world must learn to cultivate concepts such as participation, immanence, panentheism, reciprocity, and mutual indwelling. As the distance between sacred naturalism and fully immanent theism grows smaller, the old dichotomies disappear.

An agent-centered biology invites theologians into an ancient-future space of reflection. There is no reason why God, having shed "his" imperialistic trappings, cannot be invited back into this space of agential-life-lived-together—the space of *conviviality*, as other authors in this collection have named it.

Clearly, the Whitehead-inspired process theologies belong in this space, since they affirm a noncoercive divine lure that is equally operative on every agent, whatever his or her level of complexity. Process theologians have generally not included Gaia among the "actual occasions" worthy of lure; we hope we have helped to rectify that wrong.

More generally, this chapter invites a teeming profusion of green theologies, corresponding to the manifold ways that the immanent Divine may be conceived as luring, tugging, and binding the whole range of interlocking biological agents. Theologians are only beginning to set aside the anthropocentrism of the Hebrew and Greek cultures and describe an entire biosphere awash in divine presence at every level. If the co-participation of divine and finite agency is possible at the human level,[30] imagine the profusion of interagency when participation is extended across the entire spectrum, from cells to Gaia! Here, truly, is *perichoresis* (mutual indwelling), now extended beyond the intradivine persons to the interdivine entangling of divine and finite agencies.

At the very end of *On the Origin of Species* Darwin asks us "to contemplate an entangled bank, clothed with many plants of many kinds, with birds singing on the bushes, with various insects flitting about, and with worms crawling through the damp earth." We have yet to see a generation of theologians courageous enough to imagine God as truly filling and being at home in this entangled space, so that the interests of the myriad entangled agents are truly present within the divine—a theology of Supreme Entanglement. But we are drawing closer.

It is fitting that we should conclude with the words of Lynn Margulis. Her work and life symbolized the push beyond acceptable boundaries: Well-funded for her work on symbiosis, she was later blacklisted in scientific circles for her advocacy of symbiogenesis and the Gaia hypothesis. In the midst of this shunning she called for biologists and others "really to listen to the rest of life," since "as just one melody in the living Opera we are repetitious and persistent."[31]

We predict that the empirical evidence will eventually catch up with her conception. In the meantime, we offer it as a vision of what we believe is the road that lies ahead:

> We may think ourselves creative and original but in those talents we are not alone. Admit it or not, we are only a single theme of the orchestrated life-form. With its glorious nonhuman past and its uncertain but provocative future, this life, our life, is embedded now, as it always has been, in the rest of Earth's sentient symphony. . . . Life is open to the universe and to itself.[32]

NOTES

Although the entire article counts as coauthored, Elizabeth Singleton receives primary authorial credit for the final two sections of the paper. Both authors thank Catherine Keller and Mary-Jane Rubenstein for their extensive comments and suggestions for improvement. We have treated their many interpolations with the seriousness they deserved, viz., by incorporating as many as we could into the published version of the chapter. (The chapter thus performs its thesis: Truly, no author is an island.)

1. Immanuel Kant, *Critique of Judgment* (1987), 221.
2. We draw here from Stuart Kauffman and Philip Clayton, "On Emergence, Agency, and Organization," *Philosophy and Biology* 21, no. 4 (2006): 501–21.
3. Donna Haraway, "Situated Knowledges: The Science Question in Feminism as a Site of Discourse on the Privilege of Partial Perspective," *Feminist Studies* 14, no. 3 (1988): 575–99.
4. Friedrich Wilhelm Nietzsche, *Beyond Good and Evil: Prelude to a Philosophy of the Future*, trans. and ed. Marion Faber (Oxford: Oxford University Press, 1998), paragraphs 42–44.
5. See Georg Henrik von Wright, *Explanation and Understanding* (Ithaca, N.Y.: Cornell University Press, 1971); Roderick M. Chisholm, *Person and Object: A Metaphysical Study* (London: G. Allen & Unwin, 1976); Timothy O'Connor, "Dualist and Agent-Causal Theories," in *Oxford Handbook on Free Will*, ed. Robert Kane (Oxford: Oxford University Press, 2001), 337–55.
6. David W. Deamer, *First Life: Discovering the Connections between Stars, Cells, and How Life Began* (Berkeley: University of California Press, 2011); Steen Rasmussen et al., eds., *Protocells: Bridging Nonliving and Living Matter* (Cambridge, Mass.: MIT Press, 2009).
7. The imperfection of all copies/repetitions is of course a major poststructuralist trope, perhaps most powerfully expressed in Derrida's "Différance" article, published in *Théorie d'ensemble* (Paris: Editions Seuil, 1968); see also Jacques Derrida, *Writing and Difference* (Chicago: University of Chicago Press, 1978). Note the resonance with Deleuze's concept of repetition as the habitation of difference in Gilles Deleuze, *Difference and Repetition*, trans. Paul Patton (New York: Columbia University Press, 1994).
8. Philip Clayton, *In Quest of Freedom: The Emergence of Spirit in the Natural World* (Göttingen: Vandenhoeck & Ruprecht, 2009), esp. chaps. 2–3.
9. Jesper Hoffmeyer, *The Semiotics of Nature: An Examination into the Signs of Life and the Life of Signs* (Scranton, Pa.: University of Scranton Press, 2008), 98. Mentioned in the quotation is Stanley N. Salthe, *Development and Evolution: Complexity and Change in Biology* (Cambridge, Mass.: MIT Press, 1993).
10. See Jesper Hoffmeyer, *Legacy for Living Systems* (New York: Springer, 2010).
11. Stuart Kauffman, *Investigations* (Oxford: Oxford University Press, 2000), and *At Home in the Universe* (New York: Oxford University Press, 1995).
12. Terrence W. Deacon, *Incomplete Nature: How Mind Emerged from Matter* (New York: W.W. Norton, 2012).

13. Hoffmeyer, *Semiotics of Nature*, 98.
14. Elisabet Sahtouris, *EarthDance: Living Systems in Evolution* (Lincoln, Neb.: iUniversity Press, 2000), 12.
15. Donna Haraway, *The Companion Species Manifesto: Dogs, People, and Significant Otherness* (Chicago: Prickly Paradigm, 2003); Lynn Margulis, *Symbiotic Planet: A New Look at Evolution* (New York: Basic Books, 1998); Lynn Margulis, ed., *Symbiogenesis: A New Principle of Evolution* (Cambridge, Mass.: Harvard University Press, 2010); Jane Bennett, *Vibrant Matter: A Political Ecology of Things* (Durham, N.C.: Duke University Press, 2010); Karen Barad, *Meeting the Universe Halfway: Quantum Physics and the Entanglement of Matter and Meaning* (Durham, N.C.: Duke University Press, 2007).
16. Bruno Latour, *Facing Gaia: Six Lectures on the Political Theology of Nature*, Gifford Lectures 2012–13 (forthcoming). A draft version of the lectures was for a time available on the internet. A video of the lectures is still available at http://www.bruno-latour.fr/node/487, accessed February 15, 2015.
17. James Lovelock, *Gaia : A New Look at Life on Earth* (Oxford: Oxford University Press, 1987), 9.
18. J. E. Lovelock, "Hands Up for the Gaia Hypothesis," *Nature* 344, no. 6262 (1990): 100–102.
19. We regret that humans seem able to conceive the neutral gender only in mechanistic, nonagential ways, rather than organically. It is not in principle necessary to gender an entity in order to make it alive; asexual reproduction is widespread across the biosphere (e.g., bacteria).
20. D. Raup and J. Sepkoski Jr., "Mass Extinctions in the Marine Fossil Record," *Science* 215, no. 4539 (1982): 1501–3.
21. Elizabeth Kolbert, *The Sixth Extinction: An Unnatural History* (New York: Henry Holt, 2014); see also "The Sixth Great Extinction: A Silent Extermination," *National Geographic*, March 28, 2012, online at http://voices.nationalgeographic.com/2012/03/28/the-sixth-great-extinction-a-silent-extermination/, accessed January 12, 2015.
22. Sahtouris, *EarthDance*, 4.
23. However, we can make predictions based on what we do know of Gaia. Those predictions, unfortunately, suggest a tumultuous future for life as we know it in Gaia.
24. Sahtouris, *EarthDance*, 12.
25. Lynn Margulis and Dorion Sagan, *Dazzle Gradually : Reflections on the Nature of Nature* (White River Junction, Vt.: Chelsea Green, 2007), 177.
26. Sahtouris, *EarthDance*, 24. Extensive empirical data could be cited in support of the Gaia model. For example, Jason Major writes that "researchers at the University of Maryland have discovered a way to identify and track sulfuric compounds in Earth's marine environment, opening a path to either refute or support a decades-old hypothesis that our planet can be compared to a singular, self-regulating, living organism—a.k.a. the Gaia theory." (See http://www.universetoday.com/95183/is-earth-alive-scientists-seek-sulfur-for-an-answer/#ixzz2w5ITpLVJ, accessed March 8, 2014.)
27. Sahtouris, *EarthDance*, 2.

28. Stephan Harding, *Animate Earth: Science, Intuition and Gaia* (White River Junction, Vt.: Chelsea Green, 2006), 88.
29. Brian Swimme and Thomas Berry, *The Universe Story* (San Francisco: HarperSanFrancisco, 1992).
30. For a recent defense of this model, see Philip Clayton and Steven Knapp, *The Predicament of Belief: Science, Philosophy, Faith* (Oxford: Oxford University Press, 2009).
31. Lynn Margulis and Dorion Sagan, *What Is Life?* (Berkeley: University of California Press, 1995), 246.
32. Ibid.

❧ The Matter of Religion

❧ The Matter with Pantheism: On Shepherds and Goat-Gods and Mountains and Monsters

MARY-JANE RUBENSTEIN

Matter is the stage of all sorts of changes, the field of battle of contrary causes, the subject of all corruptions and of all generations; in a word, there is no being whose nature is more inconsistent with the immutability of God.
—PIERRE BAYLE, *Historical and Critical Dictionary*

And to enrich the worship of the ONE,
A universe of gods must pass away.
—FRIEDRICH SCHILLER, "The Gods of Greece"

CALLING NAMES

In his 1695 essay on Baruch Spinoza, Pierre Bayle excoriates the philosopher for having reduced God to "matter, the vilest of all creatures."[1] Matter, after all, is passive, nonrational, and changeable—and as such, everything that the God of classical theism is *not*. In conversation with the renewed focus on materiality in feminist, political, queer, and complexity theories, I would therefore like to ask: What becomes of Spinoza's unbecoming theology if matter turns out to be other than what we thought it was?

Over against the persistent—but not exceptionless—dismissals and subjugations of matter within the Western philosophical tradition, the materialist thinkers engaged in this volume call our attention to the vibrancy, activity, and animacy of matter itself. So Donna Haraway calls our attention to "morally astute dogs"; Myra Hird touts the ingenuity of bacteria and the promiscuity of mushrooms; Jane Bennett traces the vitality of iron; Mel Chen tracks the racialized workings of lead; and Karen Barad uncovers the mutual constitution of subatomic particles, experimental screens, scientists, language, "a warm bed, and a bad cigar."[2] In short, these "materialisms" present us with sites of agency—canine morality, microbial agency, animate minerality, or, while we're at it, "divine animality"—that more proper philosophies tend to deem childish, "vile," ridiculous . . . *absurd*.³ Insofar as they provoke such offense, I would like to suggest that such turns or returns to matter might find a useful supplement, were anyone to write such a thing, on the intellectual history of name-calling.

By "name-calling," the essay I am imagining would mean that perennially rejuvenated strategy of dismissing a position, practice, being, or story by affixing to it a label universally acknowledged to be distasteful. Insofar as this discursive practice presents the loathsome position as meriting no detailed elaboration, name-calling promises to save the speaker a good deal of time; to spare the speaker any possible humiliation by association; and to dissuade anyone else from adopting it, lest the speaker suffer the same sort of ridicule.⁴

Historically, some of the most notorious categories of automatic philosophical dismissal have included "schools" to which no one actually belongs ("idolatry," "nihilism," "hedonism"), "camps" resurrected and re-demolished decades after their last members have moved on ("relativism," "postmodernism," "social constructivism"), "heresies" named by their opponents ("gnosticism," "paganism," "the big bang," "Obamacare"), and "absurdities" of all flavors, whose ranks have historically been populated by such diverse teachings as heliocentrism, the hypostatic union, reincarnation, and actual infinities. Granted, many such nasty names have been strategically and even cheerfully reappropriated by the people they initially ridiculed. In these cases, the work of reappropriation demands a systematic internal evaluation of the term's boundaries and contours. "Of the mess of practices and opinions ascribed to us," the nascent group must ask, "which do we actually endorse, which should we abandon or revise, and which have been invented just to make fun of us?" Along the way, those who seek to rehabilitate a denigrated term must also come to terms with the reasons for its snarky dismissal, asking, "What is it about 'gnostics' or 'pagans' or an initial singularity or 'socialism' or 'queers' that so many people find abhorrent? Is there something perhaps *threatening* about these names we've been called?"

In this spirit, I would like to nominate a candidate for conceptual analysis and rehabilitation. As we explore the complex entanglements of materiality and theology, I think it might be useful to reflect critically on the position that most straightforwardly aligns them, which is to say, *pantheism*. And yet critical reflection is marvelously difficult when the position under consideration has been the target of such sustained and systematic name-calling among philosophers and theologians—no matter how variant and even deviant their politics, methods, and commitments. This history makes it a challenge even to define pantheism, which arguably has as many strains as critics. Etymologically, the word names the identification of *theos*, or God, with *pan*, or "all"—but does this "all" mean "all things as a unity" or "all things" in their plurality? It is perhaps because of this unexamined, constitutive equivocity that, in the words of Philip Clayton, "no philosophically adequate form of pantheism has been developed in modern Western philosophy."[5] And yet the position—if one can even call it that—is perennially and almost universally rejected.

The most common bases for this nearly exceptionless rejection are pantheism's identification of God with matter, which allegedly compromises divine agency, and its identification, therefore, of God with the world, which allegedly compromises divine freedom and infinity. But developments during the last century in nonlinear biology, quantum mechanics, complexity theory, and inflationary cosmologies have shown us in a profusion of ways that matter is not simply passive, and that "the world" is neither determined nor, in any simple sense, finite. It is my hunch, then, that many objections to "pantheism" stem from a misconstrual of materiality and worldliness alike. This chapter, therefore, traces a conceptually viable, and perhaps even theologically compelling, "pantheism" by reanimating the *"pan."* Its efforts take the form of a multilocational trek from contemporary theologies back through the hills of Arcadia, the stable in Bethlehem, and a New Jersey suburb, seeking out those pantheistic strains that center on the following affirmation: that the "vibrant," "intra-active," material "world of becoming" is itself divine[6]—that the ever-emergent universe is what we mean when we say the word "God."

PAN(ICKED)THEOLOGY

Strikingly, no matter who "we" are, pantheism tends to be the position we are trying to avoid. It is the cliff over which we will not go, the slope down which we must prevent ourselves from slipping, whether by means of Thomistic analogy, a high-octane apophatics, various blustery atheisms, or any number of carefully calibrated pan*en*theisms. To be sure, there is nothing wrong with any of these positions; there are plenty of reasons one might reject the identification of God with the material world. What I find perplexing is not the rejection itself, but the *haste* with which it is usually performed. As Grace Jantzen has remarked, "In many quarters, if a proposal is seen as pantheistic or leading to pantheistic consequences, that is deemed sufficient reason to repudiate it, often with considerable vitriol."[7] Ninian Smart attributes such vitriol to a "horror of pantheism in traditional Western theology."[8] Indeed, this horror finds dramatic enactment nearly everywhere one searches for it—even before the term "pantheism" existed.[9] One formidable exemplar of such *horror pantheismus* is Pierre Bayle's essay on Spinoza. As is well known, Spinoza dismantled Descartes's dualism between spiritual and physical substances, calling his single, ideal-material substance "God, or Nature" (*Deus sive Natura*).

Disgusted by this conflation of divinity and materiality, Bayle finds Spinoza's "singularity of substance" so repugnant as to call it "the most *monstrous* hypothesis that could be imagined, the most absurd, and the most diametrically opposed to the most evident notions of our minds."[10]

Certainly, Bayle was not the first to express such rancor in the face of this teaching; Spinoza was infamously excommunicated from his Jewish community in Amsterdam for having maintained "that God has a body," namely, the body of the world itself.[11] Perhaps needless to say, this position flagrantly violated the monotheistic insistence on God's incorporeality. For the heresy of giving God a body (which is to say, the world), Spinoza was therefore expelled bodily from the synagogue "with the anathema with which Joshua anathematized Jericho"; to wit,

> Cursed be he by day, and cursed be he by night, cursed be he when he lieth down, and cursed be he when he riseth up; cursed be he when he goeth out and cursed be he when he cometh in; the Lord will not pardon him; the wrath and fury of the Lord will be kindled against this man . . . and the Lord will destroy his name from under the heavens; and, to his undoing, the Lord will cut him off from all the tribes of Israel.[12]

In keeping with this genealogical break, the local council imposed a social quarantine, as well: "We ordain that no one may communicate with him verbally or in writing, nor show him any favour . . . nor be within four cubits of him, nor read anything composed or written by him."[13] On this matter, at least, the orthodoxies of Christianity and Judaism could agree; after all, the church had been cursing, uprooting, and even murdering proto-pantheists for centuries.[14]

Granted, identifying God with a material creation is a highly unorthodox move. But what is it about pantheism that fuels *this degree* of horror? What is it that prompts the elders' multidimensional anathematical topography (cursed be he by day, by night; when he's up, down, in, and out); that cuts the proponent off from all relation (perhaps to prevent infection?); and that constitutes not just an error, but an unforgivable one? Whence stems the *horror religiosus* that not only excommunicates Spinoza, but in the hands of Christian hierarchs condemns John Scotus Eriugena, executes the followers of Almaric of Bena, burns Giordano Bruno at the stake, incinerates Marguerite Porete, suspects even Jonathan Edwards of heresy, and would have obliterated Eckhart if he hadn't died first?[15]

> *panic, n.: (of fear): groundless . . . ; strong, infectious, unreasoning feeling of any kind.*[16]

Among the orthodox, the para-orthodox, and the guardians of rationality, pantheism often inspires something like panic: a strong revulsion with no discernible source, accompanied by an effort to dismiss it at all costs. In recent years, the most vocal critic of this generalized panic has been the late feminist philosopher of religion, Grace Jantzen. In the late 1990s, she began to argue that all the oppressive dualisms structuring Western thought were held in place by the fundamental difference between God and "the physical universe."[17] For Jantzen, the (binary) ontological distinction between God and creation establishes God's *mastery* over creation, securing in turn the supremacy of everything associated with this God (spirit, masculinity, reason, light, humanity) over everything else (matter, femininity, passion, darkness, animo-vegeto-minerality).

Admittedly, this is a well-rehearsed set of hierarchies, which feminist thinkers of both secular and sacred varieties have struggled for decades to dismantle. But as far as Jantzen is concerned, the only way to collapse the structure is to go for its root, which is to say the opposition between God and the world. "If pantheism were seriously to be entertained," she ventures, "the whole western symbolic ... would be brought into question. Pantheism rejects the split between spirit and matter, light and darkness, and the rest; it thereby also rejects the hierarchies based on these splits."[18] While affirming the spirit of this critique, one might take issue with the absolute priority Jantzen gives to the God/world opposition, which other feminist thinkers have exposed as the product of ancient patriarchies and perennial racisms.[19] It might therefore be more helpful to see these vectors of power as rhizomatically entangled than as arboreally rooted:[20] The integrity or destruction of each would depend on the integrity or destruction of the others. And for Jantzen, the position that promises to unearth the whole thicket of oppressions is pantheism.

Throughout her argument, Jantzen is careful to explain she is not working from a "realist" (or "anti-realist") stance; rather, she is working at the level of the *symbolic*. In other words, Jantzen is not saying that God is (or is not) the universe or that the universe is (or is not) divine; rather, she is trying to recode "divinity" as a concept, whether or not an "entity" called God "exists." Understandably, many feminisms, materialisms, posthumanisms, and queer theories prefer to sidestep the God-issue, having had more than enough of the Guy in the Sky. From Jantzen's perspective, however, this circumvention inadvertently leaves the God intact *conceptually* as a disembodied, omnipotent, anthropomorphic Father. Even atheists reinscribe the concept of the very God they don't believe in; for example, "theists and atheists tacitly agree on the masculinized nature of the God whose existence they dispute. Thus whether it is held that there is a God or not, the *concept* of the divine serves to valorize disembodied power and rationality."[21] And of course, the concept of the divine is the most highly valued concept we have.

For the sake of our threatened planet, in the face of our waning biodiversity, and in solidarity with those living and nonliving beings whom the Father-aligned continue to master, colonize, denigrate, and destroy, Jantzen suggests that feminist philosophers begin deliberately to project a pantheist God—a God who *is* the material universe in all its multiplicity. In her words, "If we took for granted that divinity—that which is most to be respected and valued—*means* mutuality, bodiliness, diversity, and materiality, then whether or not we believed that such a concept of God was instantiated . . . the implications for our thought and lives would be incalculable."[22] That was in 1998. Jantzen's *Becoming Divine* continues to be circulated and widely taught, and yet I don't know of anyone who has taken up its call to a feminist pantheism. Nor has anyone really argued against it.[23] A voice crying in the wilderness at the end of the second Christian millennium and then radio silence.

Silence, dismissals, and anathemas aside, there are plenty of serious objections to pantheism. The most common of these are its purported atheism, immanentism, deification of evil, denial of freedom, and denial of difference. And although each of these would in a lengthier essay merit its own elaboration and response, it seems to me the last of these is most pressing, especially insofar as it gathers so many of the rest toward it. The systematic theologian Colin Gunton weaves these objections around a common concern for "difference" when he argues, "For there to be freedom, there must be space. In terms of the relation between God and the universe, this entails an ontological otherness between God and the world. . . . Atheism and . . . materialism are in effect identical with pantheism, for all of them swallow up the many into the one, and so turn the many into mere functions of the one."[24] In short, the argument is that if there is no difference between God and the world, there can be no difference *at all*, inasmuch as the ontological distinction grounds all other distinctions. And if there is no difference, then none of the parties involved is sufficiently autonomous to be "free." So if we antiracist, genderqueer, postcolonial ecophiles know what is good for us—that is, if we want things like difference, diversity, multiplicity—we'd better hang onto the ontological distinction.

At this point, however, one might ask whether the only options out there are a two-column hierarchy on the one hand and a "denial of difference" on the other. One might even go so far as to ask whether the "two" and the "one" are really such different positions to begin with. After all, the metaphysical framework that stems from God-versus-world—opposing in turn form and matter, male and female, eternity and time, colonizer and colonized, etc.—does not establish the second as genuinely different from the first, so much as a derivation, deviation, and/or bad copy of it. One might think here of Judith Butler's analysis of lesbianism as a purported imitation of heterosexuality, or of Homi Bhabha's "colonial mimicry," which produces non-Europeans as "almost the same, *but not quite.*"[25] The oppositional logic of classical metaphysics does not, then, give us two; it gives us one, and a falling-short of that one. Nor, one might add, does this binary scheme secure the "freedom" of both terms; rather, it secures the freedom of the historically dominant term at the expense of its subjugated other.[26] And so the real concern over pantheism is not the collapse of some abstract notion of "difference"; it is the collapse of one particularly insistent and damaging way of *configuring* difference—one that gathers each instance of "difference" into a category benevolently overseen by a single metaphysical life-partner.

Again, there may be good reasons to reject pantheism's identification of God and the universe, whether from a realist or symbolic perspective, whether for sacred or secular reasons. But, to rely a bit more on Jantzen, "insofar as 'pantheism' is treated as a swear word, greeted with dismay and repudiation," it becomes an object not of critical evaluation but of *fear*. Specifically, she argues, "the fear of pantheism bespeaks a perceived if unconscious threat to the masculinist symbolic of the west."[27] Jantzen detects this panicked masculinity in the surprisingly recurrent language of pantheism's "swallowing," "consuming," and "assimilating" all otherwise "free" beings into some dark abyss, some ridiculous "night in which all cows are black."[28] "From a psychoanalytic perspective," Jantzen ventures, "one could speculate about what dread of the (m)other and the maternal womb lurks just below the surface of this fear of pantheism; what exactly is the abyss, this horror of great undifferentiated darkness into which at all costs 'we' must not be sucked?"[29]

While Jantzen's diagnosis might seem to rely on a rhetorical sleight of hand—or a set of Lacan-geeky pun(ctuation)s—a quick survey of recent rejections of pantheism does confirm the presence beneath them of something like gender-panic. For example, the evangelical theologian William Lane Craig defends the ontological distinction against pantheism (and panentheism) with the following illustration: "In marriage the antithesis of two persons is *aufgehoben* as husband and wife come together in a deep unity even as their distinctness as persons is preserved. In the same way, the opposition between infinite and finite, God and world, is *aufgehoben* in that God is intimately related to the world in various ways even as the ontological distinctness between God and the world is preserved."[30] Pantheism's demolition of the ontological distinction between God and world therefore amounts to a demolition of the sexual distinction between man and woman, the first of whom is aligned with infinity and God, while the second gets finitude and world. Reaffirming this alignment, Craig explains that God "embraces . . . his creatures . . . just as a husband embraces his wife."³¹ So we'd better hang onto that difference; otherwise who knows who might embrace whom and what un*aufgehoben*able differences might emerge.

In his defense of global capitalism as the economic vehicle for a truly global Christianity, theologian Stephen H. Webb rejects the planetary viability of a "sacred earth" cosmology. "Judaism, Islam, and Christianity," he cautions, "are unlikely to dismantle their notions of divine transcendence in order to embrace an earth goddess."³² In this declaration, at the risk of pointing out the obvious, Webb is linking the demise of divine transcendence to the emergence of divine femininity. This femininity is furthermore tied to the earth—the mother is matter, of course, and as such, is reduced to "resources" for human (read: male) development—*and* earthly femininity is tinged with the mild sexuality of an "embrace" that sounds strikingly like Craig's. Fascinatingly, at the other end of the theological spectrum, we find Richard Dawkins calling pantheism a "sexed-up atheism," so wherever you stand, pantheism is not only "absurd" but also feminized—and dangerously seductive.³³ Taking these positions together, then, what Bayle would call the "monstrosity" of pantheism—the thing that inspires such panic—amounts to a complicated hybridity of divinity, femininity, materiality, and sex, undesirable (which is to say all-too-desirable) to theists and atheists alike. And this, I suggest, is the real matter with pantheism: It threatens the Western symbolic not just with a (m)other-womb, but with a wider and more complex range of monstrosities—with parts combined that ought to be kept separate and boundaries crossed that ought to be maintained.

Of course, it all depends on what you mean by pantheism.

So far in this essay, I have opted to define the pantheist position as one that identifies God with the universe in all its vibrant materiality, emergent complexities, and intra-constituted agencies. In so doing, I have ascribed both temporality and multiplicity to the hypothetical pantheist God-world: The universe is constantly evolving, accelerating, producing new forms not incipient in the old; and as such, the universe—what the pantheist means by divinity—is not a monist totality but rather an irreducible multiplicity. To be clear, "multiplicity" here is different from numerical plurality; the idea is not that "there are a lot of things, the sum of which is God." Such a position would be impossible for anyone who takes matter seriously because (1) the world is open, evolving, and relationally self-exceeding, and therefore not a "sum," and (2) "things" themselves are open, evolving, and relationally self-exceeding, and therefore both more and less than "things." "At any moment," Jane Bennett writes, "what is at work . . . is an animal-vegetable-mineral sonority cluster."³⁴ Such hybridity is taking place whether we are speaking about cells, bacteria, the "human" genome, water, air, a cloned sheep, or a "collapsed" wave function: Each of them is composed of a mutating band of others. If, with Karen Barad, we add discursivity into the mix,[35] then our multiple-universe becomes an untotalizable and shape-shifting product of narrative-theoretical-material assemblages that are neither reducible to, nor constitutive of, "oneness." And this provisionally unified but constantly evolving multiplicity is what the "pantheism" at hand would call divine. Such a pantheism therefore stands in direct opposition to those more common pantheisms, whose fundamental assertion is "that everything . . . constitutes a *unity* and that this all-inclusive unity is divine."[36] Depending on one's starting point, "pantheism" divinizes either a messy multiplicity or a smoothed-out whole, and this particular expedition is foraging for the godly mess.

To be sure, nearly everyone who encounters such a tension between the many and the one will try to assert their *identity*, a project that begins in earnest with Plotinus and carries on at least through Hegel. Spinoza himself maintained both that there was only one substance (*Deus sive Natura*) *and* that this substance had an infinite number of attributes.[37] As Arthur Lovejoy demonstrates, however, this dual assertion landed Spinoza right back in "the peculiar paradox of neo-Platonism," which is to say the affirmation of "a being which should include the *esse* of all things without possessing the attributes of those things."[38] In other words, Spinoza's God-Nature for Lovejoy contains the abstract "being" of everything, but not the concrete *stuff* of everything:[39] The One Substance contains some ethereal version of "me," but not Polish-Italianness, recalcitrant hair, or a deep love of Gershwin. At the end of the day, then, Lovejoy insists that the pantheist will have to come down on one side or the other and assert the primacy of oneness or the primacy of multiplicity, for "if one says that God is both an absolutely simple Monad and a complex Whole of Things, the mind is certain sooner or later to cry out: *Which*, when all is said and done, am I to think of God? For honestly and vividly think of him in both ways, I cannot."[40] Now, an "I" who hopes to side with Lovejoy's "complexity of things" might want to abandon his language of "wholeness"—as well as the language of "himness"—but the distinction between the monad and the complex is helpful. And as we have already glimpsed, this difference may boil down to an irreducible etymological duplicity in "pan-theism" itself: At the end of the day, does the *pan* signify a unified "one," or a multiple "all things," each of which is "itself" multiple?

These two different meanings of "pan" map onto the distinction William James makes in *A Pluralistic Universe* between "monistic" pantheism on the one hand and "pluralistic" pantheism on the other.[41] For the monist, James tells us, the world is one "tremendous unity," in which "everything is present to *everything* else in one vast instantaneous co-implicated completeness."[42] For the pluralist, by contrast, the things of the world are "in some respects connected, [and] in other respects independent, so that they are not members of one all-inclusive individual fact."⁴³ Of course, James is a pragmatist, so as William Connolly reminds us, he knows he cannot say which of these visions is ultimately "true," or if it even makes sense to speak that way.⁴⁴ But James sides with pluralism for a host of ethical, political, and psychological reasons: If we affirm an inherent plurality rather than a primordial unity, then "evil" calls for a practical response rather than a speculative explanation; differences of opinion are signs of health rather than pathology; and our everyday experiences amount to "intimacy" with the universe itself.⁴⁵ This is perhaps James's most novel critique of the monist tradition: Presumably, the pantheist locates the divine in and as the world in order to gain intimacy with it. But if the world-as-divine bears none of the characteristics of the only world we ever experience (its desires and mistakes, its passions and pains, its earthworms and Gershwin), then the monist places himself even farther than the ordinary theist from God.⁴⁶ So James opts for pluralism, which makes of the universe what he calls a *multiverse*: a loosely coherent chain of complex connections that's never quite all-in-all.⁴⁷

Disappointingly, however, James's vision of divinity does not match his vision of cosmology, even though the two ought to be coextensive. Even though James begins by affirming pantheism as "the only [opinion] quite worthy of arresting our attention," he ends by splitting "world" and "God" into a rich, multiversal plurality on the one hand and a single, disembodied, anthropomorphic, male divinity on the other—a limited force that works alongside other limited agents in the pluralistic universe.⁴⁸ To be sure, there are plenty of reasons to affirm such a minimal theology—including concerns for theodicy, relationality, and creaturely freedom—the issue is simply that this diminished humanoid divinity does not match the complex, entangled vibrancy of the material world with which James's fully pluralist pantheism would ostensibly identify "God." Where, then, might we find a pantheism that takes refuge neither in the individual nor in the human nor in a spirituality divorced from the material? How might we set forth a *pan* that measures up to the hybridity, complexity, and multiform vegetal-mineral-animality of the multiverse?

ALL WE LIKE SHEEP

In the spirit of materialisms old and new, it might be useful to continue our search for a multiversal *pan* in a particular nodule of space-time, with a particular *object* (or two). It is 1988 on the New Jersey coastline; I am eleven; and outside my bedroom are two lawn sheep made of wool and wood, which I had begged my mother to get me, having experienced a sudden and overpowering *need* for a "flock." A quick Google search of newspaper articles from the late '80s suggests that the call I had heard was less likely the voice of the Good

Shepherd than the *mysterium tremendum* of capitalism, which creates the most baffling needs in order to fulfill them, including "stuffed animals on sticks . . . the latest craze in lawn ornaments."[49] The "craze" apparently extended all the way down to Florida, where one columnist voiced her suspicion that "lawn sheep fulfill some inner need, some secret yearning, on the part of homeowners who place them singly, in pairs, even in flocks, on their lawns."[50] She ventured that this "inner need" might have to do with bucolic longing amid suburban cul-de-sacs, or a consumerist desire for interspecies connection without the excreta. Whatever the motivations, so many sheep dotted our landscape in such quick order that the kids in our neighborhood started "tipping" them the way they'd heard rural kids tip cows, or pulling them out of the ground and relocating them on patches of grass near the convenience store, the car wash, and their high schools. One of the members of our Lutheran church was a local police officer, who said the precinct was overrun with recovered puffs of wool on sticks, and had absolutely no idea what to do with them.

"And there were in that same region shepherds, keeping watch over their flocks by night" (Lk 2:8). Not me this time, but my little brother Kenan, who at the same Lutheran church was lounging along with some high schoolers three times his size on bales of hay that my mother had talked the last local farmer into donating for our annual Live Nativity. It was Advent, of course, and it was freezing, but we staged the thing outside so that passing cars might stop at the sight of our floodlit stable—a plywood shed my mom hid behind so she could push out the angels and kings on time. For years, there were no actual sheep; the audience would just see the rumpled shepherds and *imagine* they had flocks. In 1988, we rounded out the operation with my lawn sheep, the poor things thrown all of a sudden into the fullness of time.

Unfortunately, the lawn sheep had a short run onstage; just one year later, our pastor's wife met a woman at the 7–11 who told her she kept a few farm animals in her suburban backyard and would be glad to lend them to us for the Nativity. So that year, my mother, my brother, two angels, and I crossed a four-lane highway in white albs and brown cinctures to retrieve two *real* sheep named Daisy and Baby, along with one goat with the weight of the world in his name: Mondo. Having had no experience with farm animals, it was hard for our gaggle of middle-schoolers to prevent Daisy from eating a boxwood on the way back. But it was worth the wrath of that unsuspecting homeowner to see the crowd that stopped to stare at the two live sheep and one live goat on the church lawn that year, casually tethered to the same snoozing shepherds, the smallest of whom tried to suppress a giggle as he whispered to the figure behind the stable, *"Mom! Mondo's eating my knees!"*

The angel, the Mary, the Joseph, the stall—we're so accustomed to the scene that we don't often stop to ask, "What's going on with the sheep?" Why did Luke give us shepherds watching their flocks by night, rather than bakers baking their bread or cobblers cobbling their boots? One source notes that by surrounding the scene with sheep, Luke connects the newborn king to David, who tends the flocks of his father Jesse (1 Sm 16:11).[51] Luke is concerned to establish Jesus in David's lineage through his father Joseph (quite a feat considering the virgin mom), and so Luke has the baby born in David's hometown of Bethlehem, with sheep to drive the point home. Another commentary connects the scene to the "bad shepherds" of Ezekiel 34, who symbolize the bad kings of Israel and whom the text contrasts with Israel's true shepherd, God.[52] Finally, both sources suggest that Luke is playing on the Greek and Roman motif of a royal infant discovered by shepherds—calling to mind Oedipus, Paris, Romulus, and Remus.[53] Along this line of thought, if Greeks and Romans see sheep in the story of a strange birth, they're likely to lend a mythic significance to the baby.

Connections between Christ and the classical world were fascinating to Renaissance authors in particular, who, thanks to a hilarious string of hermeneutical bungles to be detailed momentarily,[54] saw the shepherds of Luke as heralds of the Good Shepherd whom they greeted, and the Good Shepherd Himself as the consummate form of a much older shepherd-god.

> Pan, n.: allusively. A person with responsibility for shepherds and flocks; a chief shepherd (occas. applied to Jesus Christ).[55]

Half-man, half-goat, the Greek god of shepherds and goatherds originated in Arcadia, "where divine theriomorphism is well attested."[56] Herodotus tells us that the cult of Pan began to spread after the Battle of Marathon, when the goat-god appeared to Phidippides to say that if the Athenians worshipped him, he would terrify the barbarians and secure the victory of Athens.[57] A cave was quickly built atop the Acropolis—Pan is worshipped not in temples but in the wombspaces of grottoes—in which devotees would dance and sing, becoming fitfully possessed by their "noise-loving," pipe-playing deity, who liked to spring from nowhere and strike terror in the hearts of travelers.[58]

> panic, n.: "originally and chiefly used allusively with reference to a feeling of sudden terror, which was attributed by the ancient Greeks to the influence of the God Pan."[59]

Physiologically and functionally, Pan is a monstrously hard god to classify. Having "the horns, ears, and legs of a goat" with the torso and head of a man,[60] and being moreover a god, he is an irreducible hybridity—a collision of elements that any sane theology would keep separate. In his goat-half alone, he is already what Sharon Coggan calls "liminal": A goat is "not entirely tame, yet . . . not entirely wild"[61]—the kind of animal who might chew a kid's knees in a Christmas pageant. Of course, Pan is not just a goat; his triune ani-man-god-ness means he is also a shepherd or goatherd, as well as the *guardian* of shepherds and goatherds—along with their charges. Even bees were said to be under Pan's oversight, in his role as protector of flocks.[62] Ironically, however, Pan is also known as a hunter—as the god who ensures a successful kill—and is in this vein known as Pan Lykaios, or "Wolf-Pan," deadly enemy of flocks.[63] He is commonly dressed in the skin of a lynx or a fawn (wolfgoat in deercat clothing?), and his twin brother is said to be neither a goat nor a sheep nor a bee nor a wolf, but a *bear*: Arcas, ancestor of the Arcadians.[64] So Pan is what Haraway might call a "contact zone": a cross-species concatenation of "world-making entanglements,"[65] within which he is both singular predator and flockish prey, both protector and pruner of the multitude.

In addition to protecting and hunting, Pan is also known to pursue. "Plainly a lusty god," he is usually portrayed with an oversized phallus, looking to seduce anything that moves.[66] He is usually unsuccessful, rebuffed by forest nymphs and shepherd boys alike, and called by the name "Pan Duserous": "lusty, but 'Unlucky in love.'"[67] In this regard, he can be both mournful and vengeful: When the chaste nymph Syrinx refused him, ran to a riverbank, and was turned into reeds, Pan's cries made such a haunting sound across her new-found vegetality that he cut the reeds to make them (her) into his eponymous flute, the syrinx. These rejections aside, he is said to have had a tryst with Aphrodite, a fairly long-term arrangement with the muse Eupheme, and a fling with "every one of the Maenads"; so this queer god's interests range from boys to goddesses to women, and—lest we forget his other half—he is also known as "Mounter of the Goats."[68]

pan, n., adj.: an abbreviation for pansexual.[69]

Topographically, Pan is just as overdetermined. "Always an outsider to the world of Mount Olympus," Pan inhabits less sacred mountains, the "sure-footed" goat at home in all high, "rugged, rocky places."[70] But he also shows up in the subterranean caves where he is worshipped and where he sleeps as soon as the sun rises. Pan oversees pastures, of course, but also inhabits forests, where he both strikes terror in the hearts of unsuspecting passersby *and* delights his devotees with all-night dance parties set to his nymph-flute.[71] So this awesome, awful deity dwells within mountains and caves, fields and wood, vegetation and minerality—and by the way, he was said to have the power to "rescue sailors on a becalmed ship," so there seems to be nowhere he isn't.[72]

Pan, int.: international radio signal, esp. by ships and aircraft, to alert authorities that the vessel or aircraft requires assistance . . . a step below mayday.[73]

Perhaps unsurprisingly, this multilocational misfit—this hypersexual hybrid with multiple personalities—has no clear origin story, there being "no fewer than fourteen different versions of his parentage."[74] His father is most often said to be Hermes, messenger of the gods, whose patrilineage establishes Pan—at least for Plato—as the incarnation of "speech."[75] Other accounts name Pan's father as Zeus or Apollo.[76] And while his mother is usually said to be one of any number of nymphs, she is sometimes said to be the human Penelope, who in this version of the story did not wait those twenty years for Odysseus to come home; rather, she conceived Pan with one of the gods, or with one of her suitors.[77] In the more vanilla *Homeric Hymns*, Pan is the child of Hermes and the nymph Dryope, daughter of Dryops, a mortal whose sheep Hermes had tended (here again, sheep establish a royal lineage). The poet sings, "Dryope bore Hermes a dear son, marvelous to behold: / goat-footed, horned, full of noise and sweet laughter."[78] But as nymphs, shepherd-boys, and barbarians will do for centuries, Dryope jumps up in terror and flees at the sight of the goat-baby with his "rough, full-bearded face" (line 39). Hermes, by contrast, is delighted with his child and, swaddling him "in the thick fur of mountain hare," flies the strange thing to Olympus to show him off (line 42). The hymn tells us: "All the gods were delighted / in their hearts, but especially Bacchic Dionysos. / 'Pan' they named him, because he delighted them 'all'" (lines 45–47).

Thus with the *Homeric Hymns* (7th century BCE) commences a rich and strange tradition of associating Pan (*Pán*) with "the all" (*tò pân*), the closest term the Greeks have to "universe." Some of the bawdier sources perform this elision by saying Penelope was unfaithful to Odysseus not just with one god or suitor, but with them *"all,"* "and that from this intercourse was born Pan."[79] Whatever his lineage, however, this anti-Oedipal monster-god begins in Roman times to be seen "as a universal god, or god of Nature," "the pantheistic divinity," "the All."[80]

As Porphyry (ca. 234–305 CE) explains, "They made Pan the symbol of the universe and gave him horns as symbol of the sun and moon and the fawn skin as emblem of the stars in heaven, or of the *variety of the universe*."[81] We should note that Porphyry attributes "variety" to his *tò pân*, whereas most other Neoplatonist and Scholastic authors will follow Plato and Aristotle in asserting its *oneness*.[82] And so in the universalizing of the goat-god, we see a "Pan" of manifold hybridities, transgressed boundaries, and material multiplicities collide head-on with a "pan" which, depending on how you configure your universe, either means the "variety" of all things or all-things-as-one.

Christian apologetic sources go on to conflate *and* toggle between these two *pans*, depending on which strategy serves them best. Most notably, Eusebius of Caesarea (ca. 260–340) devotes two sections of his *Preparatio Evangelica* to a strange story in Plutarch that announces "the death of Pan." Opening on a boat piloted by an Egyptian man named Thamus, the story recalls the passengers' hearing a voice from the shore of Paxi calling, "Thamus, Thamus, Thamus; the Great Pan is dead!" Astonished that the voice would know his name, the captain agrees to pass the news onto the next island they reach—news whose delivery elicits "a loud lamentation, not of one but of many, mingled with amazement."[83] Once the ship returns to Rome, the captain files a report with Emperor Tiberius, who commissions an investigation, which concludes that the deceased in question was, in fact, "Pan the son of Hermes and Penelope."[84]

For a century now, many classicists have argued that the whole story was based on a misunderstanding that went over Plutarch's head.[85] Eusebius, however, takes the tale as a historical report of the death of Pan, who stands metonymically for "all" the pagan gods. Noting that the account takes place during the reign of Tiberius, Eusebius reminds his reader that these were the days of Christ's "sojourn among men," during which he "[rid] human life from demons of every kind."[86] For Eusebius then, the death of Pan is coincident with the life of Christ, who rids the world of "'All' the Greek gods, that is . . . all the evil demons."[87] And so the Lamb of God overcomes the goat-god, who goes on to become not just one evil spirit among many in the Christian imagination, but the demon of demons himself. Singling him out for his unbridled sexuality, Christian mythology parleys the "horns, hooves, shaggy fur, and outsized phallus" of Pan into the paradigmatic "image of Satan."[88]

Strikingly, however, the author who is most noted for his portrait of Satan *also* wrangles Pan into a forerunner of Christ. Calling us back to those pastures outside Bethlehem, John Milton imagines,

> The shepherds on the lawn
> Or ere the point of dawn
> Sat simply chatting in a rustic row
> Full little thought they then
> That the mighty Pan
> Was kindly come to live with them below.[89]

A half-human shepherd god of "all," Christ becomes for Milton the *true* Pan.

Ironically, François Rabelais had given voice to this similarity a century earlier through his "absurd" character Pantagruel, who interprets Plutarch's "death of Pan" as an account of the crucifixion.[90] So named by his father, who imagined him "thirsting after *the all*,"[91] Panta-gruel defends his bizarre conflation with extraordinary rhetorical flourish. The death of Pan can be interpreted as the death of Christ, he explains, "for in Greek [Christ] can rightly be called *Pan*, seeing that he is our All, all that we are, all that we live, all that we have, all that we hope, is in him, of him, by him."[92] The hapless scholar goes on to remind us that both Pan and Christ are shepherds, and that at the moment of the crucifixion, "plaints, sighs, tumultuous cries and lamentations [rang] throughout the entire machine of the Universe: Heaven, earth, sea, and Hell."[93] This, then, was the source of the cries off those Grecian shores. Reversing the Eusebian interpretation, Pantagruel presents the "death of Pan" *not* as the death of the pagan gods exorcised by Christ, but as the death of the exorcist himself: "For that Most-good, Most-great Pan, our Only Servator, died in Jerusalem during the reign in Rome of Tiberius Caesar."[94]

As classicist Wilfred Schoff illustrates, and to his great consternation, this exegetical absurdity becomes "noble verse" when Milton misses the joke and imports the whole set of associations into his Nativity Ode.[95] From there, the conflation of Christ, Pan, and all-ness becomes commonplace: Edmund Spenser reminds us that "The great Pan is Christ, the very God of all shepherds," whose death coincides with "the death of Pan";[96] Ben Johnson writes that "PAN is our All, by him we breathe, we live, / We move, we are";[97] and Elizabeth Barrett Browning tunes into that moment "When One in Sion / Hung for love's sake on the cross" to hear forests, fields, mountains, and seas cry out in agonized uniformity that "Pan, Pan, is dead."[98]

MULTIPANTHEOLOGY

The historian Robin Lane Fox tells us that of all the pagan gods whom Christianity excised, "no presence has been more haunting than Pan's."[99] This ought not to be surprising; after all, Pan has never been a great respecter of boundaries. If any deity were to cross a maze of ontospatiotemporal divides to trouble our sleep, it would likely be this polyamorous polymorph. Listening closely, we can pick up strains of him, alluring and terrifying not only the poets but humans of all sorts—especially when their exceptionalism seems threatened. Whether or not the humans in question acknowledge his presence, Pan has shown up in debates over zoological nomenclature,[100] in eco-activist struggles,[101] and, I submit, in panicked dismissals of pantheism—among Christians above all. But why are Christians so exercised about pantheism? And why do they have to keep warding it off?

On the one hand, Christians can claim total safety from pantheism, taking refuge in their lifeless earth, their human privilege, and their genetic heterolineage—all held in place by an extracosmic Father. On the other hand,

Christians also have *Christ*: in his human form an anti-Oedipal half-breed[102] who shows up amid sheep and goats, and in his divine form the principle of creation itself—the word through which the world is worlded. Considering the cosmogonic function of the logos, and assured of an infinite cosmos, Giordano Bruno concludes in the late sixteenth century that the cosmos *is* the word of God; that creation is the incarnation, which is to say the whole universe is God-in-the-universe.[103] Pan-carnation.[104] It is perhaps because the move is so easy to make for a Christian that it is so perennially attractive—or maybe it's the other way around: Pantheism is attractive because Christianity always teeters on the verge of it. But either way, as James reminds us, "Orthodox theology has had to wage a steady fight . . . against the various forms of pantheistic heresy" only because people are so steadily drawn to those forms.[105] In the case of Bruno, at least, his pantheist attraction was so threatening that "orthodox theology" burned it along with him on an Ash Wednesday in the middle of Rome at the turn of the seventeenth century.

At this late hour, I should make it clear: I am not calling for a post-Christian retrieval of the cult of Pan. Aside from believing such a return to be impossible, I find it undesirable; however protocyborgian and speciesqueer, an ithyphallic goat-man is not a god into whom I'd suggest we pile our theo-erotic energies. Neither am I advocating a (re-)turn to any number of other pantheisms; rather I am simply trying to figure out what such a thing would mean in the first place and why it has traditionally been so difficult to consider it as a coherent position. In particular, I have tried to uncover some of the sources of the aggressive and automatic dismissals of pantheism, sources that reliably amount to crossed boundaries, mixed-up categories, and monstrous combinations that usually have something to do with sex and gender. For this reason, it seems to me that the pantheism that truly threatens the Western symbolic would not be the "all is one" variety; after all, the "one" is just the "two" being honest with itself. The most threatening, and therefore most promising, pantheism would rather be the mixed-up, chimeric variety, whose *theos* is neither self-identical nor absolute, but a mobile and multiply located concatenation of pan-species intra-carnation. And one particularly salient, but evanescent, node of such intracarnational pantheism happens to be Pan himself, who crosses divisions of topography, species, function, ontology, time, space, culture, and decency not in order to make them "all one," but rather to present us with strange new sites of divinity. In such a *multipantheology*—this provisional name promiscuously mingling its many and one, its Latin and Greek—divinity would be not static but evolving: discovered, sustained, killed off, resurrected, and multiplied between and among temporary clusters of relation. As it did in those queerly intraspecies assemblages of Arcadia and Nazareth, divinity thus construed would show up in unforeseen crossings and alliances, frightening and delighting us with glimpses of the other worlds and gods that might yet emerge within the world—or God—we're *in*.

NOTES

My profound thanks to Winfield Goodwin, Lori Gruen, and Catherine Keller for their help in researching, writing, and editing this chapter.

1. Pierre Bayle, *Historical and Critical Dictionary: Selections*, trans. Richard H. Popin (Indianapolis: Hackett, 1991), 291.

2. Ivone Gebara, *Longing for Running Water: Ecofeminism and Liberation* (Minneapolis: Fortress Press, 1999), 160; Myra J. Hird, "Naturally Queer," *Feminist Theory* 5 (2004): 85–89; "Animal Transex," *Australian Feminist Studies* 21, no. 49 (2006): 41; Jane Bennett, *Vibrant Matter: A Political Ecology of Things* (Durham, N.C.: Duke University Press, 2010), 58–60; Mel Chen, *Animacies: Biopolitics, Racial Mattering, and Queer Affect* (Durham, N.C.: Duke University Press, 2012), 159–88; Karen Barad, *Meeting the Universe Halfway: Quantum Physics and the Entanglement of Matter and Meaning* (Durham, N.C.: Duke University Press, 2007), 161–68.

3. Perhaps no one gives clearer expression to the childishness of what the nineteenth century would call "animism" than Edward Tylor, who writes that "he who recollects when there was still personality to him in posts and sticks, chairs, and toys, may well understand how the infant philosophy of mankind could extend the notion of vitality to what modern science recognizes as lifeless things." Although Tylor finds much of "animism" to be charming, he is particularly repelled by the absurdity of what Derrida would call divinanimality: "To the modern educated world, few phenomena of the lower civilization seem more pitiable than the spectacle of a man worshipping a beast" (Edward Burnett Tylor, *Religion in Primitive Culture*, vol. 2 [New York: Harper and Row, 1958], 62, 315). For a rich set of explorations of varied concatenations of divinity and animality, see Stephen H. Moore and Laurel Kearns, eds., *Divinanimality: Animal Theory, Creaturely Theology*, Transdisiplinary Theological Colloquia (New York: Fordham University Press, 2014).

4. Although I don't know of anyone who has conducted a lengthy study of philosophical name-calling, there is one particularly illustrative paragraph-fragment in Schopenhauer: "If you are confronted with an assertion, there is a short way of getting rid of it . . . by putting it into some odious category. . . . You can say, for instance, 'That is Manichaeism,' or 'It is Arianism,' or 'Pelagianism,' or 'Idealism,' or 'Spinozism,' or 'Pantheism,' . . . and so on. In making an objection of this kind, you take it for granted (1) that the assertion in question is identical with, or is at least contained in, the category cited—that is to say, you cry out, 'Oh, I have learned that before'; and (2) that the system referred to has been entirely refuted, and does not contain a word of truth" (Arthur Schopenhauer, *Parerga and Paralipomena: A Collection of Philosophical Essays*, trans. T. Bailey Saunders [New York: Cosimo, 2007], 34).

5. Philip Clayton, *The Problem of God in Modern Thought* (Grand Rapids, Mich.: Eerdmans, 2000), 389. In his study of gnostic and pantheist revivals among twentieth-century Jewish theologians in particular, Benjamin Lazier notes the wide "ref-

erential promiscuity" of the term as a polemical catchall (Benjamin Lazier, *God Interrupted: Heresy and the European Imagination between the World Wars* [Princeton, N.J.: Princeton University Press, 2012], 73). Similarly, Michael Levine laments the lack "at any time" of a systematic study of pantheism, saying that the term tends to be deployed merely "as a term of 'theological abuse'"—or what I have coded "name-calling" (Michael P. Levine, *Pantheism: A Non-Theistic Concept of Deity* [New York: Routledge, 1994], ix, 17n2). Both Lazier and Levine operate with monistic definitions of pantheism—that is to say, their "pan"s amount to all-inclusive wholes. By contrast, this study is on the trail of what James would call a "pluralistic" pantheism, what I prefer to call a pantheism of multiplicity.

6. I have taken these terms, respectively, from Bennett, *Vibrant Matter*; Barad, *Meeting the Universe Halfway*; and William Connolly, *A World of Becoming* (Durham, N.C.: Duke University Press, 2011).

7. Grace Jantzen, *Becoming Divine: Towards a Feminist Philosophy of Religion* (Bloomington: Indiana University Press, 1999), 267. Similarly, Douglas Hedley notes that pantheism is often "a mere term of abuse" (Douglas Hedley, "Pantheism, Trinitarian Theism and the Idea of Unity: Reflections on the Christian Concept of God," *Religious Studies* 32, no. 1 [1996]: 75). See the nearly identical language in Levine, *Pantheism: A Non-Theistic Concept of Deity*, 93.

8. Ninian Smart, "God's Body," *Union Seminary Quarterly Review* 37 (1981–82): 51.

9. The term was coined in 1705 by John Toland, whose *Christianity Not Mysterious* (1696) was burned in Dublin for its unregenerate deism. In 1720, Toland went on to publish a book that, in the words of Paul Harrison, "dreamed of a network of Pantheist gentleman's clubs," which Toland called "Socratic-societies" (Paul Harrison, *Elements of Pantheism: Religious Reverence of Nature and the Universe* [Coral Springs, Florida: Llumina Press, 2004], 29). After a brief declaration of hatred for plurality ("the Multitude . . . is a Proof of what is worst," the *Pantheisticon* declares, "All things in the world are One, and One in all things. What is all in all things is God, and God is eternal, has not been created, and will never die" (John Toland, *Pantheisticon: or, The Form of Celebrating the Socratic-Society* [1720] [Charleston, S.C.: Nabu Press, 2010], 3, 15). Later in the work, Toland imagines this declaration as the opening invocation for the elitist meetings of his Socratic societies (Toland, *Pantheisticon*, 70). Toland's radical monism is the position most often associated with pantheism, and the one I am looking to unsettle here.

10. Bayle, *Dictionary*, 296–97, emphasis added.

11. Peter Melville Logan, *Victorian Fetishism: Intellectuals and Primitives*, Studies in the Long Nineteenth Century (Albany: State University of New York Press, 2010), 9. For a process-inspired, pan-*en*-theistic retrieval of this notion, see Sallie McFague, *The Body of God: An Ecological Theology* (Minneapolis: Fortress Press, 1993).

12. Cited in Logan, *Victorian Fetishism*, 9–10.

13. Cited in Roger Scruton, *Spinoza: A Very Short Introduction* (New York: Oxford University Press, 2002), 10.

14. See n. 15, below.

15. In 1225, Pope Honorius III condemned Eriugena's proto-pantheism as "pullulating with worms of heretical perversity," and dismissed the whole position as "Irish porridge" (Alasdair MacIntyre, "Pantheism," in *Encyclopedia of Philosophy*, 2nd ed., ed. Donald M. Borchert [Detroit: Macmillan Reference USA, 2006]). In these short words alone, we can detect a mess of fears of crossed boundaries concerning sex (perversity), species (worms, who are of course the "lowest" of animals), race (the Irish ridiculed by the noble Roman), and food, whose preparation is traditionally the purview of women. On Almaric, see Harrison, *Elements of Pantheism*, 26. On Bruno, see Ingrid D. Rowland, *Giordano Bruno: Philosopher/Heretic* (New York: Farrar, Straus and Giroux, 2008). On Porete, see M. G. Sargent, "The Annihilation of Marguerite Porete," *Viator* 28 (1997): 253–79. On Edwards, see William Mander, "Pantheism," in *The Stanford Encyclopedia of Philosophy*, ed. Edward N. Zalta (http://plato.stanford.edu/archives/win2012/entries/pantheism/, Winter 2012). On Eckhart, see Edmund Colledge and Bernard McGinn, eds., *Meister Eckhart: The Essential Sermons, Commentaries, Treatises, and Defense*, Classics of Western Spirituality (Mahwah, N.J.: Paulist Press, 1981), 5–61.
16. "Panic," in *Oxford English Dictionary* (online) (Oxford: Oxford University Press, 2014). See also Jantzen, *Becoming Divine*, 267.
17. Jantzen, *Becoming Divine*, 267. See also Grace Jantzen, "Feminism and Pantheism," *Monist* 80, no. 2 (1997): 266–85.
18. Jantzen, *Becoming Divine*, 267.
19. Just a few examples of such feminist theologians include Mary Daly, *Gyn/Ecology: The Metaethics of Radical Feminism* (New York: Beacon, 1990); Rosemary Radford Ruether, *Gaia and God: An Ecofeminist Theology of Earth Healing* (San Francisco: HarperSanFrancisco, 1992); Gebara, *Longing for Running Water*.
20. For the distinction between the tree and the rhizome, see Gilles Deleuze and Félix Guattari, *A Thousand Plateaus*, trans. Brian Massumi (Minneapolis: University of Minnesota Press, 1987), 3–25.
21. Jantzen, *Becoming Divine*, 10.
22. Ibid., 269.
23. It is remarkably difficult to demonstrate the presence of an absence, but I am struck in particular by the single, unelaborated reference to "feminist pantheism" in Pamela Anderson's review of the book (Pamela Sue Anderson, "Janzten, Grace, Becoming Divine: Towards a Feminist Philosophy of Religion," *Theology and Sexuality* 13 (September 1, 2000): 121–25; the absence of any reference to pantheism in Anderson's epistolary response to the same text (Pamela Sue Anderson and Grace Jantzen, "Correspondence with Grace Jantzen," *Feminist Theology* 9, no. 25 [2000]: 112–19); and the absence of any reference to pantheism in an academic obituary for Jantzen that focuses mainly on *Becoming Divine* ("Janzten, Grace Marion (1948–2006)," *Feminist Theology* 15, no. 1 [2006]: 121–23). Even Laurel Schneider's excellent review of the book, which reads Jantzen's constructive project quite closely and critically, does not mention pantheism: Laurel Schneider, "Becoming Divine: Towards a Feminist Philosophy of Religion," *Journal of the American Academy of Religion* 70, no. 3 (2002): 644–47.

24. Gunton in Jantzen, *Becoming Divine*, 273.
25. Judith Butler, "Imitation and Gender Insubordination," in *The Lesbian and Gay Studies Reader*, ed. Henry Abelove, Michele Aina Barale, and David M. Halperin (New York: Routledge, 1993); Homi Bhabha, *The Location of Culture*, Routledge Classics (New York: Routledge, 2004), 122.
26. Jantzen makes this point in *Becoming Divine*, 273.
27. Jantzen, "Feminism and Pantheism," 272.
28. G. W. F. Hegel, *Phenomenology of Spirit*, trans. A. V. Miller (New York: Oxford University Press, 1977), 9.
29. Jantzen, "Feminism and Pantheism," 281.
30. William Lane Craig, "Pantheists in Spite of Themselves," in *For Faith and Clarity: Philosophical Contributions to Christian Theology*, ed. James K. Beilby (Ada, Mich.: Baker Academic, 2006), 142–43.
31. Ibid., 144.
32. Stephen Webb, *American Providence: A Nation with a Mission* (New York: Continuum, 2006), 129–30.
33. Richard Dawkins, *The God Delusion* (New York: Mariner Books, 2008), 40.
34. Bennett, *Vibrant Matter*, 23.
35. "Discursive practices are ongoing agential intra-actions of the world" (Barad, *Meeting the Universe Halfway*, 173).
36. MacIntyre, "Pantheism," 98, emphasis added.
37. "Proposition XI: God or a substance consisting of infinite attributes . . . necessarily exists"; "Proposition XIV: Except God no substance can exist or be conceived" (Benedict Spinoza, *Ethics*, trans. G. H. R. Parkinson [New York: Oxford University Press, 2000], 9, 12).
38. Arthur O. Lovejoy, "The Dialectic of Bruno and Spinoza," in *The Summum Bonum*, ed. Evander Bradley McGilvary (Berkeley, Calif.: University Press, 1904), 152.
39. As Lovejoy points out, such concrete "stuff," which is to say Spinoza's attributes and modes, are "differences, not in the substance itself, but in relation to the finite intelligence which contemplates it"; in effect, "stuff" is ultimately illusory (ibid., 143). Lovejoy concedes that Spinoza does not *want* to reduce phenomena to illusions, but with the vast majority of Spinoza's readers, concludes that he cannot secure the relation between the singularity of substance and the multiplicity of attributes, and ends up prioritizing the former at the expense of the latter (ibid.). Cf. Clayton, *Problem of God in Modern Thought*, 398.
40. Lovejoy, "Dialectic of Bruno and Spinoza," 165.
41. William James, *A Pluralistic Universe* (Lincoln: University of Nebraska Press, 1996), 31. See also the discussion of monism versus pluralism in James in Mary-Jane Rubenstein, *Worlds without End: The Many Lives of the Multiverse* (New York: Columbia University Press, 2014), 4–5, 227.
42. James, *Pluralistic Universe*, 37, 322.
43. Ibid., 55.
44. See William Connolly, *Pluralism* (Durham, N.C.: Duke University Press, 2005), 80.
45. See James, *Pluralistic Universe*, 117, 33.

46. "As absolute," James explains, "or *sub specie eternitatis*, or *quatenus infinitus est*, the world repels our sympathy because it has no history" (James, *Pluralistic Universe*, 47). To paraphrase Martin Heidegger, no one would "play music and dance" before Spinoza's single substance—*or* its infinite modes (Martin Heidegger, *Identity and Difference*, trans. Joan Stambaugh [Chicago: University of Chicago Press, 2002], 72).
47. James, *Pluralistic Universe*, 325. For a treatment of the evolution of this term in James's work, see Rubenstein, *Worlds without End*, 3–5, 85, 189–90.
48. James, *Pluralistic Universe*, 30, 111; see Connolly, *Pluralism*, 85.
49. Steve Lopez, "Ornaments? Go Figure," *Philadelphia Inquirer*, March 18, 1988.
50. Diane White, "Lawn Sheep Threatening Plastic Birds," *Ocala Star-Banner*, September 11, 1989.
51. Joseph A. Fitzmyer, ed., *The Gospel according to Luke I–IX*, vol. 28, The Anchor Bible (New York: Doubleday, 1981), 395. For finding this source and the one that follows, I am indebted to my colleague Ron Cameron.
52. François Bovon, *Luke 1: A Commentary on the Gospel of Luke 1:1–9:50*, trans. Christine M. Thomas (Minneapolis: Fortress, 2002), 86.
53. See ibid., 87; Fitzmyer, *Gospel according to Luke I–IX*, 395.
54. See Wilfred H. Schoff, "Tammuz, Pan and Christ," *Open Court* 26, no. 9 (1912): 513–32.
55. "Pan," in *Oxford English Dictionary* (online).
56. "Pan," in *Oxford Classical Dictionary*, 3rd ed., ed. Simon Hornblower and Antony Spawforth (Oxford: Oxford University Press, 2006).
57. See Schoff, "Tammuz, Pan and Christ," 517.
58. "Hymn to Pan," in *The Homeric Hymns: A Translation, with Introduction and Notes*, ed. Diane J. Rayor (Berkeley: University of California Press, 2004), line 2.
59. "Pan," in *Oxford English Dictionary* (online). Robin Lane Fox tells us that "in the early fourth century, Iamblichus still referred to 'those seized by Pan' as a distinguishable class among people who had made contact with the gods" (Robin Lane Fox, *Pagans and Christians* [New York: Alfred A. Knopf, 1987], 131).
60. "Pan," in *Oxford English Dictionary* (online).
61. Sharon Lynn Coggan, "Pandaemonia: A Study of Eusebius' Recasting of Plutarch's Story of the 'Death of Great Pan'" (PhD diss., Syracuse University, 1992), 87.
62. "Pan," in *Harper's Dictionary of Classical Literature and Antiquities*, ed. Harry Thurston Peck (New York: Cooper Square, 1963).
63. "Pan," in *Oxford Classical Dictionary*, 3rd ed. See also Coggan, "Pandaemonia," 86.
64. Coggan, "Pandaemonia," 95, 86–87.
65. Donna Haraway, *When Species Meet*, Posthumanities (Minneapolis: University of Minnesota Press, 2007), 4.
66. W. R. Irwin, "The Survival of Pan," *Publications of the Modern Language Association of America* 76, no. 3 (1961): 161.
67. Coggan, "Pandaemonia," 88.
68. Ibid.
69. "Pan, 3," in *Urban Dictionary*: http://www.urbandictionary.com/define.php?term=pan.

70. Coggan, "Pandaemonia," 88, 85.
71. "Pan," in Peck, *Harper's Dictionary of Classical Literature and Antiquities*.
72. Lane Fox, *Pagans and Christians*, 130.
73. "Pan," in *Oxford English Dictionary* (online).
74. "Pan," in *Oxford Classical Dictionary*, 3rd ed.
75. "It is reasonable . . . that Pan is the double-natured son of Hermes [since] . . . speech makes all things [*pân*] known and always makes them circulate and move about and is twofold, true and false. . . . The true part is smooth and divine and dwells aloft among the gods, but falsehood dwells below among common men, is rough like the tragic goat. . . . The Pan, who declares and always moves all, is rightly called goat-herd, being the double-natured son of Hermes, smooth in his upper parts, rough and goat-like in his lower parts. And Pan, if he is the son of Hermes, is either speech or the brother of speech" (Plato, "Cratylus," in *Complete Works*, ed. John M. Cooper and D. S. Hutchinson [Indianapolis: Hackett, 1997], 408b–d).
76. "Pan," in Benjamin Hederich, *Gründliches mythologisches Lexikon* (http://woerterbuchnetz.de/Hederich/, 1770).
77. While Penelope is usually hailed as the epitome of marital virtue, some Arcadian legends have it otherwise. See the entries "Pan" and "Penelopé" in Peck, *Harper's Dictionary of Classical Literature and Antiquities*, as well as "Penelope" in *Oxford Classical Dictionary*, 3rd ed.
78. Plutarch, "The Obsolescence of Oracles," in *Moralia* (Cambridge, Mass.: Harvard University Press, 1936), 419e.
79. "Pan," in Peck, *Harper's Dictionary of Classical Literature and Antiquities*.
80. "Pan," in *Oxford English Dictionary* (online); Schoff, "Tammuz, Pan and Christ," 517; "Pan," in *Oxford Classical Dictionary*, 3rd ed.
81. See Porphyry, *On Images*, trans. Edwin Hamilton Gifford (http://classics.mit.edu/Porphyry/images.html), fragment 9.
82. See Rubenstein, *Worlds without End*, chap. 1.
83. Plutarch, *On the Obsolescence of the Oracles*, in Coggan, "Pandaemonia," ii.
84. Ibid., 1.
85. In a nutshell, the Romans on the ship misheard the Greeks on the island. The latter were celebrating the annual death of their agricultural deity, Tammuz, and cried out ritualistically, "Tammuz, Tammuz, Tammuz, the very great is dead (πανμεγας τεθνηκε)." Mistaking the god's name for his own, the captain Thamus and his passengers believed the news was just, "πανμεγας τεθνηκε," which their nonnative Greek parleyed into Παν ο μεγας τεθνκε ("Pan the great is dead"). See Schoff, "Tammuz, Pan and Christ," 521.
86. Eusebius in Coggan, "Pandaemonia," 1. As Coggan demonstrates, it was Eusebius who shifted the meaning of the Greek *daimon* from "divine being" to "evil spirit" (see Coggan, "Pandaemonia," 2–3).
87. Ibid., ii. According to Robin Lane Fox, this "Pan pun was premature, for his was the one reported death of Tiberius's reign which nobody believed. Cults of Pan continued in the very heart and identity of cities, the 'Pan hill' in the middle of Alexandria

or the grottoes and springs of Caesarea Panias, where the god's presence persisted on the city's third-century coinage" (Lane Fox, *Pagans and Christians*, 130).

88. Jeffrey Burton Russell, *The Prince of Darkness: Radical Evil and the Power of Good in History* (Ithaca, N.Y.: Cornell University Press, 1988), 17. See also R. Lowe Thompson, *The History of the Devil: The Horned God of the West* (London: Kegan Paul, Trench, Trubner, 1929); Patricia Merivale, *Pan the Goat God: His Myth in Modern Times* (Cambridge, Mass.: Harvard University Press, 1969); James Hillman and W. H. Roscher, *Pan and the Nightmare* (Irving, Tex.: Spring Publications, 1979); Schoff, "Tammuz, Pan and Christ," 517.

89. John Milton, "On the Morning of Christ's Nativity," in *The Complete Poetry and Essential Prose of John Milton*, ed. William Kerrigan, John Rumrich, and M. Fallon (New York: Modern Library, 2007), 8.85. Cf. Milton's reference to "universal Pan" in *Paradise Lost*, in *The Complete Poetry and Essential Prose of John Milton*, ed. William Kerrigan, John Rumrich, and M. Fallon (New York: Modern Library, 2007), 4.266.

90. Schoff, "Tammuz, Pan and Christ," 526.

91. François Rabelais, *Gargantua and Pantagruel*, trans. M. A. Screech (New York: Penguin, 2006), 24.

92. Ibid., 749.

93. Ibid.

94. Ibid., 750.

95. Schoff, "Tammuz, Pan and Christ," 527.

96. Edmund Spenser, "The Shepheardes Calendar," in *The Yale Edition of the Shorter Poems of Edmund Spenser*, ed. William A. Oram et al. (New Haven, Conn.: Yale University Press, 1989), 99–100.

97. Ben Johnson, "Pan's Anniversary; or, The Shepherd's Holiday," in *The Works of Ben Johnson*, ed. William Gifford (New York: D. Appleton,1879), Hymn II, 763.

98. Elizabeth Barrett Browning, "The Dead Pan," in *Aurora Leigh and Other Poems*, ed. John Robert Glorney Bolton and Julia Bolton Holloway (New York: Penguin, 1996), lines 183–84.

99. Lane Fox, *Pagans and Christians*, 130.

100. Because of their proximity to humans, chimpanzees were first placed in the genus *Homo*. Threatened by this categorical confusion of humanity and "ape," later taxonomists rechristened chimps with the genus name *Pan*, which "refers to the mythical Greek god of forests, flocks, and shepherds, represented with the head, chest, and arms of a man and the legs and sometimes horns and ears of a goat" (Clyde Jones et al., "Pan Troglodytes," *Mammalian Species* 529 [May 17, 1966]: 1–9). This explanation is perplexing, insofar as chimps have very little to do with goats. But they have quite a lot to do with "men," so it seems taxonomists named them after Pan because of his *half*-humanity. (I have no idea what they made of Pan's divinity with respect to chimps!) The species name is *troglodytes* (cave-dwellers), which is strange for animals who do not live in caves, but which both connects chimps to the fabled "cavemen" that "we" used to be and intensifies the connection to Pan. As late as 1985, the International Committee on Zoological Nomenclature was considering sneaking *homo* back into the *type* name, but the deciding vote was contingent on

the rejection of including *homo* anywhere in the official nomenclature of chimps (International Commission on Zoological Nomenclature, "Opinion 1368: The Generic Names *Pan* and *Panthera* [Mammalia, Carnivora]: Available as from Oken, 1816," *Bulletin of Zoological Nomenclature* 42, no. 4 [1985]: 365–70). Here, then, we see Pan invoked to shoulder the burden of liminality—to keep humanity safely separate from its too-close kin. I am indebted to Lori Gruen for having uncovered these sources, and for talking me through this particular thicket of theohumanimality.

101. The three reasons most commonly invoked for Christian opposition to the earth charter are its purported pantheism, its reference to the earth as "mother," and its support for women's reproductive rights. See Thomas Sieger Derr, "The Earth Charter and the United Nations," *Religion & Liberty* 11, no. 2 (2001), http://www.acton.org/pub/religion-liberty/volume-11-number-2/earth-charter-and-united-nations; Bron Taylor, *Dark Green Religion: Nature Spirituality and the Planetary Future* (Berkeley: University of California Press, 2010).

102. Even Haraway finds cyborgian potential in Jesus of Nazareth, noting his "odd sonship and odder kingship . . . disguises and form-changing habits" and calling him "a potential worm in the Oedipal psychoanalytics of representation" (Donna Haraway, "Ecce Homo, Ain't [Ar'n't] I a Woman, and Inappropriate/D Others: The Human in an Post-Humanist Landscape," in *The Haraway Reader* [New York: Routledge, 2004], 51).

103. Giordano Bruno, "On the Infinite Universe and Worlds," in *Giordano Bruno: His Life and Thought with Annotated Translation of His Work on the Infinite Universe and Worlds*, ed. Dorothea Singer (New York: Schuman, 1950), 97–98; Rubenstein, *Worlds without End*.

104. Catherine Keller uses this term with reference to Irigaray's "radicalization of the Christic symbol," and has herself been credited with or accused of a kind of pancarnationalism (Catherine Keller, *Face of the Deep: A Theology of Becoming* [New York: Routledge, 2003], 221; Carl S. Hughes, "'Tehomic' Christology? Tanner, Keller, and Kierkegaard on Writing Christ," *Modern Theology* 31, no. 2 [2015]: 256–83). Pancarnational or not, Keller's tehomic theology is far closer to the work of this essay than is Irigaray's sensible transcendental. I am concerned above all by Irigaray's rejection of "multiplicity," her recently reiterated insistence on the heterosexual *two*, and her confinement of divinity to humanity. "It is a question of the transcendence of an irreducible difference between two, of which the most universal paradigm lies between man and woman . . . who are naturally different . . . [whose] attraction . . . arises from instinct and becomes humanity and divinity. . . . I have stressed the necessary condition of the negative so that we are and we remain two, that is, so that the between-two does not merge into the multiplicity of community" (Luce Irigaray, *In the Beginning, She Was* [New York: Bloomsbury, 2013], 18, 21).

105. James, *Pluralistic Universe*, 28.

᭥ Material Subjects, Immaterial Bodies: Abhinavagupta's Panentheist Matter

LORILIAI BIERNACKI

The clay pot knows by means of my self and I know by means of the self of the clay pot.... Indeed, the clay pot knows by means of that absolute transcendent being and the transcendent deity knows by means of the clay pot.
—SOMĀNANDA, *Śiva Dṛṣṭi*

Scientific paradigms of the twentieth century, and specifically materialist paradigms, have not only set the parameters of what we take to be possible; they have indeed materially influenced in no small measure how we script the world around us. We can, for instance, thank contemporary scientific materialist reductionist models for an astounding map of the building blocks of matter, molecules and atoms, to wonderful effect in the chemistry of new medical drugs, travel to space, mapping of the genome. And less happily, we owe also to the material progress of scientific reductionism the overwhelming amount of plastics in our oceans. Materialist reductionism tells us that we can reduce just about everything, including chemistry and biology, to the tiny protons and electrons of physics. Beyond, however, the reductive materialism of the hard sciences, like physics and chemistry, the social sciences as well have scripted their own versions of materialism, *imitatio scientia*. With this, anthropology, religious studies, and sociology, for instance, look for explanations of human behavior through materialist reductionist models, simplifying the messy chaos of human religiosity, for example, to neater, more manageable material mechanisms like economic interest or the imperatives of brain biology.[1] Further, on the heels of the mind-boggling success of materialist reductionism, have come new births, of other, even bolder materialisms. For instance, Patricia Churchland's daring expansion into eliminativist materialism waves away the idea of self in a single gesture. The subjective sense of self in this farther-reaching materialist paradigm turns out to be no more than a simple, unfortunate illusion manufactured by an unwitting brain.[2] Meanwhile, materialist paradigms have also exerted powerful influence in political spheres. Indeed, steering the course of whole nations, Marx's historical materialism, and its offshoot dialectical materialism, has mapped the fate of millions under communist Russia and China.[3]

Despite this success, however, the materialist reductionism that has enabled the development of such remarkable advances in technology has since the early part of the twentieth century been grappling with its own incoherencies in the form of the new scientific discoveries reaching into the realm of the very, very small, the domain of quantum physics. The dominant response from the physics community for the latter half of the century has been to cast a blind eye—what David Kaiser pithily described as "shut up and calculate." However, a few physicists recently, including Karen Barad, have worked to make sense out of the incoherencies dogging matter and its entanglements.[4] The results of quantum experiments appear to demand a reincorporation of the observer within conceptual models describing what objectivity and scientific practice mean. In this context, Karen Barad has leveraged the anomalies of quantum mechanics into a coherent theory of subjectivity that links observer and observed in a mutual agency. Agential realism, as she names it, affords agency to both the observer *and* the photon whizzing across space as its trajectory is recorded on a screen by the experimenter/observer. Barad extrapolates from Niels Bohr's understanding of indeterminacy to suggest that observer and phenomenon observed, subject and object, both arise together through a mutual delineation. This is not an interaction, where a separate observer records the movements of an object, the photon, perhaps, changing the course the photon takes through the act of a gaze. Rather, she suggests we understand this as "intra-activity." Intra-activity means that there are not two separate parties, who then interact. Instead, we find that the two, observer and object, emerge in the first place through this very act of engagement, an intra-activity, an emergent entanglement that then allows both subject and object to become manifest into these two separate perspectives. Expanding this theory derived from Bohr's interpretation of quantum mechanics to human relations affords productive and suggestive new models for how humans might coexist with one another, certainly, and also with the wider natural world. And, indeed, it entails that the very meaning of matter dramatically shifts.

Barad's conception of the mutual delineation of subject and object curiously sounds remarkably close to a temporally distant tenth-century Indian proposal. About a millennium before Barad's keen analysis of quantum theory, the Hindu Kashmiri philosopher Somānanda offers—not through science, but theologically inspired—a philosophical formulation of the mutual imbrication of divinity and matter as subject and object. In the nondualist Śaivism of Somānanda's tenth-century philosophy of subjectivity, the consciousness of a mere clay pot gets inseparably entangled with that of his own subjective awareness—and with that of the highest divinity. "The clay pot knows by means of my self and I know by means of the self of the clay pot. . . . Indeed, the clay pot knows by means of that absolute transcendent being and the transcendent deity knows by means of the clay pot,"[5] Somānanda tells us. He is thus entangling the agency and observational capacity of god and humble clay pot with his own. Can the matter of mere clay be so indispensable to the all-encompassing deity, Śiva, who stands as the foundation of subjectivity for his school of thought? Actually Śiva, for Somānanda's philosophical system, represents more than simply a deity, more than the highest deity in a great chain of being. In Somānanda's nondualism Śiva stands less as an entity than an all-pervading principle, characterized in its first movement as *cit*, consciousness. It then unfolds within itself the capacities for will, knowledge, and action and, with these, time as it unfolds, down to the very stuff of matter, the clay that makes up the clay pot.

In this essay I examine this much earlier, radical reworking of the notion of matter, one that bears affinities with some quantum formulations of the mutual implication of observers and the objects they observe. I work here with a medieval tenth- and eleventh-century Indian school of philosophy, the nondual Kashmiri Śaivism school called "Pratyabhijñā," or "Recognition," which grew out of the work of Somānanda and his successors, Utpaladeva and Abhinavagupta. The Pratyabhijñā offers a response to the monist idealisms of earlier Indian philosophical schools, the Vedānta of the eighth-century Hindu philosopher Śaṅkara and the earlier Buddhist Vijñānavāda or "mind-only" schools, both of which tended toward leaving aside material reality as illusory in their nondualist formulations of subjectivity. The Pratyabhijñā proposed a dynamic reformulation for the ideas of objectivity and subjectivity, which retained a sense of matter as real, and incorporated the notion of matter as a perspectival expression of the subject-object relation. Abhinavagupta, the great grand-disciple of Somānanda in this school's lineage, lived from about 950–1025 CE. He offered a systematic articulation of this school's philosophy, and I draw extensively from one of his last works, his *Īśvara Pratyabhijñā Vivṛti Vimarśinī, Āgamādhikāra* (*Teachings on Cosmology, the Long Commentary on Recognition*), written about 1015 CE. In the context of Indian philosophy, with its pervasive and overt leanings toward the rejection of material existence as Māyā, or "illusion," Abhinavagupta is somewhat radical in his embrace of materiality. One might read his writing as part of a larger trend within Tantric philosophy to bring the body back, recognizing the importance of the body and matter in the larger scheme of enlightenment.

Abhinavagupta accomplishes his task by locating a fundamental dynamism between subjects and objects, consciousness and matter. Specifically, he suggests that subjectivity and objectivity ought to be understood as mutually implicated, co-constituting one another. For this chapter I read Somānanda's and Abhinavagupta's mutual implication of subject and object in dialogue with Karen Barad's quantum theory of agential realism derived from Bohr's quantum theory. The differences between these models preclude any easy comparison; Abhinavagupta in particular is more concerned with establishing an overarching subjectivity as a pervasive nondualist substratum than is Barad, who stresses relationality as mutual entanglement. In this sense, Abhinavagupta's panentheist transcendent self, transforming and unfolding itself into the multitude of others is the glue that binds together subjects and others. For Barad a mutual entanglement gives birth to selves and others through their mutual, complementary differentiations. Yet the points of convergence hint at compatible ontologies; specifically, the mutual adherence they share to a foundational inseparability of subjects and objects merits closer inspection. The point of this, however, goes beyond a sense of appreciation of remarkable similarities across a gap of centuries and continents in this Bohrian-based quantum perspective and this dynamic panentheism of a medieval Indian philosopher-mystic. Rather, it suggests that this medieval conception of the body may help us as we think through our own quantum reconceptualizations of materiality.

Abhinavagupta's theories of the entanglement of consciousness and matter lead to a reconsideration of the nature of the body. Not simply inert matter, the body's parameters shift in this formulation to afford the status of "body" with a mutuality of matter and consciousness; borrowing Barad's language, we might call the body thus construed a "material-discursive" occurrence.[6] That is, Abhinavagupta's conception of the body incorporates both materiality and discursivity entwined together, an entanglement perhaps most striking in its rendition as "subtle body" (*sūkṣma śarīra, puryaṣṭaka*). The subtle body is certainly a body, fundamentally made of matter, yet its links with consciousness afford it capacities beyond the physical body. Not quite visible to ordinary eyesight, even as it is still connected to the five senses and is itself sensible, it is the subtle body that transmigrates from life to life for these various Indian philosophies that share an affirmation of reincarnation. Not exactly a soul here—keep in mind the idea of a soul is anathema to Buddhist contexts, which strongly affirm the "no-soul" (*anātma*) doctrine the Buddha preached around 400 BCE. The subtle body does not quite match up to the Western notion of "soul," since the Western idea of soul with its implications of a permanent and unique individual essence is problematic for these Indian nondualist traditions, even for traditions such as Vedānta, that affirm the notion of "self" (*ātman*) as persisting beyond the death of the physical body. The subtle body is also not a cosmic self (*ātman*). It outlasts the physical body, even as it accompanies it through our daily lives, but it is subject to its own transformations. It is the body we use to dream in, to travel to distant places through yogic exercises without our more familiar corporeality, and it is the body we inhabit and in which we carry our particular *karmas*, propensities and habits while we look for new bodies in which to take birth between lives. For our purposes here, the fantastic, spiritualist conception of reincarnation need not distract us from what is useful about this concept. Untenable, we might even say scientifically distasteful, to contemporary Western sensibilities, nevertheless, an attentiveness to this unfamiliar map of the body allows us to glean unexpected and useful novel formulations of the relation between body and mind. The subtle body particularly demonstrates a permeability in its ontological characterization, entangling both materiality/body *and* consciousness as productions of ritual performance.

THE INSEPARABILITY OF KNOWLEDGE AND ACTION

What Abhinavagupta and Somānanda propose in this philosophical system of "Recognition" (Pratyabhijñā) is a way of linking consciousness with matter. This is no small matter. The pervasive Cartesianism that Barad (and Bohr) reject,[7] which places a wall between the consciousness and materiality, also surfaces in its own form in an Indian context beginning early in Indian philosophical history, with Sāṃkhya, around of the turn of the first millennium.[8] Arguably the basic template and bedrock of most Indian models of cosmology, Sāṃkhya proposes twenty-five basic categories for "what there is," ranging from the five great elements (water, earth, air, fire, and space) up to primordial, sentient spirit (*puruṣa*). Most succeeding Indian philosophical systems draw from the model Sāṃkhya offers, modifying the basic system with a host of additions and innovations to present variations on a theme. In the basic Sāṃkhya template, sentience and consciousness is a property of only one of the twenty-five categories, *puruṣa* or spirit, at the apex of the system. The other twenty-four categories present modulated expressions of a primordial and fundamentally insentient Nature (*prakṛti*), incorporating all else that we encounter materially.

The philosophical counter to his Sāṃkhya inheritance that Abhinavagupta uses to arrive at the mutual implication of matter and consciousness—the fundamental inseparability of clay pots, gods, and human awareness—might be said to boil down to a basic, even psychological, apprehension of the bodily links between awareness and actions. Abhinavagupta structures his arguments for the nondual pervasiveness of consciousness, even as it transforms into matter (that is, into pots that exist because they know themselves) around the convertibility of knowledge and action into each other. Why knowledge and action? He understands the ramifications of a split between consciousness and matter to reach its way down into a mundane bifurcation of human capacities to know and to do. Consciousness is at base about knowing; matter is at base about action, about the forces that move objects. This logic also plays out in our contemporary formulations of subjectivity; the phenomenological experience of *qualia*, unadulterated consciousness, eludes neuroscientific formulations of brain operations, to the extent that they can only be bracketed out as "epiphenomena." The capacity to know is separated from our bodies' capacities to move, to sense.

The notion that knowledge and action are linked is a fundamental premise of the Pratyabhijñā system. Utpaladeva, who writes the *Verses on Recognition* (the root text for Abhinavagupta's commentary) says, "Knowledge and actions are inseparably mutually associated with the Subject."[9] This goes against the larger metaphysical milieu of India's historical bifurcation of these two. This verse proposes to bring materiality back within the sphere of consciousness. Abhinavagupta glosses this verse with, "Consciousness, which has a form possessing attributes, with a body made of the energy of knowledge [*jñāna śakti*] accepts as its attribute the power of action [*kriyā śakti*], which has the nature of both externality and internality."[10] Here again we see the body (*vapuḥ*) return, knowledge itself characterized as a body. For Abhinavagupta the link between knowledge and action is an expansion; it involves a dynamism, a movement that operates on cosmic theological scales and in a person's mental processes. He tells us:

> From the perspective of mere consciousness lying in the core there exists the state containing both inner and outer. However, differentiation [*Māyā*'s creation] spreads out, expands into many branches, with the Lord still lying within the core. And that creation is two-fold still when the energy of the highest Śiva spreads open. When that energy manifests inwardly, naturally it is called the energy of Knowledge. However, when it expands in stages with its active, self-reflective awareness [*vimarśa*] gradually becoming more firm and fixed, then it manifests externally. This is pointed to as the energy of Activity.[11]

Abhinavagupta's theological language points to mental processes; an inwardly directed attention entails knowledge. An externally directed attention is directed by a capacity for self-reflective insight, *vimarśa*, which itself is the basis for action. Thus *vimarśa*, which literally translates as "distinguishing touch," involves a kind of self-measuring.[12] Over time, this process of internal measurement leads to a congealing of awareness, becoming firmer and more fixed, which then affords external manifestation. Barad's notion of matter as substance in its becoming as a "congealment of agency"[13] finds a curious parallel in Abhinavagupta's theological understanding of the firming up of active awareness, *vimarśa*, the capacity for a self-reflective, self constitution through a kind of self measurement to gradually become externalized action and then objects. Barad's understanding of knowledge is that "knowing is a material practice of engagement as part of the world in its differential becoming."[14] If we reflect (diffract?) this understanding through Somānanda's clay pot who knows by means of my self in a mutual emergence of self-other knowledge, we engage in the fundamental operation of consciousness. This is the sense of *vimarśa*, self-reflective measuring, touching and engaging between me and the clay pot, to generate an entanglement that is knowledge.

Even here, however, we note the prioritization of subjectivity in Abhinavagupta's delineation of the mental processes that transform knowledge into action, and then ultimately objects. Abhinavagupta links the object back to consciousness through a kind of subjective trace that remains an inward attribute of objects:

> The Subject's form, which is a unity of consciousness, contains an excess, an abundance of awareness. This is deposited into the side of the object that is going to be created. So inwardly the object has the attribute of *śakti*, energy which is none other than the form of consciousness.[15]

The object comes into being in conjunction with the subject; the excess of awareness forms the inside of the object. Still, even as Abhinavagupta insists on the essential unity of the subject and object, each one comes to the fore at the expense of the other in a complementarity that sees either a position of subjectivity predominating or a position of object predominating. Two things we should keep in mind here: For most of what we encounter in the world, we find a mix of both subjectivity and object alternating in expression. So when we humans, in the middle of this great unfolding of Nature (*prakṛti*), reflect on a pristine sense of "I-ness" (*ahantā*), particularly (but not exclusively) in an experience of wonder (*camatkāra*), we then access our inherent sense of subjectivity. However, when we reflect, for instance, instead on "this-ness" (*idantā*), considering, for instance, "I am fat" or "I am sad," we transform our inherent subjectivity into a state that blossoms into the predominance of one or other of the five classical elements (earth, water, fire, air, and ether). In the case of thinking "I am fat," we allow the earth element to predominate, and the element of water in the thought, "I am sad."[16] Perhaps even just rescripting these types of self-identifications in terms of a subjectivity aligned with the five elements might happily jolt us out of the doldrums that often arise from these kinds of thoughts. Might Abhinavagupta say that the prevalence of this kind of culturally pervasive self-deprecation arises from a self-exile from subjectivity, "I-ness" as *ahantā*? Cultural norms aside, it is also the case that Abhinavagupta often favors the perspective of subjectivity over that of object, stressing its priority and desirability. Moreover, like Bohr's cane (discussed below), we also see a complementarity in the relation between the subject perspective and the object perspective. In response to his fabricated interlocutor, a seemingly frustrated materialist, Abhinavagupta says:

> Now you may say, well if it occurs in this way, then if one can see the form of consciousness even in a clay pot, then what is there that's not false? Yes, correct, [Utpaladeva] says, "but it is appropriate to view the insentient" in this way. . . . By removing the form which is insentient, we allow the form which is sentient, the conscious, to take predominance and there we have a complete entrance into consciousness. This is moving forward and entering into its own form.[17]

Here while Abhinavagupta asserts a primary subjectivity at base, even for the clay pot, he shies away from Somānanda's stronger formulation, which ascribes a lively subjectivity to clay. In this respect, Somānanda demonstrates a pantheism in relation to Abhinavagupta's panentheism. Abhinavagupta, cautiously shrinking away from the radical agency Somānanda gives to the clay pot, cites Utpaladeva's explanation, that it is "appropriate" to view the clay pot in this way, recognizing its inherent consciousness, its subjectivity, precisely because our perspective, our way of seeing the clay pot, affects the clay pot's own ontological status. Our seeing it as sentient allows the clay pot's capacity for sentience to come to the forefront. Here we can read this in two ways. On the one hand, we might read again the one-sided human determination of what matter is: Human consciousness makes the pot conscious. On the other hand, we might also read this transformation of the ontological status of the clay pot affected by how we think of it as a stress on a mutuality. This mutuality derives from the entanglement of the person who sees the clay pot and the pot's own reality. Because our act of viewing, indeed a kind of measurement, brings forth the status of the pot, there is a mutuality in the emergence of the pot's life. Our view of the clay pot constrains the pot's own emergence into its own inherent consciousness, though here it is important to keep in mind that the human does not animate the pot, give it life through a human capacity. Rather, the clay pot already has its own innate life. In any case, the reverse of this is also true, particularly for Somānanda. Utpala, as he comments on Somānanda's assertion of the radical agency of pots, tells us that for Somānanda, "Śiva does not exist in the form of pots, etc., but rather the pots, etc., have Śiva as their form."[18] With this, Somānanda and Utpala emphasize that the model is not of a universal life principle pervading things like pots, but rather that things themselves condition the form of consciousness, as the clay pot emerges into its form as Śiva, absolute deity and principle of consciousness.

Abhinavagupta, in contrast, more conservative than Somānanda, ascribes natural limits to certain forms. Humans, for Abhinavagupta, more easily access subjectivity and can help objects access their own subjectivity by recognizing this capacity within them. I am reminded here of Jane Bennett's poetic encounter with debris, a beat-up old white bottle cap, a single discarded black work glove, a smooth stick, all vibrating with a palpable intensity of vitality for her on a vibrant June morning.[19] In this case, Abhinavagupta's description suggests that Bennett's recognition of the vitality within this litter affords the deep and vibrant life of these only seeming lifeless objects to shine forth (*ābhāsa*) in a way that allows both object and observer to tap into the wonder of subjectivity.

Abhinavagupta's remark points to a human responsibility to act generously, to allow the clay pot's inherent sentience to actually arise. One can thus glimpse an implicit ethics in Utpaladeva's "it is appropriate." We also see here the mechanism of complementarity. Either the form of insentience (jaḍarūpa) or consciousness (cidrūpa) will be predominant. As one rises, the other falls. This complementarity still nevertheless values sentience as a kind of ground zero base reality that can shift into either its *esse* as consciousness or into insentience.

THE SUBTLE BODY

Just as the mind is physical in this medieval linking of matter and consciousness, so also the body's matter is always imbued with its own incorporeal consciousness. Writing in a medieval Indian Tantric context, Abhinavagupta inherits an already sophisticated model of body as both physical and nonmaterial, including a map of a counterpart to the corporeal body, a nonphysical body, literally, a "subtle body," known as the *sūkṣma śarīra, liṅga śarīra*. These models of body develop a focused attention on nonphysical centers located at precise points in the physical body that interact with the physical body, the *cakra* system. With this also come ideas of a psychophysical force located within the physical body, the *kuṇḍalinī*, which rises through different areas in the physical body and ultimately brings with its motion enlightenment for the person whose *kuṇḍalinī* is awakened. A universally present liveliness, understood as deity, a goddess, residing at the base of the spine, this psychophysical force is understood as nonmaterial, and yet it is capable of generating material effects, such as a spontaneous whirling motions of the physical body (*ghūrṇita*) often accompanied by euphoria that the meditator feels as the *kuṇḍalinī* rises, among a host of other phenomena.

The Tantric body is therefore both physical and nonphysical. A hybrid entity, it contains both what we think of as physical and also what we would think of as mentality or immaterial being. The familiar physicality of the body is enhanced by an unseen energy pervading, not equally across the body, but differentially in various locations both generating and signaling material effects. In a Tantric context, very far from our own conceptual understandings of causality, some of these effects promise the prospect of using this access to this immaterial energy to gain supernatural powers such as reading other persons' minds, seeing into the future, among others.[20] Moreover, through a discursive performativity entailed within ritual practices, the practitioner engages and awakens various components of a body that is in its subtle form, directing the transformations of the physicality with which it is entangled. Thus ritual performances often enact a reformulation of the body. For example, one visualizes one's body, frequently for instance in Buddhism, as the body of a deity, among a variety of other visualization practices. Similarly, a ubiquitous Hindu practice[21] involves visualizing the sequential destruction of one's physical body and its replacement with a body made of mantras, ritual formulas understood to be deities, which are then placed in various parts of the physical body, the top of the head, the throat, the heart, and so on. This type of practice suggests that the physical body is amenable to discursive transformations through language and through visualization.[22]

For practitioners, the transformations that ritual practice bring are understood not so much as human manipulations that transform the physical body. Instead, they shift the parameters of the physical body by means of the body's already given subtle body. The subtle immaterial body has a capacity to express (connect with?)[23] the plethora of vibrant energies, and in some cases, nonmaterial beings that populate our space. If this sounds like rampant animism, it is. I suggest however, that it may be possible to think beyond the panoply of a multiplicity of spirits, to a notion of performative discursivity as a functional component always intimately connected with the physical body. Of course, this is fundamentally different from the type of bodily habits that Judith Butler points to as generating a subjectivity through discursive interpellations of culture and practice, since this discursivity accords a subjectivity to material effects of the body not governed by the human mind. Yet the inclusion of discursivity as capable of generating subjectivity is suggestive. Ultimately, I think it comes down to Abhinavagupta's suggestion that we saw earlier, that by recognizing the capacity for subjectivity in all the seeming insentient life and matter around us, we afford it the capacity to enter into its form as consciousness. The baseline understanding of the body is a recognition of its own always already present and potentially emerging subjectivity.

Like the clay pot, seemingly lifeless, but with its own reservoir of subjecthood in the wings, needing only to be unleashed, the body has a life of its own that impels the subjectivity of our thoughts.

So how exactly does Abhinavagupta describe the subtle body? Abhinavagupta tells us that the subtle body "is like the physical body, but it does not have limitations in terms of its spatial dimensions."[24] Nor is it bound by the divisions of time into past and present, though it is still connected to time as a universal.[25] That is, it is not strictly bound to linear time, but still operates within a framework of time dividing the capacity for awareness. The subtle body is not a transcendent, time-free entity. This again entails a flexibility in terms of the capacity of the subtle body to travel to future and past. The subtle body is, moreover, composed of eight components. Dubbed the "City of Eight," the *puryaṣṭaka*, these eight elements include first, the five vital breaths, called *prāṇas*. These are the in-breath, the out-breath, the upward breath, the downward breath, and the breath that mixes all of these. This makes five of the eight. Along with this is the *antaḥkaraṇa*, the inner organ, subdivided into three, including the mind, the intellect, and the ego. Finally, two more are added to these six to make eight. These are the two groups of sense organs, the *buddhīndriya*, including the ear for hearing, the nose for smelling, and so on, and the *karmendriya*, the group of organs of action, which includes the hand, the foot, the sex organ.[26] This description of the subtle body is quite similar to what we see in Sāṃkhya and yoga.

Thus we see that this very subtle body is much like a physical body. It even contains, as essence or template, if not actually corporeally, hands and feet and genitalia. For a Western context, if we imagine an idea of what might survive death, we head into the territory of notions of "soul." And for Abhinavagupta, as for Sāṃkhya and yoga, it is this subtle body that survives bodily death and reincarnates life after life. Yet—if we imagine the idea of a "soul," do we often include components like feet and genitalia?

And yet this mix of breath (the five *prāṇas* here) and very subtle feet and hands and an ego and so on is somehow not bound by space and only minimally bound by time. Even if the notion of the subtle body operates merely as a sketch, only as a very subtle outline, Abhinavagupta is very clear that this subtle body too is really, without any doubt, unambiguously a body. He tells us,

> The City of Eight does in fact have the nature of a body, because the primary elements, fire, earth etc., inhere in it.[27] Also, here, in order to remove delusion, [Utpaladeva] uses the word "body" to describe this extremely subtle state, the subtle body, *puryaṣṭaka* precisely to instigate the reader to voice doubts about the nature of this body and the applicability

that the word "body" with its physical implications entails for this subtle existence.[28]

This idea of a subtle body, a spirit body, with hands and so on, this body with spatially contoured appendages, which is however, not confined spatially, affords a polyvalent notion of the body. With this Abhinavagupta is especially at pains to reinscribe the idea of consciousness as never very far from the notion of bodies. That is, there is no space for a free-floating consciousness. There is always a body, matter of some sort emerging and interacting with awareness. Action and knowledge always coexist, entwined, unfolding mutually. The evocative, if accidental, resonances with Karen Barad's model, as she tells us, for instance, that "knowing is a material practice"[29] and that "practices of knowing and being are not insoluble. They are mutually implicated"[30] are certainly suggestive. I should note again that this notion of the subtle body is not parallel to a Western notion of soul. Although the subtle body is indeed that which transmigrates, it does not in this system carry the same notion of incorruptibility that is part of the idea of the Western soul. Rather it functions as a middle term between the physical entity that dies and some ultimate component of self—for Abhinavagupta this ultimate component would be a notion of divine subjectivity, whereas for other thinkers, such as Śaṅkara, this entails the subtle body's disappearance altogether. What then does it mean to ritually perform the refinement of this subtle body, to imagine filling it with mantras and deities? With this discursive reworking of the body, does the body talk back? Or does the subtle body come to life only if we perform it, if we habituate ourselves to a subtle materiality that accompanies the visually present physicality we engage as we move through the world?

If we understand the body in terms of a co-related discursive materiality, it affords us a polyvalent notion of the body, a body performatively enacted, yet carrying its own implicit animation, a life of the body, focused predominantly on the breaths that animate it, not so much the mind, mentality, though including this, but an animism that extends itself corporeally, a discursivity that suggests demands that the body not simply be the tool of our minds, but a materiality with its own emerging, if latent trajectories.

THE CARTESIAN "CUT"

At this point I will tie in Abhinavagupta's conceptions of the body and subtle body with the work of contemporary theorists, especially Barad, who uses the insights of quantum mechanics to reformulate ways of thinking about materiality. I should start by pointing out that the pervasive human exceptionalism ardently criticized by posthumanism finds no real foothold in this Indian context. The assertion of the human as the privileged site of agency is not really a consideration for the vast ontological landscapes of Indian cosmology. In a world already populated with a host of uncanny others—from ghosts and tree spirits (*bhūta* and *yakṣa*) to subtle bodied beings, whose main material form is sound, to talking geese and talking bulls, even to apparently garrulous inanimate elements—a fire that talks, to demons and to gods, who even when they try to mimic humans cannot help but stand an inch or so off the ground, eyes unblinking, unsweating,[31] down to worms with their own desires—the Indian context is happily bereft of a dominant anthropocentrism. Abhinavagupta, for instance, writes, "When those [beings] belonging to *Māyā*—down to an insect—when they do their own deeds, that which is to be done first stirs in the heart,"[32] signaling the perspective and agency of a conscious insect. His predecessor Somānanda takes this dispersal of agency to even greater democratic lengths, writing: "Knowing itself as the agent, the clay pot performs its own action. If it were not aware of its own agency, the clay pot would not be present."[33] Thus the Western focus on the human, which has historically enabled a seemingly "natural" division between the human and nonhuman, and its corollaries, culture and nature, is conspicuously absent in this Indian context.[34]

Yet, lest we think the source of our postmodern woes lies in our Western inability to encounter the face of the nonhuman, there is still, even in this radically nonanthropocentric paradigm, the divisions of subjects and objects. So although the human/nonhuman division is not ever a central or logical contention, nevertheless Descartes's division of "what there is" into *res cogitans* and *res extensa*, these two domains of subjects and objects, does find a counterpart quite early in Indian tradition in Sāṃkhya, dating at least from the beginning of the third century BCE and likely much earlier.[35] In Īśvarakṛṣṇa's *Sāṃkhya Kārikā*, the *res cogitans* finds expression in only the very first principle, that of *puruṣa*, the "primordial person."[36] All the rest, the succeeding twenty-four principles from Nature (prakṛti) down to earth, constitute the *res extensa*, everything else.

I should also stress that Indian philosophy and soteriology, particularly in many of its nondualist versions, has tended to favor the *res cogitans* component, the *puruṣa*, as the "real." That is, *puruṣa*, consciousness and spirit, is that which fundamentally exists in these several Indian philosophical systems, whereas matter is viewed with hermeneutical suspicion. Matter and bodies in this view are the product of an unreal *Māyā*, illusions churned out epiphenomenally from *prakṛti*. We might contrast this with twenty-first-century neuroscience, which takes the opposite cosmological position, suggesting that matter is what is "real," while the perception of consciousness is mere epiphenomenon.[37]

However, even as Indian cosmologies tend to favor the *res cogitans*, the parameters of this consciousness tend also to differ radically from our own

Cartesian parameters. For example, Descartes's cogitating mind, the thinking voice that says "I think, therefore . . . ," stands on the side of consciousness; it is the *cogitans* in the Western bifurcation of consciousness and matter. In contrast, in the Sāṁkhyan Indian perspective, mind would not make the cut for consciousness. Rather, *manas* translated as "mind," falls squarely in the camp of *prakṛti*, Nature, in all its insentient materiality. Indian systems then, following Sāṁkhya, place the Cartesian "cut" much higher up the scale, so that intentionality (*ahaṁkāra*), insight (*buddhi*), and cogitation (*manas*) all feature as materiality.

The question of where to place the cut between consciousness and matter, and the corollary subject and object, is important for contemporary understandings of what matter is. It is important for Barad's quantum interpretation, and not only for hers. Henry Stapp, another quantum physicist concerned with formulating the role of the observer in quantum theory, offers the proposal that the cut between subject and object might be understood through quantum findings as a rethinking of Heisenberg's cut. Heisenberg initially proposed the idea of the quantum "cut" as a way out of the queer metaphysics entailed by the quantum responses of matter to observers. He posited a hypothetical dividing line between the observer or measuring apparatus in a quantum experiment and the wave function being observed. Above the line, the macroscopic paradigm of classical physics is operative; below this line of the Heisenberg "cut," the weird unpredictability of microscopic quantum systems operates as quantum wave function. Heisenberg proposed this cut as a way of addressing the collapse of the wave function in the Copenhagen interpretation of quantum mechanics. An arbitrary, theoretical construct, Heisenberg's cut nevertheless tries to manage and explain why classical objects behave in the predictable way they do and why quantum objects defy classical physics laws. Like Barad, Stapp points out the artificiality of Heisenberg's macro/micro distinction, noting that quantum theory works on both levels. Moreover, he suggests that the Heisenberg cut applies not only to matter, but also to the multifold nature of reality itself. Further, Stapp draws from von Neumann's proposal that the Heisenberg cut might apply to the distinction between the physical and the mental, rather than to the macro of classical physics and the micro of quantum mechanics. With this, Henry Stapp reinterprets the Heisenberg cut as a boundary line between the perspectives of the subject and the object.[38] That is, the cut he hypothesizes is a perspectival demarcation between consciousness and matter, and specifically, consciousness and brain. Thus, Stapp suggests moving the cut out of matter altogether, using it instead as what facilitates the connectivity between matter and mind, and thus affording a reintroduction of the observer as mind (or better, consciousness and subjectivity). Stapp's location of the cut here does not, however, reinstate a

Cartesian wall between matter and mind. Rather, he uses the artificiality of Heisenberg's macro/micro distinction to point to the elephant in the room: the exclusion of mind from the equation altogether. With this, Stapp presents a possible solution to the materialist dilemma of how to account for the apparent nonmateriality of consciousness without simply whisking the notion of consciousness away as mere epiphenomenon of the gray matter inside our skulls. The mechanism he ultimately proposes for bridging between the physical and the mental is the quantum Zeno effect.[39]

The quantum Zeno effect is an empirically proven phenomenon that demonstrates that a rapid succession of measurements or observations has a capacity to influence the effects we see in the physical world. The quantum Zeno effect tells us that a rapid succession of questions—questions that originate from the observer, from the side of consciousness—will statistically alter the outcome, through some odd disposition in the order of quantum selection that somehow favors a rapid, repeated mental probing, beyond what we might expect as a statistical outcome of quantum probabilities. It bears repeating that these measurements, what Stapp terms "mental probing actions,"[40] originate on the side of consciousness, not matter, yet based on proven quantum effects, they determine the material outcomes.

Following Bohr, Karen Barad approaches the idea of Heisenberg's cut in an altogether different way from the way Stapp approaches it. She points to the contingent and dynamic nature of the cut between subject and object, telling us, "as Bohr says, there is no inherently determinate Cartesian cut,"[41] and

> the specification of the conditions necessary for an unambiguous account of quantum phenomena is tantamount *to the introduction of a constructed, agentially enacted, materially conditioned and embodied, contingent Bohrian cut between an object and the agencies of observation.*[42]

That is, the location of the cut is not predetermined. Its position changes with the material conditions we use to ascertain it, following on the materially embodied apparatus that is used to make measurements. If we understand the idea of the cut as the possibility of demarcating the differential relationship between subjects and objects, in this case the observer and the object measured, then Barad's dynamic formulation forces us to rethink the stability of the demarcation. To point out the contingent nature of the cut between subject and object, she draws on Bohr's lively metaphor of a person with a cane in a dark room that he used to explain the notion of complementarity:

> One need only remember here the sensation, often cited by psychologists, which everyone has experienced when attempting to orient himself

in a dark room with a stick. When the stick is held loosely, it appears to the sense of touch to be an object. When, however, it is held firmly, we lose the sensation that it is a foreign body, and the impression of touch becomes immediately localized at the point where the stick is touching the body under investigation.[43]

Whether this object, the cane, takes on the position of subject or the position of object depends here on whether it functions as the measuring instrument for the human or as an object the human holds in his or her hands and examines. Barad highlights the "complementarity," or mutual exclusivity, entailed by each of these usages. Here function determines status, whether subject or object, in mutually exclusive states. The same material structure, the cane, alternately performs both these functions. The cut between these two demarcations of status, that is, its *identity* as subject or object, depends on an emergent functionality.

Abhinavagupta also highlights this feature of the lability of subject and object, even as he frames this mobility of subject and object as marked through grammatical interpellation, rather than Barad's and Bohr's "unambiguous and reproducible marks on bodies."[44] Quoting a maxim of his school, Abhinavagupta draws from grammar to illustrate how objectivity and subjectivity are mobile perspectives. He says, "'Everything in fact has the nature of all things.' Even the lifeless [grammatical] third person, [the "it"], if it sheds its lifeless form can take on the first and second person forms [the grammatical I and you]. For example, 'Listen, o stones' and 'Of mountains, I am Meru.'"[45] In other words, apparently lifeless stones and a mountain in the distance do not inherently, by virtue of their seeming insentience, remain confined to the category of object. Rather, as we saw earlier with Abhinavagupta's clay pot that accesses its sentience in tandem with our awareness of it and with Jane Bennett's vibrant entanglement with debris, if we engage a stone as a "you" instead of a distant third person, "it," we open a way to a mutual and vital subjectivity. We should here keep in sight the language of agency Abhinavagupta uses for the insentient stone;[46] the stone itself, as Abhinavagupta tells us in the quote above, "sheds its lifeless form." Although appearing insentient (*jaḍa*),[47] it is the agent of its own change. Moreover, the cut between insentient matter and sentient consciousness is mobile, always available for repositioning, even for something like a stone.

Abhinavagupta's idea that "everything in fact has the nature of all things" points to a fundamental nondualism that emerges dynamically into correlated, even entangled components. Within a context of interrelation, we might venture a comparison to Barad's "intra-action." As we saw in the epigraph for this chapter, Somānanda articulates the interrelationality of clay pot, human, and god in a mutual emergence that looks on the surface to be an unapologetic pantheistic animism: "The clay pot knows by means of my self and I know by means of the self of the clay pot. . . . Indeed, the clay pot knows by means of that absolute transcendent being and the transcendent deity knows by means of the clay pot."[48] Human knowledge is on a par with the knowingness of both gods and clay pots. Further, knowledge itself is generated through a relationality between these. Insofar as the knowledge of self and god and clay pot arise through each other, Somānanda's model approaches the intra-activity that Barad points. My sense of knowing is contingent on my interaction with the clay pot; a mutuality of self-awareness arises specifically through the engagement and articulation of our subjectivities through each other. These two are co-constituted. The matrix for this mutual emergence is precisely the potentiality of a subjectivity always in the wings, as a fundamental component constituting both matter *and* consciousness. It is this prospect of a universal subjectivity, a ubiquitous and democratic citizenship of the self that makes the recovery, the recognition (*pratyabhijñā*) of inherent divinity (might we substitute for a twenty-first-century Western materialism the notion of an inherent matter-worthiness?) always available, even for something which to our modern eyes seems so hopelessly matter-bound as a clay pot.

So from this maxim of "everything in fact has the nature of all things," we arrive at a nondualism premised on a fundamental inseparability of the subject and object. Barad also rests her mutual inseparability on a nondualism. She emphasizes, "Phenomena constitute a nondualistic whole"[49] and "Phenomena are the ontological inseparability of intra-acting agencies."[50] For Barad the apparatus, which includes the machinery used to record the measures taken (the location of the electron, for instance), entails "boundary drawing practices."[51] For Abhinavagupta, the notion of apparatus is less coherent; the apparatus in his schema might perhaps be understood as the unfolding of consciousness in its various and relative reflections of *ahantā*—the state of being subject—and *idantā*, the state of being object. So although I emphasize their commonality in embracing the inseparability of subject and object, it is important not to minimize the difference of apparatus between these two nondualisms. Abhinavagupta's nondualism entails a dynamic unfolding of a singular reality; subjectivity unfolds itself into matter and objects. It never loses its innate subjectivity; yet it forgets this subjectivity in the position of being object. The Pratyabhijñā "Recognition" hints about a way of recognizing an always available vitality of subjecthood.

One other related substantive difference is, in fact, the notion of substance. Whereas Barad emphasizes *relation* over relata,[52] Abhinavagupta retains a primordial substantiality in the notion of consciousness (*cid*) which is at the heart of both objects and subjects, even if he redefines this substance of consciousness to enact a fundamental dynamism within it.

Despite these differences, we might nevertheless profitably redirect and extend some components of these medieval insights to afford ourselves in the twenty-first century other options for thinking about matter and the body. The body is by all counts pervasive in Abhinavagupta's conception of consciousness. He tells us, "The highest principle, the essence of consciousness alone is the body [*vapuḥ*] of all beings,"[53] and "the highest principle is indeed the body of all things."[54] The image of body derives from a pervasive philosophical instantiation of consciousness qua materiality, where the substance of consciousness unfolds as the materiality of the world, "on the wall of itself the picture of all entities,"[55] generating a kind of space-time fabric out of itself. That is, like Barad's notion that entities are "material demarcations not *in* but *of* space,"[56] here, entities arise out of and move within the fabric that is consciousness itself. Deriving from the *Spandakārikā*, a root text for this philosophical school, the image entails the unfolding of consciousness into materiality that is the world, matter, bodies, and differentiation itself.[57]

NOTES

1. See Latour's critique of criticism, which applies particularly in this vein to the social sciences, in Bruno Latour, "Why Has Critique Run out of Steam? From Matters of Fact to Matters of Concern" *Critical Inquiry* 30, no. 2 (2004): 235.
2. Here, literally boggling the mind. Patricia Churchland, *Touching a Nerve: The Self as Brain* (New York: W. W. Norton, 2013). Daniel Dennett's work offers a similar extension of materialism to write out the notion of subjective self.
3. Not to mention its negative reverberations in the United States and Europe.
4. David Kaiser, *How the Hippies Saved Physics: Science, Counterculture and the Quantum Revival* (New York: W. W. Norton, 2012), 53. Special thanks to Mary-Jane Rubenstein for connecting this point to Kaiser's work and for her especially thoughtful and helpful suggestions throughout this article.
5. Somānanda, *Śivadṛṣṭi*, 5.106–7. For the first translation into English of this complex Sanskrit text, see John Nemec's excellent translation and study: John Nemec, *Ubiquitous Śiva: Somānanda's Śivadṛṣṭi and His Tantric Interlocutors* (New York: Oxford University Press, 2012).
6. Karen Barad, *Meeting the Universe Halfway: Quantum Physics and the Entanglement of Matter and Meaning* (Durham, N.C.: Duke University Press 2007), 335.
7. Ibid., 125, 130–31.
8. One early classical expression of Sāṃkhya is Īśvarakṛṣṇa's roughly third century CE Sāṃkhya Kārikā, however various permutations of Sāṃkhya were in circulation probably as early as the fourth century BCE. See Gerald Larson, *Classical Sāṃkhya: An Interpretation of Its History and Meaning* (Delhi: Motilal Banarsidass, 1969).

9. *Īśvara Pratyabhijñā kārikā*: 3.1. Utpaladeva was Abhinavagupta's predecessor and teacher of Abhinavagupta's teachers.
10. Abhinavagupta, *Īśvara Pratyabhijñā Vivṛti Vimarśinī* (IPVV), 3 vols., ed. Paṇḍit Madhusudan Kaul Shāstrī (Delhi: Akay Reprints, 1985), 3:257: jñāna śakti vapuṣā dharmirūpaḥ san tām antarbahirātmatām kramikāvabhāsām kriyā śaktim dharmatvena svīkurute iti.
11. Ibid., 3:262.
12. Ibid., 3:258. I translate vimarśa as "active awareness" to keep the inherent connection with action that vimarśa entails; however, Mark Dyczkowski, encountering the term in other texts of this school translates it as self-reflection.
13. Barad, *Meeting the Universe Halfway*, 184.
14. Ibid., 91.
15. IPVV, 3:258.
16. Ibid., 3:280.
17. Ibid., 3:315: Nanu evaṃ ghaṭe 'pi yadi cidrūpatāṃ kaścit paśyet, tarhi kim abhrānta eva asau. Om ity āha "jaḍasya tu" iti "Yukta eva" iti ucita evety arthaḥ. . . . ayaṃ samyag iti jaḍarūpatānyakkriyayā cidrūpatāprādhānyena āveśaḥ ā samantāt svarūpapraveśaḥ ity arthaḥ.
18. Nemec, *Śivadṛṣṭi*, 270.
19. Jane Bennett, *Vibrant Matter: A Political Ecology of Things* (Durham, N.C.: Duke University Press 2010), 4.
20. These are the *siddhi*-s, magical powers, historically one of the primary reasons a Tantric practitioner takes on particular ritual practices.
21. Bhūta śuddhi
22. See Gavin Flood, *The Tantric Body: The Secret Tradition of Hindu Religion* (London: I. B. Taurus, 2006).
23. The philosophical cosmology of a particular system determines the answer to this.
24. IPVV, 3:306.
25. Ibid.
26. Ibid., 3:334.
27. Via the connection of the subtle elements to the sense organs.
28. IPVV, 3:306.
29. Barad, *Meeting the Universe Halfway*, 342.
30. Ibid., 185.
31. The twin gods of the morning and evening star, the Aświns, try to pass as the human suitor Nala in the Nala and Damayantī love story in the Mahābhārata. Here we also encounter a talking goose, as we do a talking bull, various talking birds and a fire that talks as well in Satyakāma's story, Chandogya Upaniṣad, 4.4.1–4.9.3 in *Upaniṣads*, trans. from the Original Sanskrit by Patrick Olivelle (New York: Oxford University Press, 1996), 130–33.
32. IPVV, 3:260.
33. *Śiva Dṛṣṭi* 5.16: jānan kartāramātmānaṃ ghaṭaḥ kuryāt svakāṃ kriyām/ajñāte svātmakartṛtve na ghaṭaḥ sampravartate. See also John Nemec's fine critical edi-

tion and translation of the first three chapters of this seminal text, John Nemec, *Ubiquitous Śiva* (New York: Oxford University Press, 2012).
34. In fact, the medieval Śaiva context for instance, does not even employ the species-dependent formula for categorization. Rather, the classification here falls into seven levels of beings: (1) sakala, (2) pralayākala, (3) vijñānākala, (4) mantra, (5) mantreśvara, (6) mahāmantreśvara, (7) Śiva. Some systems add more or different categories—Vidyeśvaras, for instance, in some classifications. The system of classification derives from ontological priority based on degrees of manifestation of subjectivity. Thus, a being on the level of mantra (some gods, for instance) displays a greater sense of awareness that what exists is not different from one's own self than the entity at the level of vijñānākala, who understands the essential unity of reality but not to the degree to be able to act in the world as a god (at the level of mantra) might. These distinctions can cut across species; a particular animal, a crow or a cow, might attain a level of vijñānākala (literally, the one who has unbroken awareness) or mantreśvara, whereas a human at the level of sakala (literally, containing internal divisions) might not. These differences are ontological and appear in some cases to specifically pertain to that entity's capacity for greater or lesser effective action in the world. The notion of species in this context is, in contrast, more ephemeral, perhaps due to the idea in Patanjali, *Yoga Sūtra*, 2.13, that species can change from birth to birth dependent on karma.
35. I have argued elsewhere that it is precisely this distinction of the thinking mind as already on the side of matter fundamentally that David Chalmers points out in his discussion of the difference between the phenomenal and the psychological components of consciousness. In Loriliai Biernacki, "A Cognitive Science View of Abhinavagupta's Understanding of Consciousness," *Religions* 5, no. 3 (2014): 767–79.
36. *Sāṃkhya Kārikā*, 19–20. Larson, *Classical Sāṃkhya*, 265. The *"puruṣa"* is as we might expect always translated as male gender; I am taking license here to universalize it to person rather than "male person." Descartes's references to the *res cogitans* and *res extensa* come from his Second Meditation, II.08 in René Descartes, *Meditations on First Philosophy*, ed. Stanley Tweyman (New York: Routledge 1993), 34–40.
37. For instance, Daniel Dennett, *Consciousness Explained* (Boston: Little, Brown, 1991).
38. Henry Stapp, "A Quantum Mechanical Theory of the Mind/Brain Connection," in *Beyond Physicalism: Toward Reconciliation of Science and Spirituality*, ed. Ed Kelly, Paul Marshall, and Adam Crabtree (New York: Rowman and Littlefield, 2015), 166–69.
39. Ibid., 168, 173.
40. Stapp uses the term "mental probing actions" as a corollary to what von Neumann terms an "abstract ego." For Stapp the mental probing of a consciousness maintains "a quantum dynamical linkage to its associated physical brain" (ibid., 165, 172–73, 181).
41. Barad, *Meeting the Universe Halfway*, 114.
42. Ibid., 115.

43. Bohr quoted ibid., 154, used also by Merleau-Ponty with a different sense, Barad points out.
44. Ibid., 320. Barad uses this criteria from Bohr as the definition of objectivity. For instance, she quotes Bohr on 119: "'Objective' means reproducible and unambiguously communicable—in the sense that 'permanent marks' . . . [are] left on bodies which define the experimental conditions" (taken from Bohr in Henry J. Folse, *The Philosophy of Niels Bohr: The Framework of Complementarity* (New York: North Holland Physics Publishing, 1985), 15.
45. Abhinavagupta, *Parātrīśikavivaraṇa*, downloaded from GRETIL: Gottingen Register of Electronic Texts in Indian Languages: http://gretil.sub.uni-goettingen.de/gretil/1_sanskr/6_sastra/3_phil/saiva/partrvpu.htm, 212: sarvaṃ hi sarvātmakam iti. narātmāno jaḍā api tyaktatatpūrvarūpāḥ śāktaśaivarūpabhājo bhavanti, śṛṇuta grāvāṇaḥ [cf. Mahābhāṣya 3.1.1; cf. Vākyapadīya 3 Puruṣasamuddeśa 2], meruḥ śikhariṇām ahaṃ bhavāmi (Bhagavadgītā 10.23).
46. The Sanskrit makes this apparent: literally, "Those who have the essence of nara (man/human—here referring to the third person, the "it" of the stone), having abandoned that earlier form, they become beings who partake of the śākta-second person and Śaiva-first person forms." No external agency is invoked here.
47. Here Abhinavagupta's and Somānanda's insistence on the innate consciousness and sentience of the stone is the exception to a long tradition in India that begins with Sāṃkhya demarcating between matter and consciousness. As noted earlier, Sāṃkhya tends to consider only consciousness sentient, and even mind and intellect fall on the side of insentient nature, not to mention stones. Abhinavagupta, Somānanda, and Utpaladeva are radical in their ascription of consciousness even to matter.
48. Somānanda, *Śiva Dṛṣṭi* 5.105–7.
49. Barad, *Meeting the Universe Halfway*, 205.
50. Ibid., 206.
51. Ibid.
52. Ibid., 139.
53. IPVV, 3:257: Tad eva ca sarvabhāvānāṃ pāramārthikam vapuḥ.
54. Ibid.: Śivatattvaṃ hi sarva padārthānām vapuḥ.
55. Ibid.: Tad bhitta pṛṣṭhe ca sarva bhāva citra nirbhāsa.
56. Barad, *Meeting the Universe Halfway*, 181.
57. This derives from the Spanda Karikā opening metaphor, where consciousness paints the picture of the world out of its own self, SpK Nirṇaya:1.1: svasvarūpātsvatantra svacchasvātmasvabhittau kalayati dharaṇītaḥ śivāntaṃ sadā yā.

❧ Theophanic Materiality: Political Ecology, Inhuman Touch, and the Art of Andy Goldsworthy

JACOB J. ERICKSON

But making anything you have explored time,
and exploring time you have created the world,
even if it is only a little cairn of broken bricks
at the end of a rainbow.
—MARY RUEFLE, "Rumors of Earth," in *Trances of the Blast*

My writing was seduced by the matter at hand—I was lured by our new and gathering senses of materialism, this "new materialism"[1] which lures us to recall rather old senses as well. The history of Christian thought, one might say, is a history of the possibilities of materiality—at least the materiality of glory, sin, and flesh or divinity enfleshed, incarnadine. Even early Gnosticisms, in their own way, are theologies, apparatuses, reconfiguring theories of the capacities of matter.[2] The question becomes not whether Christian theology shapes material imagination, but how the materials of theology have fixed, fixated on, or shaped material possibility in economic and ecological forms. If our current moment indicates anything—situated in the Anthropocene,[3] overwhelmed by "atmospheric powers" of our own instigation,[4] with new knowledge and a knowing ignorance of quanta—it is that ecotheology must rethink materiality on a planet where matter indeterminately shifts, swirls, performs, flows, and is reconfigured with intensity and vibrancy. On a "tough new planet," as Bill McKibben calls our ungrounded ground, we tenuous earthlings may feel—deeply, affectively—that we need to take the matter we are seriously, anew, and shift the grounds of our thought to adapt, adapt with, and adapt to the new textures of this earth.[5]

Still, in feeling out those new textures, theology quite often rubs the wrong way. There have been glorious glimpses of material reflection in recent ecotheology, process thought, and other liberal theologies, as, for example, when

Paul Tillich notes, gesturing toward Henri Bergson and other vitalists in the third volume of his *Systematic Theology*, that "the genesis of stars and rocks, their growth as well as their decay, must be called a life process," and then notes, in a sort of understated lament, that "the religious significance of the inorganic is immense, but it is rarely considered by theology."[6] Still, as the story goes, theology has considered inorganic matter lifeless, passive, to be shaped by an omnipotent creator, in the masculinity of Adamic dominion, in an *imago dei*, riddled with human exceptionalism.[7] In its own brief patchwork life, then, this chapter pursues the queerly constructive task of rethinking the strange entanglements of divinity and matter in our new key. Seduced simultaneously by the so-called land art of Andy Goldsworthy, the new materialisms of Karen Barad and Jane Bennett, and the theophany traditions of Christian thought, I offer the concept of "theophanic materiality," in which divine energy is entangled in the performance of indeterminate material agencies. That is to say, I'm attempting to create a conceptual possibility for the queer intimacy of divinity and earth. To do so might effect a reimagined response to our current ecological crises.

THE ENCHANTMENT OF CAIRNS

Again, I was seduced by the matter at hand. As I ascended Mount Kilimanjaro and contemplated this writing, my eyes kept diverting themselves to the side of the path, to piles of rocks of all odd shapes and sizes, constructed on outcroppings, on the top of the mountains, in the middle of paths, on top of boulders, falling apart, covered in moss, perched on by crows. Cairns, that is—cairns that mark important rituals, places, views, hopes, or the haunting remembrances of people who were once there. You may recall small piles of stone that you have seen on the side of the road, or climbing up a mountain, or made in remembrance of someone. You may have piled stone upon stone yourself, attracted to their shapes and sizes. Some were built for plain fun; some were made to commemorate some sort of event or hope. They capture our attention with their eeriness, their communality, or their opaque histories.

In a multiplicity of cultural contexts, piles of stones take on various forms of significance. As S. Brent Plate describes in *A History of Religion in 5½ Objects: Bringing the Spiritual to Its Senses*, "From the Gaelic, a cairn is simply a pile of rocks. Sometimes they have religious connotations, sometimes not."[8] But genealogically, the practice of making cairns connects to particular, deep, and rich theological traditions. Timothy Joyce observes that in Celtic Christianity moving earth and piling rocks took on special significance in navigating the thin boundaries between "natural" and "spiritual" worlds. Joyce writes, "A common ritual practice was that of the *cairn*. Celts would draw a circle in the earth around themselves with a finger as they prayed to the Trinity to encircle and protect them."⁹ Earth can become a site of invocation or ethical attention. Cairns can live as simple piles of stone or as ritually encircling bodies, as an ambient practice of safety or hope. Their stony and precarious architecture (perhaps too improvisational for that term) becomes, on occasion, a fragile practice of prayer or praise *in media res*.

Thus Plate redirects the search for the historical cairn to a pointed, contemporary observation. He writes, "Crucial here is the fact that a cairn is made up of more than one stone, and they indicate some sort of human expression, somewhere between nature and culture, usually situated in the realm of ritual."¹⁰ The domain of the cairn is an odd one, indeed, spanning the oft-constructed walls between human being and ecological context. The stones reconfigure place and reconstitute ecological space. And yet the stones themselves never fully conform to human vision or order. They make no claim to permanence; they are precarious and unique. Their rocky bodies resist neat lines; they break against each other; they erode and wear over time; and they fall over. Left to their own handiwork—without industrial technological manipulation—human beings can configure the stones only in ways appropriate to shape or texture or jagged brokenness. Cairns force common caricatures of "nature" and "culture" to coincide, collapse, or begin to break apart. Unlike a monument or temple, these stones strive for neither permanence nor the holy façade of an exclusively human world.

As a ritual spanning nature and culture, cairns offer possibilities of ecological discernment and wise navigation. Cairns help mark or create brief views in space and time that one may not usually experience. Cairns may serve as markers for how others have made navigable paths through tough terrain. Cairns may lure others to touch and see their ecologies differently. One may take and remove stones at will. And some cairns may lure us along, even as they remain mysteries to us, testaments to memories or lives other than our own. Some cairns will fall over when they've been forgotten; animals will knock some over; and new cairns emerge in similar areas with differences in position and shape.

Although they may offer possibilities and potentialities, cairns do not necessarily dictate a forced path or view. They offer possible views and paths, relational aids in imaginatively navigating landscape. They may call for respect or spiritual contemplation. When the practices of cairns become prayer or praise, they also seduce others across time and space to possibly reconfigure their care, sensuality, touch, and sight of a locale and therefore the universe. That is, Plate notes, "They mark our presence in a physical place for people and ages to come. Facebook and Twitter accounts are filled with updates on people in transition. And yet, in a real sense, humans have been checking in and sending status updates for millennia."[11] The piling of ordinary stones is a form of social media and multicommunicative ecological touch.

Taking note of a cairn—in the midst of a city, in the midst of a hike, on the side of the road, on the side of a mountain—is enchanting. The stones can hold attention, shift focus, cause us to wonder, urge us to topple or add to them. They simultaneously attempt solidity and expose the fragile movements and hauntings of a place. Plate concludes of stones, "At least—if we are listening—they are telling us something about life itself: performing for us, whispering, standing up and down, killing, drawing us near or pushing us away, pointing us here and there, reviving, ticking down the moments of eternity, allowing us to stand in awe and contemplation."[12] In attuning our senses to something new or raising possibilities, imagination, or memory, cairns exhibit the circulations of human-ecological relationality. Cairns, for all their stoniness, expose human and nonhuman ecology as a fluid, open process.

A TOUCH OF ENERGY: ON GOLDSWORTHY AND COLLABORATIVE ATTUNEMENT

Alongside their human circulations, everything of the cairns I witnessed, the piles of stones—moss, crows, and the occasionally discarded food wrapping on the side of Kilimanjaro—enchanted me not unlike the work glove, the oak pollen, the dead rat, the bottle cap, and the stick of wood that "shimmied back and forth between debris and thing" for Jane Bennett.[13] All of this matter hardly lifeless, instead vibrant.[14] These cairns sent my imagination off or opened it doubly in wonder, curious to the rocks' stellar lives and futures as social and ecosocial media. Who made them? Why? Do they mark anything significant? Were they made playfully? Did human beings make them at all? What kind of response do they call from me, if any? How do their lives extend far beyond my own? They touched my imagination and lured me to new ways of touching and seeing my own ecology. And the witness of these stones and questions invoked a mountainside conversation, animated with mysterious, transfiguring energy, with my fellow climbers over the work of the contemporary artist Andy Goldsworthy, whose land art is quite often inspired by cairns of this very sort.

In popular culture, most people—if they have heard of him—know the name of Andy Goldsworthy because of a 2001 documentary, *Rivers and Tides*, which followed Goldsworthy as he went about his work near his home in Penpont, Scotland, and in a variety of ecological settings around the globe.[15] Since the 1970s, Goldsworthy has been creating works of art that utilize local and found materials. His work is often identified within the (slightly) longer and nebulous tradition of "land art" or "earthworks" that emerged in the 1960s.[16] As Kathryn B. Alexander describes it, "Its founders, who include Robert Smithson, Nancy Holt, Richard Long, and Michael Heizer, wanted to free landscape from painting and schmaltzy sentimentalism. Instead of creating decorative placebos for urban angst, land artists sought to aggravate that angst by displaying what had become of the natural world in an industrialized and consumerist age."[17] These artists, also influenced by feminist and other political artists, sought to create ecologically engaged pieces that challenged artistic convention and traditional relationships between human culture and nature.[18]

In its emergence and life today, "Land art as a genre includes forms such as landscape art, earth art, nature art, ecological art, installations, gardening, garden architecture, even flower arranging."[19] Artists vary in their understanding of creativity and relationship to ecology and matter, and various artists fine-tune their creativity in very different ways. With the documentary, Goldsworthy's artistic process is fascinating to watch. He attempts to tune into the lives and rhythms of a place. He becomes entranced by the elements, creatures, and the ecological materials of that place. With those materials calling for his attention, he then attempts a tenuous act of creation in "collaboration," as he calls it, within that place. He usually accomplishes this work with matter and his bare hands or breath. He stacks stones and sticks into cairns; he weaves leaves into patterns of color; he throws sticks and snow into the air; he laces the wool of sheep around rock; he melts delicate icicles together with his fingers into the most glorious of forms. He works with these materials to weave together the most beautiful of moments—usually for just a brief photograph before the piece collapses or melts or is washed away by a river.

Thus photography is precisely how Goldsworthy communicates. Photography is the "language" by which Goldsworthy speaks about his artistic collaborations, how his professors in art school knew what Goldsworthy was up to, and how the world "sees" much of Goldsworthy's work—especially the fragile, transient, and collapsed pieces.[20] To bring these pieces indoors would ruin their artistic vibrancy, contrast, and context. As the process theologian Jay McDaniel notes of photography, "This act of being with things, by means of photography, is a mysterious thing. On the one hand it frames things and has a controlling side. It is an act of power on the part of the photographer. But on the other hand it is a way of helping the things of the world show themselves, present themselves, not only to the photographer but to the world."[21] Photography is an art form filled with its own peculiar tensions, limitations, and possibilities. Of course, photographs communicate only so much to a viewer's senses—quite often a single photograph hides just as much of its subject as it reveals—and possesses its own constructive vision.[22] Still, for Goldsworthy, the language of photography moves much of his particular ecological vision out into larger audiences.

We might meditate on a number of Goldsworthy's photographs for the fluid energy of matter, their fragile and resilient transformations, and all of the collaborative energies that flow into making a single moment. Take, for example, a piece from 1987 subtitled "Bricks and river mud; rising tide" done at the south bank of the River Thames in London.[23] Goldsworthy stacks bricks found on the banks in circles and up into a kind of tower or cairn. The bricks hold together with mud and at the top Goldsworthy has constructed, out of the mud, a deep, dark hole at the center that lures your attention. But that dark hole is impermanent. The picture itself is of the tide of the Thames coming in later, beginning to wash away the mud, and then the cairn, and then the hole—all materials disappearing again, reconfiguring, into the rhythms and flows of the place.

In his 1994 collection of photos, *Stone*, Goldsworthy reflects on making cairns, stating that he "had to forget my idea of nature and learn again that stone is hard and in so doing found that it is also soft."[24] Working with stone forced Goldsworthy to recognize the hard and fluid properties stone takes across space and time. Everything is, in some sense, fluid and in flux. Later in that same book, he argues, "Everything is fluid, even the land, it just flows at a very slow rate."[25] Even stones, which quite often give illusions of changelessness and permanence, flow, erode, and change with time. Even the territories of the earth, which quite often feel timeless and stable, change.

Alongside these reflections on fluidity and fragility, consider other works such as 1987's "Rowan leaves around a hole,"[26] "Pebbles around a hole"[27] (from that same year), or 2007's "Wool laid on a sheet of ice lifted from nearby pool placed on river stone,"[28] which, on much smaller poetic scales, gather leaves, stones, and wool into carefully constructed piles and uses their various shadings and colors to produce a similar kind of focused effect. The hole in the center seduces you in, causes you to hover over and dive into the deep, the multiplicity of stones, the leaves, and lines of wool frothing on the surface. And again, these materials fade or are washed away.

Although these pieces can be initially interpreted as a kind of *axis mundi*, a monument, or central honorary location, Goldsworthy attempts to gesture toward the mysterious energy of the material earth in the edges and the unforeseen, and his pieces seduce the viewer into the flow of those material realities. "The black of a hole is like the flame of a fire," Goldsworthy writes. "The flame makes the energy of fire visible. The black is the earth's flame—its energy."[29] The material collaboration of the earth eludes in mysterious, multiple constitutions, relations. The cavities in these pieces exhibit a kind of infinite depth. As Miranda Strickland-Constable writes, "This gives the artist a sense of the energy contained in the earth, and energy is one of his favourite concepts. It appears in the way he uses colours, manipulating them to bring out unexpected brilliances; it is the intensity of a found colour, rather than its individual hue which fires his enthusiasm."[30] And that depth and those intense hues unsettle him; they seduce him to work with that energy to create beautiful, captivating (though not fully captured) art.

I am drawn to these works because they seem to point toward a kind of figuration of materiality in Goldsworthy's imagination—of a kind of collaborative attunement to the indeterminate energies of earth. Matter is not simply a passive or inactive substance to be shaped by human hands. Ecological matters perform agencies all their own that collaborate with, captivate, reconstitute, or frustrate human agencies. We might call it a kind of "vibrant" or "enchanted" materialism, following Bennett. Goldsworthy's art makes the "energy of the fire visible" precisely by working creatively with its brilliant darkness, precisely by working with a multiplicity of materials that erode, wash away, and illumine material flows. Even rocks and stones are fluid as he gets to know the geohistories of a place. Earth is a vibrant place. Goldsworthy wants to discern and work with those vibrations.

These works reflect intimacy and collaboration with a fluid materiality, without any pretense to full or omnipotent control over that materiality. Goldsworthy's art points to a maxim similarly articulated by the artist Cornelia Parker: Matter is transformed.[31] And Goldsworthy participates in that creative transformation fully. Indeed, James Putnam notes in his introduction to Goldsworthy's collection *Enclosure*, that "Goldsworthy regards himself to be as much a receiver as a transmitter of creativity, since he 'tunes in'" to an intimate knowledge of place—intimate, part of, not omniscient, not discretely separate.[32] That is, Goldsworthy doesn't swoop into a place, make something he alone deems pretty, and then disappear. He discerns the creativity of the place, regards matter, precisely *as he is enmeshed in the place*. It is an art of "intra-action." Goldsworthy recognizes that he is precisely ecological matter working with other ecological matters to produce intimate expressions of beauty, energy, and relation. Such intra-active intimacy uncovers multiplicity. "Though he might talk about the need to feel connected to a place, the work he does with his hands has meant that place is always plural, that his connections are never exclusive or definitive, settled or permanent."[33] "Place" is not about simple location—it is a flow of material processes ever under transformation. His art, just as cairns, exposes a multiplicity of views, navigational, or ritual potentialities in landscape.

Goldsworthy's work is about the openness of creativity, then, the creative beauty of forms perpetually perishing.[34] Viewers of the 2001 documentary will note the amount of "failure," or "collapse," Goldsworthy experiences in the process of making. Stone structures collapse in the middle of a project, sticks fall apart, and ice melts, shatters, and breaks. Goldsworthy's collaboration with and as materiality is a kind of chaotic indeterminacy—no form is fixed, permanent, perfect. Instead, in each photograph, Goldsworthy notes, "There is an intensity about a work at its peak that I hope is expressed in the image. Process and decay are implicit in that moment."[35] And elsewhere, "The very thing that brings the work to life is the thing that will cause its death."[36]

It seems then that intimately working with matter, for Goldsworthy, is about attunement, flow, indeterminacy, energy, unknowing. In writing woven into Goldsworthy's own *Hand to Earth*, the writer John Fowles indicates that he would place Goldsworthy in "a much older tradition in our culture, that of the nature mystic—if it were not that that suggests a dreamy remoteness from actual life."[37] Goldsworthy himself expresses some discomfort with the term "mystical" when he argues, "My work is spiritual but not mystical, art that continues the long tradition of making art in order to see, feel and understand nature."[38] He fears that the term "mystical" might risk disembodying and de-mattering the mysterious animated energies of the earth. We also might observe, however, that "spiritual" might work with a similar disembodying effect.

Despite discomfort, a more earthen version of "nature mystic" may not be too far off. Goldsworthy is getting some recent attention from theologians and scholars of religion such as Jeffrey L. Kosky, Mark C. Taylor, and Kathryn B. Alexander. Kosky argues that artists like Goldsworthy may help us—those of us motivated by our own "disenchantment with modern disenchantment"—think a bit more mysteriously without returning to old theological orthodoxies that rely on an omnipotent creator. These works of art are "places where we might encounter mystery and wonder."[39] In Goldsworthy's work, the unforeseen in collaboration, the creative novelty, or "creative unsecuring" in time, as Kosky calls it. Goldsworthy knows he cannot make the earth in his own image. Instead, he works with matter in a kind of mystical or apophatic attempt to speak of earth's processes—bare to the touch, seductive, and collaborative—yet always, in some ways, unknowable for those very reasons.

Mark C. Taylor argues that Goldsworthy's attentiveness to ecological energies and place might actually work as a rich political critique of our contemporary moment. He argues that "Goldsworthy's art provides a corrective to the excesses of finance capitalism and proposes a vision that addresses many of the critical environmental issues we are facing."[40] Goldsworthy's land art serves as a stunning counter to the unbridled consumerism of popular art and the global excesses of neoliberal capitalism. For Taylor, Goldsworthy's collaboration with ecological materials, his light touch, his attentiveness to the fragility of matter and the transformation of the earth offer a subversive practice of attention to the subtle rhythms of an earth in peril.

THEOPHANY, DIVINE ENERGY

Goldsworthy's attention to earthy energies and rhythms certainly cannot be called theistic. And yet they do evoke a set of theologically energized discourses in my mind. Christian theology, in its most orthodox keys, usually constructs a sense of materiality along a string of various doctrines—creation, incarnation, salvation.[41] Here, however, I want to think of the touch of divine energy, the theophany traditions that occasionally appear in all those messy biblical imaginations and theologies of creation—especially of the Eastern Orthodox sort.

"Theophany" is a very conceptual or theological term gesturing toward the occurrence of divinity, incandescent in places or things in places. The original Greek term *theophania*, a hybrid of *theos* (God) and *phaneros* (appearance or manifestation), does not occur in the Bible. The term is, however, later widely applied as a category to those stories where the Divine Other appears in places, special times, visions, and discernments of Word, Spirit, Wisdom, and personal experience. One thinks of the biblical stories of Moses and the burning bush, mountains that invoke Moses' shining face on Sinai, stones bursting with water, Elijah's encounter with silence, clouds or the transfigurations in the gospels. Stephen Farris notes that these appearances of the divine "must be understood in light of a near paradox." He observes, "The divine presence is made known but the stories are often replete with imagery of disruptions to nature that emphasize the otherness of God."[42] One might be able say that theophany is precisely a mysterious attunement to divine energies and rhythms in a host of strange ecologies. Quite often these encounters usually are unsettling in their strangeness of call and locale. They often demand a call for justice or, for Moses, liberation.

Classical reflections on the occurrence of theophanies from the biblical to the Cappadocian Fathers to Gregory Palamas to more contemporary theologians have a way of touching on the materiality of earthy places as well. These theologians often make a distinction between the "essence" and "energies" of God—God in Godself and God variously relating to creation as Creator and End. Although the essence of God is unknowable, we might discern the edges of divinity's working immanently in the world in wisdom, goodness, justice, and the energetic list goes on and varies in emphasis from theologian to theologian.[43]

Gregory of Nyssa, in his discussion of the burning bush in *The Life of Moses*, notes that the crackling encounter of divinity, prophet, branches, flame, and bare feet, occurs in a material key. He writes, "Lest one think that the radiance did not come from a material substance, this light did not shine from some luminary among the stars, but came from an earthly bush and surpassed the heavenly luminaries in brilliance."[44] He continues by praising "the Radiance which shines upon us through this thorny flesh."[45] Divinity occurs in thorn and thickets.

The ninth-century theologian John the Scot Eriugena went so far in his *Periphyseon* to say that in some sense each and every creature is a theophany of God, of God multiplying, running out into and as creation.[46] Eriugena writes, "Therefore it is through bodies in bodies, not through Himself, that He shall be seen. . . . For 'God shall be all in all'—as if the Scripture said plainly: God alone shall be manifest in all things."[47] Divinity, manifest in bodies, entangling all to all. He continues that God "runs *throughout all things* and never stays but by His running fills out all things. . . . For He is at rest unchangingly in Himself, never departing from the stability of His Nature; yet He sets Himself in Motion through all things in order that those things which essentially subsist by Him may be."[48] The whole of physical reality contains billions of creatures made manifest through divinity, billions of mysterious burning bushes—an apophatic cosmology returning to the divine. As William Franke notes, "Not only God but *all* things are unknowable in their essences, since they have their being and essence only from God, who is unknowable."[49] All things, despite perceived opposition, burn and coincide in their life and origin in the divine, all in all. Sadly, the reception of his theology, the intimacy of the cosmos and God, got him posthumously declared a pantheist and heretic.

The contemporary Eastern Orthodox theologian Kallistos Ware, riffing on the theophany traditions of Christianity, might poetically exhibit a similar meditation. Ware argues that with a spiritual attention to the earth, "To contemplate nature, then . . . is to discover, not so much through our discursive reason as through our spiritual intellect, that the whole universe is a cosmic Burning Bush, filled with the divine Fire yet not consumed."[50] The practice of seeing the cosmos in spiritual intellect may reveal the burning of divine energies throughout all of creation, sustaining and enlivening creation with spirit back to an end in God.

Although Gregory and these others let their meditations of divinity flow to materiality for brief moments, they quickly flow back to the omnipotence and oneness of their nonmaterial divinity.[51] Matter becomes a passive mode or mediation of divine knowledge, incidental, trumped by the occurrence of divinity. And even while their thought might unravel at its tensions between that oneness of God and the multiplicity of creation, the light of knowledge, wisdom, might seem to take precedence in many places in their writing over the mystery of the encounter.[52] As David Bradshaw observes about the concept of theophany in some Eastern Orthodox thought, "Nature is theophany, but only because it points beyond itself to its source."[53]

Still, theophanic thought points toward the energy of divinity as it flows in material places, of divine energy occurring in and as material place in ways that glow with luminous darkness at its edges. And it is here, fusing Goldsworthy's attention to earthy energy with theophanic attention to divine energy that we might begin to reconstruct earth and divinity as irreducibly entangled and animated. According to Donna Bowman and Clayton Crockett, "It makes sense to view energy as material and spiritual at the same time rather than to dualistically oppose them."[54] Or, to think of the occurrence of divinity with the occurrence of matter-energy as well, we might think of energy as Catherine Keller does, "with possibly more fidelity both to science and to the spirit of our shared life—as the rhythm of interactivity." To think divine and matter-energy as pulsing to a rhythm of interactivity is to think about them as vulnerable and resilient together, simultaneously. Keller continues, "Energy signifies the pulsation of life: life as boundless vitality, life that exceeds the distinction between organic and inorganic. In theological terms, energy connotes not only the efforts of work but the effortlessness of grace as well."[55]

THEOPHANIC EARTH, QUEER INTIMACY

If Goldsworthy's effortful and effortless attention to place actually uncovers the radical indeterminacy, energy, agency, and flow of matter, we might revise an ecologically attentive concept of theophany to do something creatively similar. We might approach the theophany tradition with the kind of light collaborative touch that Goldsworthy uses in his art. To rethink the concept of theophany as the fluid, rhythmic site of intra-active collaboration—the rivers and tides of energies crackling, touching, and transforming, creatively unsecuring one another (to borrow a metaphor of Goldsworthy's artistic process), a ritual of making cairns—is to think of the theophanic earth as an indeterminate, performative, fluid place, where new kinds of divine and earthy relations are always possible. As Taylor remarks, "For Goldsworthy, nature is bursting with energy and explodes like a fern's uncoiling frond reaching for the sun."[56] What might it mean to consider the bursting energy of that vision constructively read with the luminous burning bush of theophany traditions?

My hunch here is that we might think of the divine energies indicated under theophany as entangled—amazed and undone by the earth in a kind of queer intimacy. That is, to think of theophany as the transfiguring flow of divine energies in creation, reconfiguring the divine again, is also to think of the flow of earthy energies in the divine, refiguring the earth again. I might call it a theophanic interactivity.[57] Or I might simply say, as Goldsworthy does, that "we all touch nature and we are a part of this process of interaction and change, we rely on each other." And I might, finally, suggest that what Goldsworthy is really talking about is not "interaction" (in the sense of a relation of preexistent entities) but "intra-activity"[58]—a touching relationality we all make up.[59] Perhaps, in Mayra Rivera's turn of phrase, "the touch of transcendence."[60]

This material touch of our relations, our theories, our theologies, our poetics, our arts, Karen Barad might remind us, creatively transfigures worlds. And it does so in very queer fashion, for our touch is a matter of matter touching matter, "self-touching is an encounter with the infinite alterity of the self. Matter is an enfolding, an involution, it cannot help touching itself, and in this self-touching it comes in contact with the infinite alterity that it is."[61] (Oh, the inhumanity!) Theologically, that might mean that if divine energies pulse with the rhythmic intra-action of matter, promiscuously entangled or enfolded in the rhythms of theophany, then earth and divinity pulse with infinite and infinitesimal alterities and intimacies.[62] Perhaps the holy, moving ground is as much in awe of the touch of our bare feet. Truly, to borrow Barad's language, "Polymorphous perversity raised to an infinite power: talk about a queer intimacy."[63] That talk about intimacy whispers at the edges of alterity, at the flame of black holes and bushes. That talk unravels at the many ragged edges of our writing and speech. That talk, so touching in its range of verbose responses to the call of material and divine things.

Politically, such a theology and such an art might evoke a protest, as Taylor urges, of the growth of neoliberal capital endangering planetary life. As Clayton Crockett, Jeffrey W. Robbins, and Michael W. Wilson argue in their own call for a new materialism, "Art must become fugitive and multiple—*plastic/plastique*—disappearing and refusing to participate when tactically necessary—sabotaging, attacking, and occupying sites of power to redistribute resources and attention when possible."[64] Although not obviously a tactics of sabotage, a theophanic materiality inspired by the metaphors and practice of Goldsworthy's process is fugitive and multiple. Perhaps a theophanic materiality performs a subtle kind of sabotage by pulling attention from the dangerous flows of global capitalism to the flows of the tactile, fluid earth, to the vibrancies of everyday matters, to the shimmering glory of imperiled creatures. How might we create cairns—conceptual, material, relational—that redirect our attention to addressing ecological crises with memory and hope?

In further notes on vibrant matter, with attention to the culture of hoarding, Bennett writes, "A less verbose practice (performance art, photography, painting, music, dance) is probably better suited to the task of acknowledging the call of things."[65] The loud mystery of tactility is what drew me around, up, and down the mountain and to Goldsworthy. That mystery lured me to compose this cairn of words to meditate on the less wordy practices of land art. And those less verbose practices are probably better suited to acknowledge the call of divinity in place as well. They acknowledge the fluid materiality of place, the fluid materiality of divinity, and the ways they appear, mysteriously entangled. They might let the indeterminacy of earth, terrifying at first in its transience, be the setting of thought and might urge new figurations of intimacy with divinity and earth. Or with, as Bennett and others now note, "the earth we are." And that self-touch, *adam* from *adamah*, human from *humus*, from and to dust, various theophanic transfigurations is a very intimate thing indeed. A theophanic conviviality, a darkly luminous, inhuman life together, perhaps. We are all partly, after all, a little cairn of broken bricks at the end of a rainbow—for *this* touch of dust . . . and *this* . . . and *this* . . .

NOTES

1. I am referring here to a growing multiplicity of works cohering with similar themes of the vibrancy, performativity, or agency of matter, the complicated refiguring of nature and culture, the new porous boundaries of humanity and animality. These may include works from Jane Bennett and Karen Barad (to be discussed elsewhere in this chapter), but also a number of other promising works. The following list is suggestive only, not exhaustive. See, for example, Stacy Alaimo, *Bodily Natures: Science, Environment, and the Material Self* (Bloomington: Indiana University Press, 2010); Diana Coole and Samantha Frost, eds., *New Materialisms: Ontology, Agency, and Politics* (Durham, N.C.: Duke University Press, 2010); Mel Y. Chen, *Animacies: Biopolitics, Racial Mattering, and Queer Affect* (Durham, N.C.: Duke University Press, 2012); Rick Dolphijn and Iris van der Tuin, *New Materialism: Interviews, and Cartographies* (Ann Arbor, Mich.: Open Humanities Press, 2012); the writings of Elizabeth Grosz, more recently, *Becoming Undone: Darwinian Reflections on Life, Politics, and Art* (Durham, N.C.: Duke University Press, 2011); and Timothy Morton, *Hyperobjects: Philosophy and Ecology after the End of the World* (Minneapolis: University of Minnesota Press, 2013.)
2. Patricia Cox Miller points us toward the material infatuations of the fourth century, for example. (See her *The Corporeal Imagination: Signifying the Holy in Late Ancient Christianity* (Philadelphia: University of Pennsylvania Press, 2009.) Even Luther meditated on the sacramental indwelling of Christ in matter, in elements, noting that "I might find him in stone, in fire, in water." Even though Luther then hedges his bets, pastorally, "He is present everywhere, but he does not wish that you grope for him everywhere." See "The Sacrament of the Body and Blood of Christ—Against the Fanatics," in *Luther's Works*, vol. 37: *Word and Sacrament III*, ed. Robert H. Fischer (Philadelphia: Fortress Press, 1961).
3. The term "Anthropocene" was coined in its current usage by the Nobel Laureate and atmospheric chemist Paul Crutzen to refer to a time of unequivocal, long-lasting, and pervasive human manipulation of the planet and its atmospheric composition. Human influence on the planet is so pervasive, he argues, that that it implies a new geological epoch. See P. J. Crutzen and E. F. Stoermer's piece, "The 'Anthropocene,'" *Global Change Newsletter* 41 (2000): 17–18. http://www.igbp.net/download/18.316f18321323470177580001401/1376383088452/NL41.pdf.

4. I borrow this turn of phrase primarily from Willis Jenkins, *The Future of Ethics: Sustainability, Social Justice, and Religious Creativity* (Washington, D.C.: Georgetown University Press, 2013).
5. See Bill McKibben, *Eaarth: Making Life on a Tough New Planet* (New York: Times Books, 2010).
6. Paul Tillich, *Systematic Theology*, vol. 3 (Chicago: University of Chicago Press, 1963), 12–18.
7. See my earlier critique and revision of this concept in "The Apophatic Animal: Toward a Negative Zootheological *Imago Dei*," in *Divinanimality: Animal Theory, Creaturely Theology*, ed. Stephen Moore (New York: Fordham University Press, 2014).
8. S. Brent Plate, *A History of Religion in 5 ½ Objects: Bringing the Spiritual to Its Senses* (Boston: Beacon Press, 2014), 35.
9. Timothy Joyce, *Celtic Christianity: A Sacred Tradition, a Vision of Hope* (Maryknoll, N.Y.: Orbis Books, 1998), 26.
10. Plate, *Religion*, 35.
11. Ibid.
12. Ibid., 59.
13. Jane Bennett, *Vibrant Matter: A Political Ecology of Things* (Durham, N.C.: Duke University Press, 2010), 4. Or, perhaps, to risk saying too much, they indirectly invoked Gordon Lathrop's earlier liturgical theological sense of "holy things," where things take on meaning in their assembly and juxtapositions. He argues, "Even the assembly itself, the place of those juxtapositions, may be regarded as a *thing*, in the archaic old Norse and old English sense—that is, a gathering of people with a purpose, an assembly of the free and responsible ones." See his *Holy Things: A Liturgical Theology* (Minneapolis: Fortress Press, 1993), 10.
14. As Bennett writes in her preface that her "hunch is that the image of dead or thoroughly instrumentalized matter feeds human hubris and our earth-destroying fantasies of conquest and consumption." See *Vibrant Matter*, ix.
15. The documentary made by Thomas Riedelsheimer met with well-deserved praise. *Rivers and Tides: Andy Goldsworthy Working with Time*, DVD (Mediopolis Film with Skyline Productions, 2004).
16. Richard Long, incidentally, lectured at Preston Polytechnic, the art college Goldsworthy attended in Lancaster. For some historical context see the vast literature on land and environmental art, such as Jeffrey Kastner and Brian Wallis, *Land and Environmental Art* (New York: Phaidon Press, 2010). Other important contemporary artists resonating with this style include Martin Hill, Christo and Jeanne-Claude, Graham Goddard, Olafur Eliasson, Cornelia Parker, James Turrell, Isamu Noguchi, and Walter De Maria. More, the body and performance art of Marina Abramović adds much to the conversational matrix of art, ecological consciousness, and creaturely energy. The list is far too long for this short one to be comprehensive, of course.
17. Kathryn B. Alexander, *Saving Beauty: A Theological Aesthetics of Nature* (Minneapolis: Fortress Press, 2014), 103. Sadly, I wasn't aware of Alexander's reflection on Goldsworthy's art until after this piece had been written, and so only am able to refer-

ence her well-constructed work (and some similar conclusions about the ways in which Goldsworthy's work challenges strict dualisms of "nature" and "culture").
18. Important to note is that the origins and definitions of this tradition, like those of most traditions or styles or schools, are contested.
19. Alexander, *Beauty*, 103.
20. This becomes very evident later on in *Rivers and Tides*.
21. See the reflections of the process theologian Jay McDaniel, "How Photography Becomes a Prayer," at *Jesus, Jazz, and Buddhism*. Located http://www.jesusjazz buddhism.org/how-photography-becomes-prayer.html Last Accessed: February 2, 2015.
22. Had I more space here, I would reflect longer on how photography is an intra-active collaboration, an apparatus for relating and thinking about matter.
23. See the image at Andy Goldsworthy, *Hand to Earth* (New York: Abrams, 1990), 32–33.
24. Andy Goldsworthy, *Stone* (New York: Abrams, 1994), 6.
25. Ibid., 65. See Mark C. Taylor's reflection on this process, especially in chapter 5 of *Refiguring the Spiritual: Beuys, Barney, Turrell, Goldsworthy* (New York: Columbia University Press, 2012).
26. Andy Goldsworthy, *A Collaboration with Nature* (New York: Abrams, 1990), 10. The piece is actually fully subtitled, "Rowan leaves laid around; hole collecting the last few leaves; nearly finished; dog ran into hole; started again; made in the shade on a windy, sunny day." Constructed at the Yorkshire Sculpture Park, West Bretton, October 25 1987.
27. Ibid., 11. Made at Kiinagashima-Cho, Japan, December 7, 1987.
28. Andy Goldsworthy, *Enclosure* (New York: Abrams, 2007), 27.
29. Goldsworthy, *Hand to Earth*, 24.
30. Miranda Strickland-Constable, "Beginnings," in Goldsworthy, *Hand to Earth*, 19.
31. Such is the piece of Parker's that captivates on the cover of Jane Bennett's *Vibrant Matter*.
32. Goldsworthy, *Enclosure*, 13.
33. Jeffrey Kosky, *Arts of Wonder: Enchanting Secularity—Walter De Maria, Diller + Scofidio, James Turrell, Andy Goldsworthy* (Chicago: University of Chicago Press, 2013), 145.
34. There are interesting resonances between Goldsworthy's artistic sense of perishing and Whitehead's cosmological sense of perishing, of failure—for another time.
35. Goldsworthy, *Hand to Earth*, 9
36. See *Rivers and Tides*. I might bring up a further point here, that Goldsworthy often speaks of time in very complicated ways. Often, that talk sounds quite linear—he works in a place regarding its past, present, and future. Often, he'll speak of time itself as one of the "tools" he works with. In collaborating with nature, it might be that Goldsworthy's sense of temporality is in fact too linear and simple. In collaborating with the multiplicity of place, Goldsworthy may in fact be reconfiguring and re-creating time itself as well—that is, he might be collaborating in the creation of both material space and time: "space-time." See Andy Goldsworthy, *Time* (New York: Abrams), 2000.

37. Goldsworthy, *Hand to Earth*, 160.
38. Quoted in Alexander, *Beauty*, 101.
39. Kosky, *Arts of Wonder*, xii–xiii.
40. Taylor, *Refiguring the Spiritual*, 157.
41. And I have elsewhere argued for an intracarnational perspective, a carnivalesque panentheism with the queer pairing of Karen Barad and Martin Luther. My "Indecent Ecologies: Karen Barad, Naturecultural Performativity, and Queer Ecotheology" (to be published in a collection of essays on queer ecology edited by Whitney A. Bauman with Punctum Books).
42. Stephen Farris, "Theophany in the NT," in *The New Interpreter's Dictionary of the Bible*, vol. 5, ed. Katharine Doob Sakenfeld (Nashville, Tenn.: Abingdon Press, 2009), 565. We might also note that the term "theophany" is quite often interchanged with that of "epiphany." I want to reserve the possibility that the two terms can do differing theological, collaborative work. That is for another time, however.
43. For further distinctions on this theological construct see David Bradshaw, "The Concept of Divine Energies," *Philosophy and Theology* 18, no. 1 (2006): 93–120.
44. Gregory of Nyssa, *The Life of Moses*, trans. Abraham J. Malherbe and Everett Ferguson, Classics of Western Spirituality (New York: Paulist Press, 1978), 59.
45. Ibid., 60.
46. Most accessible in the United States, though outdated now, is Myra L. Uhlfelder's translation of the *Periphyseon: On the Division of Nature* (Eugene, Ore.: Wipf and Stock, 1976). Two translations are now standard, however, for scholarly study: I. P. Sheldon-Williams and John O'Meara, Dublin Institute for Advanced Studies Book I (1968), Book II (1972), Book III (1981), ed. Sheldon-Williams and Ludwig Bieler; Book IV (1995), ed. Édouard Jeauneau. For more on the complexities of Eriugena's complex understanding of Theophany and argument of the Word of Wisdom running into creation see Michael A. Sells, *Mystical Languages of Unsaying* (Chicago: University of Chicago Press, 1994), 34–62. For more recent constructive and critical work, see Eugene Thacker, *After Life* (Chicago: University of Chicago Press, 2010).
47. John the Scot Eriugena, *Periphyseon*, Book I, trans. I. P. Sheldon-Williams. 57. See sections 450C–451A.
48. Ibid., 61. Sections 452C–D.
49. William Franke, *On What Cannot Be Said: Apophatic Discourses in Philosophy, Religion, Literature, and the Arts*, vol. 1: *Classic Formulations* (Notre Dame, Ind.: University of Notre Dame Press, 2001), 183.
50. Quoted in Jurretta Jordan Heckscher, "A 'Tradition' That Never Existed: Orthodox Christianity and the Failures of Environmental History," in *Toward an Ecology of Transfiguration: Orthodox Christian Perspectives on Environment, Nature, and Creation*, ed. John Chryssavgis and Bruce V. Foltz (New York: Fordham University Press, 2013), 141.
51. I would argue that this phenomenon is actually more complex in Eriugena. Again, see Sells, *Mystical Languages of Unsaying*.
52. Mary-Jane Rubenstein gestures to some of the possibilities of the deconstruction of this tension in "The Fire Each Time: Dark Energy and the Breath of Creation."

Especially when she writes, "If the energies cannot be dislodged safely from the divine essence *or* from creation, then they do seem to bind the whole lot into an eternal, synergetic dance. This is not to say that God is creation or that creation is God, but by virtue of the eternal energy that 'bridges' them, it is possible that both become, and unbecome, in mysterious relation to each other" (Mary-Jane Rubenstein, "The Fire Each Time: Dark Energy and the Breath of Creation," in *Cosmology, Ecology, and the Energy of God*, ed. Donna Bowman and Clayton Crockett [New York: Fordham University Press, 2012], 40–41).

53. He is referring, in particular, to the thought of Evagrius of Pontus. See David Bradshaw, "The *Logoi* of Beings in Greek Patristic Thought," in *Toward an Ecology of Transfiguration: Orthodox Christian Perspectives on Environment, Nature, and Creation*, ed. John Chryssavgis and Bruce V. Foltz (New York: Fordham University Press, 2013), 15.

54. Donna Bowman and Clayton Crockett. Introduction to *Cosmology, Ecology, and the Energy of God*, ed. Donna Bowman and Clayton Crockett (New York: Fordham University Press, 2012), 3.

55. Catherine Keller, "The Energy We Are: A Meditation in Seven Pulsations," in *Cosmology, Ecology, and the Energy of God*, ed. Donna Bowman and Clayton Crockett (New York: Fordham University Press, 2012), 12.

56. Taylor, *Refiguring the Spiritual*, 171.

57. Keller refers to this as a kind of "intercarnation." I might want to hold onto the possibility of an "intracarnational" or "intratransfigurational" relationality of divinity and earth.

58. I am riffing on this term of "intra-active" from the work of Karen Barad, of course. See her stunning *Meeting the Universe Halfway: Quantum Physics and the Entanglement of Matter and Meaning* (Durham, N.C.: Duke University Press, 2007).

59. Goldsworthy, *Hand to Earth*, 164.

60. Mayra Rivera, *The Touch of Transcendence: A Postcolonial Theology of God* (Louisville, Ky.: Westminster John Knox Press, 2007). For further exploration of the complicated relations of carnality, flesh, and new materialism, see also her *Poetics of the Flesh* (Durham, N.C.: Duke University Press, 2015).

61. An earlier version is located in Karen Barad, "On Touching—The Inhuman That Therefore I Am," *Differences* 23, no. 3 (2012): 206–23, whereas a "corrected" and revised version can be found online at http://womenstudies.duke.edu/uploads/media_items/on-touching-the-inhuman-that-therefore-i-am-vi–1.original.pdf (last accessed February 2, 2015).

62. Perhaps what we're working with here is a version of what Laurel Schneider calls "promiscuous incarnation." See her article of the same title in *The Embrace of Eros: Bodies, Desires, and Sexuality in Christianity*, ed. Margaret Kamitsuka (Minneapolis: Fortress Press, 2010).

63. Barad, "On Touching." Or what I'm calling theophany, then, might be something so near to what Mayra Rivera calls "glory." "A dark and luminous halo," she writes," envelops all life, but we seldom perceive it." Those halos are multiple, I think, earthly and divine, some made by Goldsworthy, some unseen—dark, luminous,

evoking an incarnational verbosity in our attempts to write them and yet pointing to the inadequacy of those words as well. See Rivera's essay "Glory: The First Passion of Theology?" in *Polydoxy: Theology of Multiplicity and Relation*, ed. Catherine Keller and Laurel C. Schneider (New York: Routledge, 2011

64. Clayton Crockett and Jeffrey W. Robbins. *Religion, Politics, and the Earth: The New Materialism* (New York: Palgrave Macmillan, 2012), 68.

65. Jane Bennett, "Powers of the Hoard: Further Notes on Material Agency," in *Animal, Vegetable, Mineral: Ethics and Objects*, ed. Jeffrey Jerome Cohen (Washington, D.C.: Oliphaunt Books, 2012), 242. I would add here the strong thematic resonances of the choreographic theories of Erin Manning and Andrea Olsen. See Erin Manning, *Always More Than One: Individuation's Dance* (Durham, N.C.: Duke University Press, 2013), and Andrea Olsen, *Body and Earth: An Experiential Guide* (Lebanon, N.H.: University Press of New England, 2002).

❧ Interdisciplinary Ethics: From Astro-Theology to Cosmo-Liberation Theology

THEODORE WALKER JR.

The Spirit of the Lord is upon me, because he has anointed me to bring good news to the poor. He has sent me to proclaim release to the captives and recovery of sight to the blind, to let the oppressed go free, to proclaim the year of the Lord's favor.
—LUKE 4:18–19 (NRSV)

The purposes of this interdisciplinary deliberation are to recover William Derham's early modern "astro-theology" and to offer a revised astro-theology, a constructive postmodern astro-theology. This postmodern revision supplements early modern astro-theology by adding a cosmology and a metaphysical realism essential to meeting the challenge of ethical realism. This is an important contribution to liberation theology. Having proposed the term "astro-liberation theology" in an earlier exploration,[1] I now propose "cosmo-liberation theology" as a better label for this deliberation.

MODERN ASTRO-THEOLOGY

Among the earliest explicitly transdisciplinary publications in the history of North Atlantic modern science, there is William Derham's *Astro-Theology; or, A Demonstration of the Being and Attributes of God, from a Survey of the Heavens* (1715).[2] First printed in London, England, in St. Paul's Church Yard, this richly ornamented book features 58 pages of introduction, 228 pages of body, and eleven illustrated figures on two pages of copper plates. Its author, William Derham (1657–1735), was an observational astronomer and an Anglican clergyman. He sought to demonstrate "the being and attributes of God" by appealing to "a survey of the heavens" using "modern" scientific methods and instruments, especially Newtonian mathematics and telescopes.[3] In Derham's person and work, modern math- and telescope-assisted astronomy plus Christian theology produced *modern* astro-theology.

Derham's early modern astro-theology differs from almost all of the subsequent literature claiming to be astro-theology. Unlike Derham, who used the term "astro-theology" to mean modern scientific astronomy plus theology, subsequent authors gave the terms "astro-theology" and "astrotheology" an array of very different meanings, including astrological, mystical, magical, occult, ritualistic, shamanistic, and even atheistic meanings.[4] Google and YouTube searches show that today Derham's literal meaning (astronomy + theology) has been almost altogether eclipsed by spooky, paranormal meanings. Hence, it is important to emphasize that we are now concerned with recovering and revising Derham's early modern astro-theology.

CONSTRUCTIVE POSTMODERN REVISION

Constructive postmodern science (meaning science instructed by Alfred North Whitehead's *Science and the Modern World* and subsequent process philosophy)[5] requires two major revisions to Derham's modern scientific astro-theology. The first is to expand the scope of our concern from astronomy to cosmology. Astronomy is about parts of the universe, especially stars and galaxies. Cosmology is about the whole universe. A revisionary astro-theology includes astronomy *and cosmology*, along with theology. The second revision is to replace the "Demonstration" in Derham's subtitle (*A Demonstration of the Being and Attributes of God*) with "Exemplification." In modern science and theology alike, "demonstration" implies empirical, factual proof. According to constructive postmodern science, however, empirical, factual data can only illustrate or exemplify metaphysical truths. Therefore, the constructive postmodern astro-theology I propose here holds that scientific astronomy and cosmology provide *exemplifications* (examples, illustrations) of theological truths, not demonstrations. Moreover, Derham's near contemporary, also an Anglican clergyman, John Wesley (1703–91) appropriated natural scientific demonstrations by Derham and others with a calculated selectivity that provides a Wesleyan "precedent" for postmodern unwillingness to embrace empirical, factual demonstrations of theological claims.[6]

Today, there are many various postmodern philosophies, including the varied theories stemming from the work of Gilles Deleuze, Michel Foucault, and Jacques Derrida. According to David Ray Griffin, however, Whitehead's postmodern philosophy is "radically different."[7] Unlike the more recent deconstructive postmodernisms, Whiteheadian postmodernism is "constructive" or "reconstructive."[8] It critically appreciates the factual data of modern science, and reconstructively supplements modern science with a critical appreciation of relations between physics (ranging from quantum physics to cosmology) and metaphysics.[9]

This radically different postmodern philosophy allows us to better evaluate and more rightly appreciate scientific contributions to theology. We now recognize that astronomical observations *illustrate* a logically necessary existential truth, which is to say a metaphysical truth: We are parts of reality existing among other parts of reality.[10] Of course, no rightly interrogated observational report of any kind (whether biological, anthropological, psychological, sociological, or historical) could fail to illustrate or at least imply that we are parts of reality existing among other parts of reality. Nevertheless, astronomy and cosmology illustrate this necessary existential truth on a uniquely vast space-time scale.

Furthermore, constructive postmodern process philosophers distinguish the set of all parts of reality called universe from the all-inclusive whole of reality, which is greater than the set or sum of all its parts, called God.[11] According to process thought, God includes and exceeds (transcends) the universe. This theology is therefore classified as "pan-en-theism," meaning that all parts of reality (*pan*) are included within (*en*) God (*theos*).[12]

Given such panentheism, we must affirm precisely one God and one universe. It is logically incoherent to affirm more than one presently actual all-inclusive whole of reality.[13] And it is equally incoherent to affirm more than one presently actual set of all parts of reality. Moreover, where the word "God" refers to the all-inclusive whole of reality, atheism is illogical and nonsensical. No sensation, no observation, no experience, and no experimental outcome denies what all of natural science unavoidably confirms: that *we are partly inclusive parts among variously inclusive parts of the all-inclusive whole of reality*.

METAPHYSICS OF NATURE, METAPHYSICS OF MORALS

According to a Whiteheadian account of the metaphysics of nature, rightly interrogating any observation or experience can only bear witness to the metaphysical-theological idea of an all-inclusive whole of reality.[14] Some observations, however, reach us more deeply than others. Astronomical observations, and the resulting cosmological hypotheses, are frequently among the most deeply inspiring witnesses. Hence, it is no surprise that a biblical psalmist sang, "The heavens are telling [necessarily exemplifying] the glory of God" (Ps 19:01 NRSV).

Moreover, there is a metaphysics of morals. In *Process and Reality: An Essay in Cosmology*, Alfred North Whitehead says, "Morality of outlook is inseparably conjoined with generality of outlook."[15] Though modern moral theory characteristically fails to explicate metaphysical necessity, constructive postmodern theory does not. In *The Divine Good: Modern Moral Theory and the Necessity of God* Franklin I. Gamwell advances the Hartshornean argument that reference to the "divine good" is metaphysically necessary for adequate moral theory.[16]

MEETING THE CHALLENGE OF REALISM

Christian ethics commands that we love our enemies (Mt 5:44), love our neighbors as ourselves (Mt 22:39), and bring such "good news to the poor" (Lk 4:18). These commands are often dismissed as naïve, utopian, sentimental, idealistic, and unrealistic. The hard-core, commonsense alternative is called "realism." This realist challenge is long-standing. For instance, in the gospel of Matthew, Jesus contrasts his command to "love your enemies and pray for those who persecute you" (5:44) with the commonsense realism known to all: "You have heard that it was said, 'You shall love your neighbor and hate your enemy'" (5:43).

Similar expressions of commonsense realism in the contemporary world include various appeals to Reinhold Niebuhr's *Moral Man and Immoral Society: A Study in Ethics and Politics*.[17] Poised between two world wars and faced with the rise of European fascism, Niebuhr held that when we compare the human individual ("man") with large social groups ("society"), we find that the rational and religious resources of the individual are much greater than those of large social groups such as tribes, states, and nations. Niebuhr reasoned that individual morality must be significantly different from "the morality of nations."[18] Accordingly, the morality appropriate to individuals is "ethics," and the morality appropriate to social groups and nations is "politics." And sometimes there is "conflict between individual and social morality."[19] Conflict between ethics and politics requires recognizing a "frank dualism in morals."[20] And although Niebuhr himself argued to the contrary in *An Interpretation of Christian Ethics* (1935) and subsequent works, some Niebuhrians have concluded that Christian ethics apply only to relations among individuals and small local groups, such as families and individual church congregations. Accordingly, the Christian ethical commandments to "love your enemies" (Mt 5:44) and to "love your neighbor as yourself" (Mt 22:39) apply only on the individual level. On the political level, Christian ethics is not realistic.

Liberation theologians disagree. We hold that *Christian ethics is realistic*. Moreover, I hold that metaphysical realism supports true Christian realism. Recall that astronomy and cosmology unavoidably illustrate the metaphysical truth that we are partly inclusive parts among variously inclusive parts of reality, that mereological analysis reveals that all parts of reality are parts of the whole of reality, and that the transcendent all-inclusive whole of reality is God. Accordingly, the command that we love our enemies and our neighbors as we love ourselves is utterly realistic, insofar as our enemies and neighbors really are as we are. They and we really are partly inclusive parts among variously inclusive parts of the all-inclusive divine whole of reality. Hence, in calling us to love enemies and neighbors as we rightly love ourselves (as parts among parts of the divine whole), Christian ethics is calling us to live in accordance with reality.

Recognizing that Christian ethics is reality-based can be liberating. Let us recall that at various times in history, there were pessimists who wrongly conceived of themselves as "realists" because they could not imagine more righteous futures such as the end of transatlantic slavery, the end of legal segregation in the southern United States, or that Nelson Mandela would become president of South Africa. Similarly, today, many of us still have difficulty conceiving that the Rev. Dr. Martin Luther King Jr. was being a realist when he prescribed nonviolence toward enemy nations, and when, like Luke 4:14–19, he prescribed the "abolition of poverty" throughout this "world-wide neighborhood."[21] When we are attentive to metaphysical realism, however, we can appeal to the reality of our existence as parts among parts of the divine whole. Recognizing this realism can help liberate us from the long-standing, energy-draining habit of wrongly identifying chronic pessimism and untamed cynicism with realism.

COSMO-LIBERATION THEOLOGY

In *Liberation and the Cosmos: Conversations with the Elders*, the theological ethicist Barbara A. Holmes prescribes that we should consult the wisdom of the elders and the ancestors, and that we should creatively engage scientific studies of the cosmos, including quantum physics and quantum cosmology.[22] She argues that cosmology is important for liberation theology. Similarly, in *Race and the Cosmos: An Invitation to View the World Differently*, Holmes proposes that we view race and other social concerns in cosmic context.[23] Holmes judges that some liberation struggles have stalled on account of failure to see and appreciate our cosmic context. Appreciating our cosmic context can be deeply inspiring. Accordingly, Holmes invites us to appreciate our relations to planets, stars, galaxies, and other macroscopic/cosmic realities, and to consider that we are composed of microscopic, cellular, subcellular, molecular, atomic, subatomic, and quantum parts, particles, wavicles, and events. Cosmic myopia can yield only seriously inadequate or wrong views of the world.

To develop more adequate liberation theologies and inspire more liberating struggles, we need to be more attentive to scientific inquiries at all levels, including astronomical and cosmological levels. We need "cosmic orientation."[24] By studying the cosmology-oriented liberation theology of Barbara A. Holmes, we can see emerging—what may now be called—*cosmo-liberation theology*.

NOTES

1. Theodore Walker Jr., "Astro-Theology in the Journal of Cosmology," *Journal of Cosmology* 19 (June 2012): 8552–55; Theodore Walker Jr., "The Liberating Role of

Astronomy in an Old Farmer's Almanac," *Journal of Cosmology* 19 (June 2012): 8556–82.
2. William Derham, *Astro-Theology; or, A Demonstration of the Being and Attributes of God from a Survey of the Heavens* (London: William Innys, 1715).
3. Ibid., 7–9.
4. Henry Moseley, *Astro-Theology* (London: A. Varnham, 1874); Edward Higginson, *Astro-Theology* (London: E. T. Whitfield, 1866); Clifford Stevens, *Astrotheology for the Cosmic Adventure* (Techny, Ill.: Divine Word, 1969); Jan Irvin, Andrew Rutajit, and Jordon Maxwell, *Astrotheology and Shamanism: Unveiling the Law of Duality in Christianity and Other Religions* (San Diego, Calif.: Book Tree, 2006); Jan Irvin and Andrew Rutajit, *Astrotheology and Shamanism: Christianity's Pagan Roots: A Revolutionary Reinterpretation of the Evidence* (Crestline, Calif.: Gnostic Media Books, 2009).
5. Alfred North Whitehead, *Science and the Modern World* (New York: Free Press, 1997 [1925]).
6. Randy Maddox, "John Wesley's Precedent for Theological Engagement with the Natural Sciences," in *Divine Grace and Emerging Creation: Wesleyan Forays in Science and Theology of Creation*, ed. Thomas Jay Oord (Eugene, Ore.: Pickwick, 2009); Theodore Walker Jr., "Postmodern Astro-Theology, Cometary Panspermia, and the Polonnaruwa Meteorite: Derham, Wesley, Whitehead, Griffin, and Cobb," *Journal of Cosmology* 22 (March 2013): 10141–51.
7. David Ray Griffin, *Whitehead's Radically Different Postmodern Philosophy: An Argument for Its Contemporary Relevance* (Albany: State University of New York Press, 2007). See also John B. Cobb, "From Crisis Theology to the Post-Modern World," *Centennial Review* 8 (Spring 1964): 209–20; "Ecology, Science, and Religion: Toward a Postmodern Worldview," in *The Reenchantment of Science: Postmodern Proposals*, ed. David Ray Griffin (Albany: State University of New York Press, 1988), 99–114.
8. Griffin, *Whitehead's Radically Different Postmodern Philosophy*, 48.
9. Concerning Whitehead, modern science, and metaphysics, see Alfred North Whitehead, *Adventures of Ideas* (New York: Free Press, 1967 [1933]); *Essays in Science and Philosophy* (New York: Philosophical Library, 1947); Victor Lowe, Charles Hartshorne, and A. H. Johnson, *Whitehead and the Modern World: Science, Metaphysics, and Civilization: Three Essays on the Thought of Alfred North Whitehead* (Boston: Beacon Press, 1950); Charles Hartshorne, *Aquinas to Whitehead: Seven Centuries of Metaphysics of Religion* (Milwaukee, Wisc.: Marquette University Press, 1976).
10. Schubert Ogden reminds us that the contingency, or nonnecessity, of human existence requires distinguishing between metaphysical aspects of specifically human existence (*metaphysica specialis*) and metaphysical aspects of existence generally (*metaphysica generalis* [strict metaphysics]). See Schubert M. Ogden, "The Criterion of Metaphysical Truth and the Senses of Metaphysics," *Process Studies* 5, no. 1 (1975): 47–48.
11. With Alfred North Whitehead, Charles Hartshorne holds that God is "the one universal individual" (Charles Hartshorne, *Reality as Social Process: Studies in Metaphysics and Religion* (Boston: Beacon Press, 1953), 176. Schubert M. Ogden's formulation of this Hartshornean doctrine is that God is "the one all-inclusive whole of reality"

(Schubert M. Ogden, "Process Theology and the Wesleyan Witness," *Perkins School of Theology Journal* 37, no. 3 [1984]: 21).
12. See Charles Hartshorne and William L. Reese, *Philosophers Speak of God* (Chicago: University of Chicago Press, 1953). Whiteheadian-Hartshornean-Ogdenian panentheism is distinct from other varieties of panentheism and "pantheism" that do not insist that there is a living, feeling, interacting universally creative divine whole of reality [the Creator] that creatively synthesizes all parts of reality [all creatures and creations]. See Charles Hartshorne, *Creative Synthesis and Philosophical Method* (La Salle, Ill.: Open Court Press, 1970). Also Whitehead scholars rightly insist that their metaphysical account of God is not adequate to religious needs for specific data concerning divine interactivities with specific creaturely populations such as ourselves. Metaphysical theology makes only generic claims about God. It says nothing specific, nothing contingent, only what is generally and universally necessarily true. As such, it is both essential and inadequate to our religious needs. In addition to knowing that God exists, we religious people care to know that God is doing "abc" and "xyz" to/for and with us.
13. Ogden, "Process Theology and the Wesleyan Witness," 21.
14. According to Hartshorne, no conceivable experience could "exemplify" the "nonexistence of God" (Charles Hartshorne, *Anselm's Discovery: A Re-Examination of the Ontological Proof for God's Existence* [LaSalle, Ill.: Open Court Press, 1965], 64.) Instead, "the occurrence of anything whatever" implies divine existence (ibid).
15. Alfred North Whitehead, *Process and Reality*: (New York: Free Press, 1979), 15.
16. Franklin I. Gamwell, *The Divine Good: Modern Moral Theory and the Necessity of God* (San Francisco: HarperSanFrancisco, 1990); Charles Hartshorne, "An Outline and Defense of the Argument for the Unity of Being in the Absolute or Divine Good" (PhD diss., Harvard University, 1923).
17. Reinhold Niebuhr, *Moral Man and Immoral Society: A Study in Ethics and Politics* (New York: Charles Scribner's Sons, 1932).
18. Ibid., chap. 4.
19. Ibid., chap. 10.
20. Ibid., 271.
21. Martin Luther King Jr., *Where Do We Go from Here: Chaos or Community?* foreword by Coretta Scott King, New Introduction by Vincent Harding (1967; Boston: Beacon Press, 2010), 174, 77, 81.
22. Barbara A. Holmes, *Liberation and the Cosmos: Conversations with the Elders* (Minneapolis: Fortress Press, 2008).
23. Barbara A. Holmes, *Race and the Cosmos: An Invitation to View the World Differently* (London: Bloomsbury T&T Clark, 2002).
24. Charles H. Long, *Alpha: The Myths of Creation* (Chico, Calif.: Scholars Press, 1963), 18.

Vascularizing the Study of Religion: Multi-Agent Figurations and Cosmopolitics

MANUEL A. VÁSQUEZ

The notion of science isolated from the rest of society [is] as meaningless as the idea of a system of arteries disconnected from the system of veins. . . . Rich vascularization . . . makes the scientific disciplines alive.
—BRUNO LATOUR, *Pandora's Hope: Essays on the Reality of Science Studies*

In this essay, I explore the implications of a nonreductive, materialist, networks approach for the study of religion. In particular, I am interested in pluralizing and expanding agency beyond the one-dimensional and domineering Cartesian-Kantian model of the sovereign, buffered subject, which has haunted not only Western secular humanism and theological discourse but also the discipline of religious studies, through what Russell McCutcheon calls the "private affair tradition," which sees authentic religion as issuing from the internal, irreducible experiences of *homo religiosus*.[1] I contend that a framework that allows for a rich exploration of the multiplicity of religious phenomena in their sociocultural and environmental contexts requires a recognition of the dynamic entanglement of multiple types of agencies. In turn, this recognition entails a re-embedding of embodied, historical human actors in vascularized and interactive ecological figurations from which they have evolved, and through and within which they carve out shared and contested spaces of livelihood.

In physiology, the term "vascularized" is used to describe living tissues, whose relative vitality, robustness, and complexity are generated and sustained by a myriad of vessels that enable multiple flows to and from these tissues. Generally, living tissues show greater development as the vessels that supply them proliferate and ramify. The science studies scholar Bruno Latour and other actor-network theorists have borrowed the term to advocate for richer accounts of the social world, approaching it as the stabilized yet contested outcome of the collaboration of heterogeneous actants, including a "multiplication of nonhumans," assembled by intersecting networks.² To ground my argument, I examine the case of a Mexican Pentecostal church in Atlanta, showing how networks enable the recognition of alterity and singularity, while foregrounding connectivity, embeddedness, and interplay.

• • •

The parking lot was full of cars, many of them pickup trucks with license plates from throughout the South, from the Carolinas to Alabama. Families decked out in their finest were converging on La Iglesia Luz del Mundo (Light of the World Church) in Marietta, on the northern outskirts of Atlanta. They were attending arguably the most important event in the church's liturgical calendar: La Santa Cena (Holy Supper). Founded in Guadalajara, Mexico, in 1926 by the apostle and prophet Aarón Joaquín González, Luz del Mundo is now the largest Mexican Pentecostal church, with temples throughout the Americas, from Chile to Canada, as well as in Europe, including in Spain, Italy, and the UK. The church also claims missionary activity and *casas de oración* (prayer houses) in Israel and Australia. In the United States the first Luz del Mundo temple was founded in Houston in the early 1960s. Today, there are temples throughout the country, beyond traditional gateway cities such as Los Angeles, Tijuana, and El Paso, which have been key nodes in long-standing migration networks from Mexico. There are now many temples in the "new immigrant destinations," from Omaha, Nebraska, and Salt Lake City, Utah, to Atlanta, Georgia.

On that steamy August afternoon, I accompanied Patricia Fortuny, a Mexican anthropologist who has written extensively on Light of the World. The visit was part of an interdisciplinary project to study transnational religious networks and flows among Brazilians, Guatemalans, and Mexicans in the New South. La Santa Cena offered a unique opportunity to observe the construction of a transnational community in the church. To celebrate Holy Supper, church members throughout the world are encouraged to travel to Guadalajara to congregate in La Luz del Mundo's colossal *templo sede* (international headquarters temple) and witness Apostle Samuel, Aarón's son and successor, lead a soaring service that lasts many hours. With seating for 15,000 persons, the templo sede is located in the midst of a complex of neighborhoods populated by church members and called Hermosa Provincia (The Beautiful Province), meant to represent a New Jerusalem. Borrowing from Erving Goffman, Fortuny describes Hermosa Provincia as a "total institution." Church members "tend to live within the same neighborhood, attend the same temple, go to school together, shop in local establishments often run by members, recreate together, and in general create their own world within the religious community."³ This model of a community anchored around the sacred *axis mundi* of the temple has been replicated in other locations in Mexico and the Americas. During the week-long celebrations of Holy Supper, the population of Hermosa Provincia swells to as many as 350,000, as pilgrims from throughout Mexico and beyond come to join the festivities, staying with relatives, friends, and at guesthouses arranged by the church.

Those unable to attend—a number that in the last few years has grown because of the increasing militarization of the U.S.–Mexico border and the general anti-immigrant climate—gather in the temples abroad to join the celebration via satellite link, thus forming a diasporic "sodality of worship and charisma," a powerful moment of collective effervescence, of *communitas* that transcends national borders.[4] As they welcomed us into the temple in Marietta, the ushers divided our research team. Patricia and other women in the team were offered veils to cover their heads and directed to the right side of the temple. I went to join the men on the other side. The atmosphere was literally charged, alternating between intense periods of joy and sorrow as the large chorus, impeccably dressed in white, wound its way through several hymns. In front of the temple, towering over the altar, was a huge screen projecting the proceedings in the templo sede. As word began to filter that Apostle Samuel was making his way into the temple in Guadalajara, emotions in Atlanta reached a crescendo. Men around me dissolved into uncontrollable sobs. Through the cries, I could hear the man next to me, tears streaming down his cheeks, saying out loud: "I am not worthy of you, Jesus; yet you have chosen to save me, to give us our beloved Apostle. I am such a sinner, but I rejoice in your compassion and love." All of a sudden, the transmission from Mexico was interrupted, and the huge screen went blank. A collective gasp rose from those in the crowd. They lifted their hands as one, eyes closed in deep prayer, and their faces showing profound concentration and distress. They wanted to touch the screen, using their bodies to gather and project the power of the Holy Spirit to make the screen restore the image. The screen, however, proved obdurate, not immediately responding to the summons of the faithful and the Holy Spirit.

A Brazilian sociologist, who had also joined us at the Santa Cena in Atlanta and who has studied Pentecostalism in Brazil, commented after the service how different the mood was at La Luz del Mundo from that of the average Pentecostal church in Brazil. "It is too depressing," he told us. "There is way too much crying, particularly among the men. In Brazil, they would be clapping and dancing more and beating the tambourine." Much has been written about Pentecostalism and gender in Latin America, about how it "domesticates males," who after conversion must be accountable to the "Lord of Lords."[5] The diasporic context, however, adds another dimension to the relationship between Pentecostalism and masculinity. La Luz del Mundo offers a protective and intimate space where undocumented Mexican and Central American men can express their vulnerabilities and anxieties, amid a dangerous and uncertain world that stigmatizes them and in the face of a state that seeks to render them visible in order to regulate their existence. The important point to make here is that there are multiple ways of being Pentecostal, characterized by an array of "body techniques" that inculcate in the bodies of the believers a certain habitus: abilities, habits, and dispositions that allow for authoritative and efficacious religious performance.[6] These body techniques circulate and act efficaciously in transnational Pentecostal networks as part of a global pneumatic culture of the spectacle, a "charismatic corpothetics," a full-body aesthetic that is disseminated globally through the Internet, TV, and DVDs.[7]

After what seemed like an endless period of consuming anxiety, the video image came back. Samuel appeared front and center, greeting the pilgrims and entering the templo sede. Cheers filled the muggy Atlanta night, cheers that proved only temporary, as the signal faded several times, producing yet more despair. This was the beginning of an intense evening, which Fortuny, borrowing from Victor Turner, has characterized as a "social drama."[8] It culminated with a sumptuous shared meal, which included not only Mexican taquitos but Guatemalan tamales and Salvadoran pupusas, reflecting the fact that the congregation in Marietta brings together immigrants from different countries in Latin America.

This vignette illustrates the complex role of transnational networks in the production of religious and cultural diversity. By linking various localities and allowing what Peggy Levitt and Nina Glick Schiller describe as "simultaneous embeddedness" in multiple spatial scales, transnational and electronic networks lead to the unsettling of established, territorialized cultural and religious habits and to the articulation of hybrid practices, identities, landscapes, and institutional arrangements.[9] Increased hybridity may result in the intensification of heterodoxy and pluralism—and sometimes simultaneously with the attempt to reaffirm pristine origins, purity, and orthodoxy. Jacqueline Hagan has shown how transnational churches help undocumented immigrants from Mexico and Central America confront the risks and uncertainties of migration to the United States, from making the difficult journey across a perilous border to facilitating the process of settlement by offering safe spaces of belonging and resources for personal strength and collective mobilization—all in response to the state's focus on enforcement and the generalized climate of insecurity and hostility that the migrants confront in their day-to-day lives.[10] Particularly for single, young, and first-time migrants, who "do not have access to either personal networks or financial resources to hire coyotes to assist them on their journey," churches and religiously based shelters have become nodes in a transnational "social infrastructure that sustains transit migration in the region."[11] These nodes enable immigrants "to connect with one another and exchange ideas and information about the journey. From deportees who are returned to the shelters by government officials who rely on the shelters, unseasoned migrants learn the ropes of crossing the border. . . . Nonkin migrant networks, forged in the shelters, buffer against the disruption and isolation, offering solidarity, social support, and protection on the road."[12]

In the long term, transnational religious networks allow migrants not "to trade in their home country membership card for an American one but to belong to several communities at once. They become part of the United States and stay part of their ancestral homes at the same time. They challenge the taken-for-granted dichotomy between either/or, United States or homeland, and assimilation versus multiculturalism by showing that it is possible to be several things simultaneously, and in fact required in a global world."[13] Transnationalism, hence, enriches diversity by challenging exclusionary notions of citizenship and belonging based on naturalized notions of the nation-state. In some quarters, this challenge may be perceived as a mortal threat to national unity, to the imagined core values of the nation, leading to nativist impulses that seek to arrest and domesticate diversity.

La Luz del Mundo was from the outset a hybrid church, blending U.S. Pentecostalism that traveled to Mexico through migrant networks in the wake of the Azusa revival, with traditional popular Mexican Catholicism, itself deeply inflected by indigenous practices and cosmologies, as the example of the Virgin of Guadalupe demonstrates. This Catholicism places devotion, pilgrimage, and commensality at the core of ritual life. In Mexico, La Luz del Mundo has led the way in the introduction of Evangelical Protestantism in a context in which national identity is inextricably tied to Catholicism. The rise of La Luz del Mundo in Jalisco represents an even more dramatic rupture in the Catholic doxa, given that the state, unlike northern Mexican states bordering the United States, which have had extensive exposure to Pentecostalism, is a recognized Catholic stronghold.

In the United States, La Luz del Mundo joins—if not in direct collaboration, at least as part of a global vision of "making disciples of all nations"—other transnational Pentecostal networks based in Latin America, including the Brazilian Igreja Universal do Reino de Deus (Universal Church of the Kingdom of God) with temples in 171 countries, and the Guatemalan Iglesia Principe de Paz, whose Ministerio Apostólico Yeshua operates across North and Central America. Together with other charismatic Christian churches from Nigeria, Ghana, and South Korea, these transnational networks are engaged in a "reverse evangelization" that, by revitalizing Christianity in the global North, particularly in Europe, is challenging the notion that secularization is an inexorable process.[14] In fact, the vitality and proliferation of South-to-North and South-to-South religious networks and flows might be signaling the emergence of a "new polycentric global cartography of religion" constituted by multidirectional networks of varying intensity, extensity, morphology, and durability.[15]

In this emerging, lumpy landscape, to study religious diversity is to ask: "Who are the key actors involved in transnational, global, and diasporic religious networks? Are they religious elites, missionaries, itinerant pastors, pilgrims, and/or religious tourists? In what spatiotemporal scales do they operate? By what specific mechanisms? Informal exchanges or formalized chains? What kind of media are involved (from familial or kinship-based networks to electronic media)? What is flowing? Is it commodities, gifts, texts, relics, saints, theodicies, (video)taped sermons, money, or bodies?"[16] How do these materialities change their valence, with commodities becoming sacred artifacts and vice versa, as they circulate among different "regimes of value"?[17] In what ways are these networks imbricated in the dynamics of a resurgent, panoptical nation-state and of a deepening neoliberal capitalism?[18]

Although La Luz del Mundo is a territorialized church centered in Hermosa Provincia, its transnational activities de-territorialize it. Borrowing from Clifford Geertz, we may understand Hermosa Provincia as a transposable "model of and model for" the sacred community and homeland in diaspora.[19] In other words, what is circulating within Luz del Mundo networks is a particular "geopiety," a way of carving out lived sacred space that interacts with other religious and secular cartographies in the societies of origin and settlement.[20] Because of Atlanta's urban cartography as a de-centered city, a city characterized not by distinctive neighborhoods but by the logic of the car, by the big highways that crisscross it, members of La Luz del Mundo have not been able to reproduce the model of the Hermosa Provincia. Nonetheless, the temple in Atlanta serves as the reference point for the faithful scattered throughout the U.S. South. It is a second-order node, which derives its legitimacy through its connections with the templo sede: the physical connections (via traveling pastors and the remittances that immigrants send to Mexico and Central America), the virtual connections, and the imagined connections. Here we see, once again, the intricate operation of transnational religious networks: Their entwinement in processes of de-territorialization does not lead to the erasure of all distinctions and borders. Rather, de-territorialization and (re)territorialization enter into a complex and tensile interplay that cannot be reduced to relations of opposition or mutual exclusion. Out of this interplay emerges a craggy "transnational social field . . . a set of multiple interlocking networks of social relations through which ideas, practices, and resources are unequally exchanged, organized, and transformed. . . . Social fields are multidimensional, encompassing structured interactions of differing forms, depth, and breadth."[21]

In La Luz del Mundo, the national, transnational, and postnational coexist in a tensile relation. On the one hand, the church is unabashedly Mexican, just like the Church of Jesus Christ of Latter-Day Saints is inextricably connected with the United States. Not unlike the LDS church, La Luz del Mundo sacralizes the national landscape in which it emerged as its mark of prophecy, charism, and authenticity. Following Nina Glick Schiller and Georges Fouron, we might even say that La Luz del Mundo represents a religiously inflected form of "long-distance nationalism."[22] Or borrowing from Thomas Tweed, that the projection of the Hermosa Provincia is not just a translocative dynamic, linking back to an imagined, idealized homeland where the extended, transnational "family" gets together to share a meal, but a supralocative phenomenon whereby divine and human times and spaces overlap.[23] On the other hand, despite its avowed Mexican identity, in the United States the church takes a pan-Latino character. It moves from being a national church in diaspora to being an ethnic church among other immigrant churches, appropriating the constructed category of "Latinos" that in the United States serves to homogenize a very diverse population into a collective identity jostling against other collective identities within the framework of American pluralism. Thus, as this case shows, transnational networks introduce alterity without excluding processes of homogenization.

Finally, in its modus operandi, La Luz del Mundo transgresses the boundaries between tradition and modernity, between "belonging through memory, nostalgia, [and] imagination" and using the latest communication technologies.[24] In fact, the communication technologies, rather than disenchanting the lifeworld, become a means through which memory, nostalgia, and imagination are reenacted in diaspora, breaking down the sharp separation between sacred and profane.

NETWORKS AND NONANTHROPOCENTRIC AGENCY

The case of La Luz del Mundo goes a long way toward showing us how a networks approach can contribute to cultural and religious diversity. Bruno Latour and other actor-network theorists point toward the need to go beyond reductive, anthropocentric understandings of agency. In my view, however, even this approach fails to capture the potency of the multiple materials, media, actors, and environments, whose dynamic entanglements make a religious event such as La Santa Cena efficacious.

In his study of the way in which pasteurization became a widespread practice in nineteenth-century France, Latour challenges the heroic view of history that centers on Louis Pasteur's genius in his struggle to protect humanity against germs. Latour argues that a dense and heterogeneous network of physicians, hygienists, government officials, scientific journals, test tubes, microscopes, sterilizing substances, and microbes interacted to stabilize a scientific view of illnesses that guaranteed the public acceptance of pasteurization. As Latour puts it, "The laboratory becomes, so to speak, the prosthesis allowing the ferment of lactic acid to speak as well as Pasteur; it allows the articulation between Pasteur and 'his' ferment, between the ferment and 'its' Pasteur."[25] In other words, agency is not an essence but a relation. It is not something that belongs intrinsically to humans as subjects standing apart from the "inert" nonhuman entities that surround and interact with them. Rather, agency is the negotiated outcome of shifting relations among various components in a particular network.

Actor-network theory (ANT) is "a semiotics of materiality. It takes the semiotic insight, that of the relationality of entities, the notion that they are produced in relations, and applies this ruthlessly to all materials—and not simply to those that are linguistic."[26] Moreover, ANT "tells us that entities achieve their form as a consequence of the relations in which they are located. But this means that it also tells us that they are *performed* in, by, and through those relations. A consequence is that everything is uncertain and reversible, at least in principle. It is never given in the order of things."[27] In other words, it is not that everything goes or that history does not matter, but that reality is fragile because it is disclosed relationally, as various elements summon each other and collaborate with each other.[28] This is why science studies must be concerned with "science's blood flow," with giving nuanced and empirically rich accounts of the "unpredictable and heterogeneous links that explain the circulatory system that keeps scientific facts alive."[29] I suggest that to capture something of the vitality of religious practices and experiences, theology and religious studies must also be concerned with religion's blood flows, with the multiple materialities and networks of humans and nonhuman entities of various kinds that enliven religious facts like the Santa Cena.

For Latour, this relational and dynamic view of agency as emerging out of variegated assemblages poses a direct challenge to the Cartesian notion of an unencumbered rational self and to "the illusion of modernity . . . that the more we grew, the more separate objectivity and subjectivity would become, thus creating a future radically different from our past."[30] Latour's networks approach allows "the 'outside world' to invade the scene, break the glassware, spill the bubbling liquid, and turn the mind into a brain, into a neuronal machine sitting inside a Darwinian animal struggling for its life."[31] This rematerialization and re-embedding of the self in nature, in turn, leads Latour to the conclusion that "objectivity and subjectivity are not opposed, they grow together and they do so irreversibly." If the gap between objectivity and subjectivity is closed, then, artifacts "deserve better. They deserve to be housed in our intellectual culture as full-fledged social actors. Do they mediate our actions? No, they are us."[32]

Certainly, during La Santa Cena, the blinking screen was one with the crowd, straining to make Apostle Samuel present. We cannot simply dismiss the agency of the screen as nothing more than a fetish, the naïve and distorted projection of the alienated consciousness of the "oppressed creature" yearning to be free, in this case, undocumented immigrants exploited by global neoliberal capitalism, a reading that, for all of its emancipatory impetus, reinscribes the Cartesian sovereign self and mind-body and spirit-matter dualisms. Nor is the screen a mere prop. The screen itself does work along with Samuel, the chorus, the taquitos, the worn-out Bibles held aloft, and the monumental architecture. Within this assemblage, the screen alters the course of events, enabling and disabling some actions, participating in the generation of certain powerful affects and forms of religious subjectivation that are essential to the efficacy of La Santa Cena.

Latour's networks approach also challenges the "denial of coevalness"[33] that is closely related with secular modernity's rejection of religion, which in the modern mind is associated with the intimate interaction and ontic shifts among humans, nonhuman animals, plants, things, and spirits.[34] Science understood these ontic shifts as the projections of primitive thinking, as naïve forms of fetishism. Apparitions, miraculous icons, and beliefs in shape-shifting shamans are all misattributions of causality that modern science shatters with facts. But Latour shows that "we have never been modern," or rather that the "moderns" also depend heavily on and interact closely with things and technologies. Thus, what we have are not facts opposed to fetishes, but "factishes," constructed yet binding facts that emerge from historical, contested yet stable "regimes of enunciation."[35] The task then is not to debunk religion, but to understand comparatively the "'conditions of felicity' of the various activities that in our cultures are able to elicit truth."

> If it were necessary to take science seriously, without giving it some sort of "social explanation," such a stand is even more necessary for religion: debunkers and iconoclasts simply would miss the point. Rather my problem concerns how to become attuned to the right conditions of felicity of those different types of "truth generators."[36]

Building on Latour and ANT, the religious studies scholar Matthew Day has called for an understanding of various religions as evolving, agentic, heterogeneous networks that may involve not just priests, prophets, monks, missionaries, immigrants, pilgrims, healers, tourists, and scholars but also texts, relics, icons, money, embodied habituses, architectural styles, and notions of honor and prestige, as well as "transcendental beings" such as gods, spirits, and impersonal forces.[37] Understanding religions as dynamic and contested but relatively stabilized multi-agentic heterogeneous assemblages would force scholars to take the following question of Latour seriously:

> Why not say that in religion what counts are the beings that make people act, just as every believer has always insisted? That would be more empirical, perhaps more scientific, more respectful, and much more economical than the invention of two impossible non-existing sites: one where the mind of the believer and the social reality are hidden behind illusions propped up by even more illusions. Besides, what is so scientific in the notion of "belief"?[38]

In other words, Latour is challenging us to break with the modernist doxa in religious studies and *"grant the gods their agency,"*[39] for "as long as they make a difference in what people do, the gods are *real actors* with *relative existence.*"[40] Thus, scholars of religion need to "start attending to the networks from which these actors emerge and the labor required to make them real and obvious."[41] Day goes on to write that a networks approach to religion can help us overcome the unproductive impasse between emic and etic perspectives and between criticism and caretaking.[42] With this approach, he suggests, we can

> kick the habit of treating those "non-obvious beings" that fill the pages of books about religion as if they were stand-ins for something else. The gods, ghosts, spirits, and prophets that speak are not ciphers for "Society" (e.g., Durkheim), "Culture" (e.g., Geertz), or "Economy" (e.g., Marx). Rather, they are one of the many non-human actors who circulate within a given network: agents who make their presence felt by sharing the labor required to gather, attach, move, motivate or bind their fellow actors together into a social aggregate.[43]

Defining "non-obvious beings" as actants within particular networks has opened new spaces in which to develop a richer anthropology of the "uncanny" and of the alternative ontologies that may accompany it.[44] For instance, studying ghosts and spirits in Buddhist festivals in Laos would involve analyzing "under which circumstances they appear to which people. This would imply a focus on the multifaceted regimes of communicability that evolve between humans and spirits."[45] After all, "ghosts can be beings with desires, with taste, with biographies. They appear in specific ways at places at a certain time; they slip into objects, they live in them, they consume things, leave material traces and demand a certain treatment as social beings."[46] Thus, the task is to trace the "tracks," "footprints," or "imprints," the signals "left in the material domain of something that in conventional ways is not graspable for most people not endowed with special capacities to do so. . . . The trace itself can point to the immanence of a being through its material manifestations."[47]

In general terms, a Latourian perspective would make the comparative study of religious diversity into a painstaking, empirically informed exercise in "ontological diplomacy," seeking not to fit the multiplicity of phenomena into procrustean beds (as in the case of Eliadean archetypes) or to locate them in self-standing, seamless cultural systems (as for Geertz). Rather, the task would be to map out the contingent, heterogeneous, praxis-driven assemblages that give rise to and sustain fragile yet efficacious modes of existence that contend with each other and/or that enter relations of cross-fertilization.[48]

Networks, then, contribute to religious diversity not just by de-territorializing and re-territorializing cultures, by mixing and remixing them at various scales, but also by enlarging the boundaries of agency and incorporating assemblages of humans and nonhumans, including spirits and gods, with varying trajectories and locations. Besides "liberating the gods" within a nonreductive naturalistic, agnostic framework,[49] networks also produce "natural-cultural entanglements," permeable "material-semiotic" zones of contact that cross-fertilize humans, nonhuman animals, and technology.[50] Diversity, thus, becomes more capacious. It ceases to be a purely human matter, to be beholden to "the fantasy of human exceptionalism . . . the premise that humanity alone is not a spatial and temporal web of interspecies dependencies."[51] Diversity then is no longer just about negotiating alterity and commonalities on the bases of race, class, gender, immigrant status, national origin, or religious affiliation. It is also about being-with, about dwelling in "mixed communities," in "multispecies knots."[52]

Posthumanist diversity is characterized by heavily vascularized networks in which humans are situated along with other species among whom they cohabitate and with whom they have coevolved. To characterize this expanded notion of diversity, Donna Haraway invokes the figure of the cyborg as a fusion of the organic and the cultural (via technology as situated discursive and nondiscursive practices). "Cyborg anthropology attempts to refigure provocatively the border relations among specific humans, other organisms, and machines. The interface between specifically located people, other organisms, and machines turns out to be an excellent field site for ethnographic inquiry in what counts as self-acting and collective empowerment."[53] In her later work, Haraway has moved from cyborgs to companion species, not only including dogs, cats, horses, and cattle, but also "organic beings such as rice, bees, tulips, and intestinal flora, all of whom make life for humans what it is—and vice versa."[54] Along the same lines, Latour speaks of an "internalized ecology," as exemplified in agriculture and the domestication of animals, which involves "the intense socialization, reeducation, and reconfiguration of plants and animals—so intense that they change shape, function, and often genetic makeup."[55]

Sketching the politics that would accompany this more capacious diversity, Latour has called for a "parliament of things," that acknowledges "the silent masses" of nonhuman entities among whom and through whom we construct the lifeworld.[56] However, the science theorist Karen Barad is skeptical about the radicalness of Latour's proposal, which seems merely to extend human privileges to nonhuman entities, rather than problematizing the construction of the human as privileged.[57]

> The main point has to do with power. How is power understood? How are the social and the political theorized? Some science studies researchers are endorsing Bruno Latour's proposal for a new parliamentary governmental structure that invites non-humans as well as humans, but what, if anything, does this proposal do to address the kinds of concerns that feminist, queer, postcolonial, (post-)Marxist, and critical race theorists and activists have brought to the table? . . . What conception of power, what model of citizenship, what immigration policy is being enacted when a new representationalist democracy is being proposed that only acknowledges two kinds of citizens and their offspring—the fully human (those who had already been granted citizenship) and the fully non-human and their hybrids.[58]

Following Barad, we might say that diversity in a networks approach is not simply about extending webs of relationality to recognize alternative identities and highlighting transnational and cross-species hybridity, but that it also involves the critical examination of persistent power asymmetries within particular assemblages and between different networks. Thus, in the case of La Luz del Mundo, the question of orthodoxy (i.e., who has authority across the transnational church networks?) and of "illegality" (who is at a particular time recognized by the state as citizen?) are central to the production of diversity.

The networked agency of nonhuman entities is also clear in Caroline Walker Bynum's study of Christian materiality in late medieval Europe. She observes that in "contrast to the modern tendency to draw sharp distinctions between animal, vegetable, and mineral or between animate and inanimate, the natural philosophers of the Middle Ages understood matter as the locus of generation and corruption."[59] This nondualistic epistemology helps us make sense of the pervasive weeping statues, moving crucifixes, healing relics, and wafers that turned into blood in traditional popular Catholicism. The sacred objects did not simply "comment on, refer to, provide signs of, or gesture toward" the transcendental, unseen power of the divine.[60] Rather in their "insistent materiality," in the simultaneous affirmation of their "thingness" and their capacity for paradoxical transmutation, they were the locus of the divine; they presented themselves as efficacious surplus to the believers. Miraculous matter asserted "life in death, eternity in change, ever-living blood in ever-decaying bread, wood, or bone. Transformation miracles were moments in which matter contradicted itself. But it could do so only in material ways."[61] This paradoxical co-implication of extravagant physicality and divine transubstantiation explains why "medieval devotees responded to the overtly tactile appeal of such objects. They kissed and fondled them—actions still apparent to us centuries later in the many cases where a key part of a manuscript illumination or statue has been rubbed smooth by the contact of devout lips and fingers."[62]

In stressing the "power of the material as material," Bynum goes beyond Latour and ANT theorists, for whom agency is still the product of delegation by humans as they interact with nonhuman entities.[63] She points to the need to appreciate fully that for medieval theologians and philosophers and particularly for rank-and-file Christians, matter was "a dynamic stratum."[64] "It was not static and inert; it was by definition potent and changeable—both that which decayed in death and that which teemed with new life."[65]

Bynum's observations about Christian materiality resonate with the work of the "new materialists," who, borrowing from thinkers such as Epicurus, Lucretius, Spinoza, Nietzsche, Bergson, Merleau-Ponty, and Deleuze and in conversation with Latour, Haraway, and Barad, move from mechanistic views of nature toward an understanding of matter as "vibrant" and autopoietic, that is, self-propelling, self-creating, and self-organizing. The political scientist Jane Bennett uses the term "thing-power" to affirm "material recalcitrance," the "capacity of things—edibles, commodities, storms, metals—not only to impede or block the will and design of humans but also to act as quasi agents or forces with trajectories, propensities, or tendencies of their own."[66] For the new materialists,

> materiality is always something more than "mere" matter: an excess, force, vitality, relationality, or difference that renders matter active, self-creative, productive, unpredictable. In sum, new materialists are rediscovering a materiality that materializes, evincing immanent modes of self-transformation that compel us to think of causation in far more complex terms; to recognize that phenomena are caught in a multitude of interlocking systems and forces and to consider anew the location and nature of capacities for agency.[67]

Thus, for new materialists, the vitality and self-generating power of matter goes together with a nondualistic, immanentist epistemology. Bodies—human and nonhuman—and the identities and practices they make possible are the products of "differential patterns of mattering," the outcomes of the nonlinear interaction of social fields, discursive matrixes, and neural and ecological networks. Accordingly, "the phenomenological task is to show how consciousness emerges from, yet remains enmeshed, in [the] material world."[68] This is precisely the task that the enactive approach to cognition undertakes.[69] For this approach, cognitive processes are not the abstract computations of a disembodied Cartesian mind. Rather, they "emerge from the nonlinear and circular causality of continuous sensorimotor interactions involving the brain, body, and environment."[70] These interactions articulate dynamic structures of experience, incorporated habituses, and the environment, which is itself transformed by the creative deployment of these habituses.

Echoing and extending the new materialism and enactive cognition, Tim Ingold introduces an ecological dimension, challenging Latour for assuming that "the materiality of the world . . . is fully comprehended in the things connected."[71] For Ingold, relationality and the distributive, delegatory properties are themselves grounded in and enabled and constrained by materiality itself, "by the material media in which living things are immersed." Ingold sees reality not as "an assemblage of bits and pieces but a tangle of threads and pathways. Let us call it meshwork, so as to distinguish it from . . . networks. My claim, then, is that action is not the result of an agency that is distributed around the network, but rather emerges from the interplay of forces that are conducted along the lines of the meshwork," the webs that are not only anchored in nature but are part of it. In that sense, "bringing things to life, is a matter not of adding to them a sprinkling of agency but of restoring to them the generative fluxes of the world of material in which they came into being and continue to subsist. This [is the] the view that things are in life rather than life in things,"[72] which latter view was the simplistic way in which anthropologists have characterized primitive religion, that is, animism, since E. B. Tylor.

With Ingold, we come to the current frontiers disclosed by the networks approach: a holistic, relational, and naturalistic but nontotalizing and nonreductive understanding of phenomena. "What poststructuralists cannot imagine is a nontotalizing structure that nonetheless acts as a whole. But this is precisely what complex adaptive networks do . . . [so that] what structuralists view as fixed forms are actually emergent figures that provide the ever-changing parameters of constraint within which individual identities and differences are constantly formed and reformed."[73] We must understand diversity as part and parcel of self-organizing, coevolving, co-determining, and interconnected living networks of varying degrees of complexity, networks that link in often nonlinear ways the ecological, neuro-somatic, cultural, social, religious, and technological dynamics to give rise to the worlds in which we dwell. Thus, fresh understandings of diversity must take the "complexity turn" and draw inspiration from the study of emergent complex adaptive systems (CAS) in fields as diverse as ecology, immunology, chemistry, artificial intelligence, economics, and computational sociology. CAS involve

> a sense of contingent openness and multiple futures, of the unpredictability of outcomes in time and space, of charity toward objects and nature, of diverse and non-linear changes in relationships, households and persons across huge distances in time and space, of the systemic nature of processes, and the growing hypercomplexity of organizations, products, technologies and socialities.[74]

The diasporic community of worship and charisma that is La Luz del Mundo is afforded and constrained not only by transnational social and religious networks monitored by the nation-state but also by electronic technologies, as well as the sacred landscapes and material culture in Hermosa Provincia that are transposed to and reenacted in Atlanta. Circulating in the network are also the visceral experiences of ecstasy and sorrow that are enabled by a charismatic habitus, the embodied structuring and structured structures that are the product of the interplay of sensorimotor, social, and ecological dynamics. And as Bynum reminds us through her work on the Eucharist, we should not neglect the power of commensality. We should not forget that the food that is prepared, shared, blessed, and consumed is a key element of the La Santa Cena's ritual efficacy. La Santa Cena is a variegated, relatively stabilized entanglement of materialities: the food, the satellite connections, the screen, and the Holy Spirit embodied in the weeping faces and outstretched hands of the virtual pilgrims to Hermosa Provincia are all part of the material infrastructure, the necessary "conditions of felicity" that render this entanglement vibrant.

A networked nonreductive materialism is not a new grand narrative, but it does have a strong elective affinity with a "cosmopolitics" that recognizes the fragility and simultaneous viability of coeval ways of presencing the world.[75] By cosmopolitics, Isabelle Stengers does not mean just a postnational and posthuman politics, but one that departs from the possibility that Western, science-driven modernity could have it all wrong. Rather than merely tolerating or accommodating the Other, we have to entertain seriously "the troubling and exhilarating feeling that things *could be different*, or at least that *they could still fail*."[76] Only this destabilizing humility can open us to alterity, to a painstaking but ever-expanding rhizomatic politics of inclusion and a vascularized pluralistic holism and naturalism that transgresses the limits of Western modernity, which has made the rational, unencumbered Cartesian self and the unified, territorialized nation-state the exclusive receptacles of legitimate identities, rights, and duties.

NOTES

1. According to McCutcheon, "private affair tradition," which has been dominant in religious studies since Schleiermacher, notwithstanding the attempts of *Religionswissenschaft* to set itself apart from theology, "presupposes that religion cannot be *explained* as a result of various historical and cultural factors and processes; instead it is argued that the deeper meaning of religion can only be deciphered and understood to make manifest in culture certain essentially religious or transcendent values and feelings" (Russell T. McCutcheon, *Critics Not Caretakers: Redescribing the Public Study of Religion* [Albany: State University of New York Press, 2001], 5).
2. Bruno Latour, "On the Partial Existence of Existing and Nonexisting Objects," in *Biographies of Scientific Objects*, ed. Lorraine Daston, 247–69 (Chicago: University of Chicago Press, 2000); Bruno Latour, *We Have Never Been Modern* (Cambridge, Mass.: Harvard University Press, 1993), 135.
3. Patricia Fortuny, "The Santa Cena of the Luz Del Mundo Church: A Case of Contemporary Transnationalism," in *Religion across Borders: Transnational Immigrant Networks*, ed. Helen Rose Ebaugh and Janet Saltzman Chafetz (Walnut Creek, Calif.: AltaMira Press, 2002), 24.
4. According to Appadurai, mass-mediated sodalities "are often transnational, even postnational, and they frequently operate beyond the boundaries of nation. [They] . . . have the additional complexity that, in them, diverse local experiences of taste, pleasure, and politics [and religion!] can crisscross one another, thus creating the possibility of convergences in translocal social action that would otherwise be hard to imagine" (Arjun Appadurai, *Modernity at Large: The Cultural Dimensions of Globalization* [Minneapolis: University of Minnesota Press, 1996], 8).
5. See, for example, Elizabeth Brusco, *The Reformation of Machismo: Evangelical Conversion and Gender in Colombia* (Austin: University of Texas Press, 1995).
6. Marcel Mauss, "Body Techniques," in *Sociology and Psychology: Essays* (London: Routedge and Kegan Paul, 1979). Fleshing out Mauss's notion, Bourdieu defines the habitus as "systems of durable, transposable dispositions, structured structures predisposed to function as structuring structures, that is, as principles which generate and organize practices and representations that can be objectively adapted to their outcomes without presupposing a conscious aiming at ends or an express mastery of the necessary operations necessary in order to attain them" (Pierre Bourdieu, *The Logic of Practice* [Stanford, Calif.: Stanford University Press, 1990], 53).
7. The anthropologist Simon Coleman uses the term "corpothetics" to highlight "how believers may not merely contemplate, but also bodily engage and elide the self with images" (Simon Coleman, "Constructing the Globe: A Charismatic Sublime?" in *Traveling Spirits: Migrants, Markets and Mobilities*, ed. Gertrud Hüwelmeier and Kristine Krause [New York: Routledge, 2009], 188). The term also points to "how charismatic travel—visualized and often vicarious—becomes a chronic

means of negotiating between embodied subjecthood and transcendent propriety" (ibid., 189).
8. See Victor Turner, *Dramas, Fields, and Metaphors: Symbolic Action in Human Society* (Ithaca, N.Y.: Cornell University Press, 1974).
9. Peggy Levitt and Nina Glick Schiller, "Conceptualizing Simultaneity: A Transnational Social Field Perspective on Society," *International Migration Review* 38, no. 3 (2004): 1002–39.
10. Jacqueline Hagan, *Migration Miracle: Faith, Hope, and Meaning on the Undocumented Journey* (Cambridge, Mass.: Harvard University Press, 2008).
11. Ibid., 162.
12. Ibid., 163.
13. Peggy Levitt, *God Needs No Passport: Immigrants and the Changing American Religious Landscape* (New York: New Press, 2007), 2.
14. Philip Jenkins, *The Next Christendom: The Coming of Global Christianity* (New York: Oxford University Press, 2011).
15. Manuel Vásquez and Cristina Rocha, "Introduction: Brazil and the New Global Cartography of Religion," in *The Diaspora of Brazilian Religions*, ed. Cristina Rocha and Manuel Vásquez (Leiden: Brill, 2013), 24.
16. Manuel A Vásquez, *More Than Belief: A Materialist Theory of Religion* (Oxford: Oxford University Press, 2011), 302.
17. Appadurai, *Modernity at Large*, 15.
18. An interesting question to ask is whether the satellite transmission of La Santa Cena represents a counter optic to the power of the state to render immigrants visible as "illegals," or whether it is itself part of the transnational panopsis of an image-driven capitalism. I wish to thank Catherine Keller for raising this question.
19. Clifford Geertz, *Interpretation of Cultures* (New York: Harper and Row, 1973).
20. John K. Wright, "Notes on Early American Geopiety," in *Human Nature and Geography: Fourteen Papers, 1925–1965* (Cambridge, Mass.: Harvard University Press, 1966); Thomas Tweed, "Diasporic Nationalism and Urban Landscape: Cuban Immigrants at a Catholic Shrine in Miami," in *The Gods of the City: Religion and the Contemporary American Urban Landscape*, ed. Robert Orsi (Bloomington: Indiana University Press, 1999), 132.
21. Levitt and Schiller, "Conceptualizing Simultaneity," 1009.
22. Nina Glick Schiller and Georges Fouron, *God Woke Up Laughing: Long-Distance Nationalism and the Search for Home* (Durham, N.C.: Duke University Press, 2001).
23. Thomas Tweed, *Our Lady of the Exile: Diasporic Religion at a Catholic Shrine in Miami* (New York: Oxford University Press, 1997).
24. Appadurai, *Modernity at Large*.
25. Bruno Latour, *On the Modern Cult of the Factish Gods* (Durham, N.C.: Duke University Press, 2010), 19–20.
26. John Law, "After ANT: Complexity, Naming and Topology," in *Actor Network Theory and After*, ed. John Law and John Hassard (Oxford: Blackwell, 1999), 4.
27. Ibid.

28. William Connolly, *The Fragility of Things: Self-Organizing Processes, Neoliberal Fantasies, and Democratic Activism* (Durham, N.C.: Duke University Press, 2013).
29. Bruno Latour, *Pandora's Hope: Essays on the Reality of Science Studies* (Cambridge, Mass.: Harvard University Press, 1999), 99.
30. Ibid., 214.
31. Ibid., 9.
32. Ibid., 214.
33. Johannes Fabian used the expression "denial of coevalness" to characterize the epistemic violence that anthropology performs to establish and legitimize its authority, setting the anthropologist in a different, more advanced, time frame than the people he or she studies (Johannes Fabian, *Time and the Other: How Anthropology Makes Its Object* [New York: Columbia University Press, 1983]). In a similar fashion, modernity sets itself as forward-looking, as concerned with a present that has broken with the past and is unfolding toward a utopian future, in contrast to the Dark Ages, the age of religion, which was dominated by tradition and the backward reference to the past, what Mircea Eliade called *illo tempore*, the time of the original revelation.
34. Charles Taylor, for example, argues that modernity is characterized by a "buffered self" in contrast to the premodern "porous self." According to him, "One of the big differences between us and our ancestors of five hundred years ago is that they lived in an 'enchanted' world, and we do not; at the very least, we live in a *much less* 'enchanted' world. We might think of this as our having 'lost' a number of beliefs and the practices which they made possible. But more, the enchanted world was one in which these forces could cross a porous boundary and shape our lives, psychic and physical. One of the big differences between us and them is that we live with a much firmer sense of the boundary between self and other. We are 'buffered' selves. We have changed" (Charles Taylor, "A Secular Age: Buffered and Porous Selves" [2008], http://blogs.ssrc.org/tif/2008/09/02/buffered-and-porous-selves. Accessed May 16, 2013).
35. Borrowing from Michel Serres, Latour also uses "quasi-objects" or "weak objects" to describe hybrids between objects and things and to acknowledge that these constructed hybrids "too act, they too do things, they too make you do things" (Bruno Latour, "Why Has Critique Run Out of Steam: From Matters of Fact to Matters of Concern," *Critical Inquiry* 30, no. 2 [2004]: 243).
36. Latour, *On the Modern Cult of the Factish Gods*, 100.
37. Matthew Day, "How to Keep It Real," *Method and Theory in the Study of Religion* 22, no. 4 (2010): 272–82.
38. Bruno Latour, *Reassembling the Social: An Introduction to Actor-Network Theory*, Clarendon Lectures in Management Studies (New York: Oxford University Press, 2007), 235.
39. Day, "How to Keep It Real," 278.
40. Ibid., 280–81.
41. Ibid., 281.

42. In anthropology, emic perspectives privilege the categories used by the practitioners themselves in seeking to understand a particular cultural pattern. In contrast, etic explanations rely on conceptual tools that may not make sense to those directly involved in the action or event. See also McCutcheon, *Critics Not Caretakers*.
43. Day, "How to Keep It Real," 278.
44. Gillian Goslinga, "Spirited Encounters: Notes on the Politics and Poetics of Representing the Uncanny in Anthropology," *Anthropological Theory* 12, no. 4 (2012): 386–406; Casper Bruun Jensen et al, "Comparative Relativism: A Symposium on an Impossibility," *Common Knowledge* 17, no. 1 (2011): 1–165.
45. Patrice Ladwig, "Ontology, Materiality and Spectral Traces: Methodological Thoughts on Studying Lao Buddhist Festivals for Ghosts and Ancestral Spirits," *Anthropological Theory* 12, no. 4 (2013): 431.
46. Ibid., 428.
47. Ibid., 431.
48. John Tresch, "Another Turn after ANT: An Interview with Bruno Latour," *Social Studies of Science* 43, no. 2 (2013): 302–13.
49. I share Latour's understanding of agnosticism: "as a way of ceasing to believe in belief" (Latour, *On the Modern Cult of the Factish Gods*, 2). The issue is not what people believe or what is in their heads, but the material "modes of existence," the efficacious and binding lifeworlds they make present through their discursive and nondiscursive practices. Latour, however, is critical of naturalism, "the faith in a single natural world, comprehensible Science—or rather, through a mistaken definition of (Western) natural science whose purpose has been to eliminate entities from the pluriverse" ("Whose Cosmos, Which Cosmopolitics? Comments on the Peace Terms of Ulrich Beck," *Common Knowledge* 10, no. 3 [2004]: 458). In his view, naturalism, "like any fundamentalist ideology, amounts to a prejudice against fabrication" (ibid., 461). I believe this critique is right on target when it comes to positivism and scientism. However, Latour assumes that there is a single, monolithic Western naturalism. Certainly, Spinoza's, Nietzsche's, Whitehead's, Merleau-Ponty's, and Quine's naturalisms are very different from those of Comte, Carnap, and Churchland. I believe that it is possible to operate with a rigorous-yet-fallible naturalism, which is keenly aware that the boundaries between what counts as "natural" and "supernatural" are constructed and negotiated within communities of interpretation. In fact, Latour's work is a prime example of that naturalism.
50. Donna Haraway, *Simians, Cyborgs, and Women: The Reinvention of Nature* (New York: Routledge, 1991).
51. Donna Haraway, *When Species Meet*, Posthumanities (Minneapolis: University of Minnesota Press, 2007), 11.
52. Ibid., 35.
53. Donna Haraway, *Modest_Witness@Second_Millennium. Femaleman_Meets_Oncomouse* (New York: Routledge, 1997), 52.
54. Ibid., 363.
55. Latour, *Pandora's Hope*, 208.
56. Latour, *We Have Never Been Modern*, 144–45.

57. Latour, *Reassembling the Social*, 208. I am also critical of Latour's stress on the integrative and collaborative (even if fragile) dimensions of networks, which he deploys polemically against critical theory. "Dispersion, destruction, and deconstruction are not the goals to be achieved but what needs to be overcome. It is much more important to check what are the new institutions, procedures, and concepts able to collect and reconnect the social" (ibid., 11). Despite his own clever deconstruction of Durkheimian-Parsonian notions of society and the social, Latour seems to reproduce Durkheim's and Parsons's concern for stability.
58. Karen Barad, *Meeting the Universe Halfway: Quantum Physics and the Entanglement of Matter and Meaning* (Durham, N.C.: Duke University Press, 2007), 58, 59.
59. Caroline Walker Bynum, *Christian Materiality: An Essay on Religion in Late Medieval Europe* (New York: Zone Books, 2011), 30.
60. Ibid., 35.
61. Ibid., 285.
62. Ibid., 65.
63. Ibid., 121.
64. Ibid., 250.
65. Ibid., 52.
66. Jane Bennett, *Vibrant Matter: A Political Ecology of Things* (Durham, N.C.: Duke University Press, 2010), viii.
67. Diana Coole and Samantha Frost, "Introducing the New Materialisms," in *New Materialisms: Ontology, Agency, and Politics*, ed. Diana Coole and Samantha Frost (Durham, N.C.: Duke University Press, 2010), 9.
68. Diana Coole, "The Inertia of Matter," ibid., 101.
69. Francisco Varela, Evan Thompson, and Eleanor Rosch, *The Embodied Mind: Cognitive Science and Human Experience* (Cambridge, Mass.: MIT Press, 1991).
70. Evan Thompson, *Mind in Life: Biology, Phenomenology, and the Sciences of the Mind* (Cambridge, Mass.: Harvard University Press, 2007), 11.
71. Tim Ingold, *Being Alive: Essays on Movement, Knowledge and Description* (New York: Routledge, 2011), 89–94.
72. Ibid., 29.
73. Mark C. Taylor, *After God* (Chicago: University of Chicago Press, 2007), 310.
74. John Urry, "The Complexity Turn," *Theory, Culture & Society* 22, no. 5 (2005): 3; Ilya Prigogine, *The End of Certainty* (New York: Free Press, 1997); Grégoire Nicolis and Ilya Prigogine, *Exploring Complexity: An Introduction* (New York: W. H. Freeman, 1989); Manuel De Landa, *A Thousand Years of Nonlinear History* (New York: Swerve, 1997).
75. Isabelle Stengers, *Cosmopolitics I* (Minneapolis: University of Minneapolis Press, 2010); Isabelle Stengers, *Cosmopolitics II* (Minneapolis: University of Minnesota Press, 2011).
76. Latour, *Reassembling the Social*, 89.

❧ Ethicopolitical Entanglements

❧ Stubborn Materiality: African American Religious Naturalism and Becoming Our Humanity

CAROL WAYNE WHITE

The same stream of life that runs through my veins night and day runs through the world and dances in rhythmic measures. It is the same life that shoots in joy through the dust of the earth in numberless blades of grass and breaks into tumultuous waves of leaves and flowers. It is the same life that is rocked in the ocean-cradle of birth and of death, in ebb and in flow. I feel my limbs are made glorious by the touch of this world of life. And my pride is from the life-throb of ages dancing in my blood this moment.
—RABINDRANATH TAGORE, *Stream of Life*

In his groundbreaking volume, *Souls of Black Folk* (1903), W. E. B. Du Bois sketched the complex unfolding of nineteenth-century African American religiosity, revealing the institutionalization of a people's hopes, fears, core values, aspirations, ethical convictions, cosmological assumptions, and grasp of death. In this and other works, Du Bois offers a compelling view of African American religiosity as an evolving, humanistic enterprise with monumental social and communal implications, capturing its vigor and accentuating its vital role in comprehending the complex psyche of the American nation. His insights into the rich and layered texture of African American religiosity, and his bold claims for its enduring legacy, are as relevant and cogent today as they were at the beginning of the twentieth century. Notwithstanding its range of worship expressions, African American religiosity has continued to evolve historically (and perhaps primarily) as a complex social and cultural mechanism that has aided African Americans in their untold struggles against various forms of injustices. This evolution is not surprising, given the harrowing experiences of Africans on the North American continent as enslaved subjects whose full humanity was often questioned or denied. Emerging from the context of slavery, the dominant African American religious tradition has often interwoven its declaration of blacks' full humanity with theological tactics and strategies that provide the necessary ontological justification and ethical reasoning. In so doing, it has consistently revealed a general structure of human desire that underscores the value of life, insisting on our efforts to bring forth conditions that would accentuate the fullness of life for all those interested in processes of transformation.

In this essay, I explore the conceptual space opened by Du Bois, focusing on this theme of human desire within African American thought that values fullness of life. The particular humanistic discourse that I foreground in anticipating this task, however, departs from a tradition of liberal humanism that I find both limiting and problematic for various reasons. First, this trajectory of liberal humanism has consistently overestimated the autonomy of human animals, positioning us outside of complex, myriad nature and rendering invisible our inextricable connection to other life forms and material processes. With its problematic forms of anthropocentricism and the reinscription of exceptional human nature, this model of liberal humanism has also advanced a facile environmentalism in which nonhuman natural processes are often accorded value according to their usefulness to humans. Concomitant with this ecological view are ethical sensibilities in which humans project our own notions of ourselves as the measure for valuing other parts of nature. According to this worldview, for example, whales and dolphins are worthy of human sympathy because they reflect our focus on cognition and autonomy, whereas other parts of nature (dung beetles and fungus) are unworthy of our regard, or not so valuable to humans.

In rejecting this dominant humanistic model, as well as its retention of a hierarchical model of nature built on the "great chain of being" concept, I prefer a materialist view that emphasizes a deeper, appreciable view of nature that constitutes the human itself. My humanistic approach in this essay has affinities with other recent developments in feminist theories, animal studies, posthumanism, and vegetal studies—all of which offer expanded (or new materialist) views of the human and the animated nonhuman world.[1] As I have argued elsewhere, contributing to ongoing efforts of decentering the human may help provide the basis for envisioning different personal and biopolitical futures.[2] Yet in reconfiguring the human within the context of African American intellectual culture, I also acknowledge a thorny but important issue: the traditional exclusivity of the category itself, or what some of us have recognized as lacunae in conceptualizing the human. The theoretical violence perceivable in how the category of the human has been constructed has been acutely recounted in feminist, liberationist, and postcolonial critiques. Their persuasive critiques stress that the normative human subject has been primarily conceptualized as, and associated with, the lived experiences of white males of European descent, into whose ranks African Americans and other minoritarian subjects have not traditionally been admitted. In short, all human subjects have not traditionally been included in what is "properly" human.

My exploration of humanistic reasoning within the context of African American thought and history thus presents me with a complex yet important task. On the one hand, the model of humanistic discourse I invoke aims at reconstructing the category of the human in such a way that it can address the tyrannical presence of white supremacy, as well as other "isms" that have denied black (and other marginalized) subjects their rightful claims to becoming human and actualizing their humanity in many ways. On the other hand, my retrieval of the human necessarily rejects the aims of traditional humanisms that have positioned humans outside of nature and eclipsed the interrelatedness of all natural processes. Whatever conceivable notion of black humanity I claim in this essay will be ontologically enmeshed and entangled with other forms of natural life.

Bearing in mind these conceptual qualifications, I argue for a radical humanism that claims a religious sensibility for itself. Identifying African American religiosity as the ingenuity of a people constantly striving to construct their humanity and eke out a meaningful existence for themselves amid harrowing circumstances, I construct a concept of "sacred humanity" and ground it in existing hagiographic and iconic African American writings. With this term "sacred humanity," I evoke the notion of humans as value-laden, social organisms in constant search of meaning, enamored of value (or the good), and instilled with a sense of purpose (telos).[3] This capacious view of humanity foregrounds African Americans as dynamic, evolving, social organisms having the capacity to transform ourselves and create nobler worlds where all sentient creatures flourish, and as aspiring lovers of life and of each other. As such, sacred humanity underscores the functional value of black religion as one of the highest aspirations of African American character: its claim on life.

Furthermore, as I discuss later in detail, the conception of humanity as sacred emerges from a naturalistic vision that emphasizes deep interconnectedness among humans and celebrates our kinship with other sentient life, accentuating a modality of existence in which transformation occurs. Utilizing the tenets of religious naturalism in conjunction with values discourse, I consider humans' awareness and appreciation of our connection to "all that is," as an expression of sacrality, or what we perceive as ultimately important and valuable.[4] Since religious naturalism does not use "supernatural" concepts or theories in comprehending humans' need for value and meaning, the realm of nature is the focus (this includes both natural processes and human culture for most religious naturalists). Religious naturalists draw on two fundamental convictions in understanding basic human quests for meaning and value: the sense of nature's richness, spectacular complexity, and fertility, and the recognition that nature is the only realm in which people live out their lives.[5] In this context, then, religious naturalism offers an eloquent rendering of sacred humanity in explaining and advancing human animals' deep, inextricable homology with the rest of the natural world. Further, the materialist spin I offer will underscore that the inexhaustible connection or entanglement with other natural processes, or with the more-than-human, constitutes the very notion of the human as such.

To advance my argument, I outline a trajectory of modern philosophical reasoning that has shifted emphasis from a transcendental divine Other, or the traditional object of faith (deity), to emphasize human subjectivity. I then assess the dominant model of human subjectivity arising in modern humanism—the prototypical tyrannical model, replete with gender, racial, class biases—that reinforced problematic differentials of Enlightenment ideals, outlining specific shortcomings suggested by both postmodern and African American cultural critiques. Next, I provide examples from select African American intellectuals whose ideas helped generate a liberationist vein of humanism in African American thought. Building on their insights, I begin a construction of sacred humanity—as the specific site for the emergence of an African American religious naturalism.

This naturalistic vision emphasizes deep human interrelationship in the world, and it accentuates humans' propensity toward life. More important, within African American religious culture, a naturalist view of humanity helps us grasp the conceptual richness of a liberationist motif within black religiosity and to celebrate its enduring legacy. With the concept of sacred humanity, I thus bring to light deeper understandings of basic convictions in African American religious expression. Posed as questions, these convictions ask: What does it means to be human and to affirm the essential value of blacks? How do we continue justifying religiously that indeed all humans share in the same ontological reality? Why do racist conceptions of blacks' humanity remain woefully impoverished and inadequate in light of current scientific views that show a deep interrelatedness among all biotic life forms?

THE EMERGENCE OF MODERN HUMANISM: ANOTHER VIEW

With a general movement from Immanuel Kant's positioning of morality as the focal point of religion, through G. W. F. Hegel's speculative idealism, to Friedrich Schleiermacher's elevation of intuition or feeling, religious thought in the West has put more and more emphasis on human ingenuity, and less on the divine as the transcendent Other. From Cartesianism through the Enlightenment to idealism and Romanticism, those attributes traditionally predicated of the divine subject have progressively been transferred to the human subject. In other words, as Mark C. Taylor has suggested, through a dialectical reversal, the creator God dies and is resurrected as the creative subject.[6] As God created the world through the Logos, so (Western) humanity now creates a "world" through conscious and unconscious projection. Like the God of classical theology, this sovereign subject relates asymmetrically only to what it constructs and is, therefore, unaffected by anything other than itself. The subject becomes the first principle (formerly identified as God) from which everything arises and to which all must be reduced or returned. Moreover, an undue emphasis on human knowledge eventually leads to a modernist project in which all objects of knowledge exist for the epistemological subject.

In areas of knowledge as diverse as science, theology, and early modern philosophy (albeit directed toward different ends), one locates the same exclusive concern for and interest in humanity. Within the general area of theology, for example, this general trend culminated (and was best represented by) a Protestant liberal movement derived from the German intellectual tradition. Although too broad and complex to outline here, twentieth-century modern liberal theology has had an interesting array of thinkers who have grappled with the role and task of theology in an increasingly complex and troubled world. Its trends have ranged from the historic-critical research of Albrecht Ritschl, through the neo-orthodoxies of Emil Brunner, Rudolf Bultmann, and Reinhold Niebuhr, to the empirical process orientations of the Chicago School of the 1930s and '40s. In the latter part of the twentieth century, such post–World War II liberal movements as the existentialism of John Macquarrie and Paul Tillich also represented the modern search for new foundations.[7]

With a shift increasingly focusing on the human subject of faith, twentieth-century liberal theology introduced a new problematic within religious scholarship: an attempt to articulate the deep structures of the experiences and consciousness of a self that is said to be representative for all humans. For example, Tillich's existential approach that speaks of human nature in general abstract terms reflects a fundamental flaw in Western reason: its universal and ahistorical tendencies. In a fashion akin to modern philosophy, with its obsession with identity and singularity, these types of theological configurations have failed to speak convincingly of the human subject in all its historical complexity. Just as the categories of Western philosophy often obliterated differences of gender and race as these shape and structure the experience and subjectivity of the self, so a dominant model of modern theological thought posited itself as the discourse of the one self-identical subject, thereby blinding adherents to (and in fact delegitimizing) the presence of otherness and difference that did not fit into its categories.[8]

These trends show that influential notions of the human and its spheres of creativity, as represented by an influential form of modernity, have often been ambiguous as well as potentially lethal in their consequences. Ironically, in its creativity and success in enacting its most cherished ideals, Western humanity brought forth material conditions and an ethos of dominion that threatened its own life as well as that of other sentient beings. Begun as a revolt against oppressive structures (political and religious), enlightened rationality fell short of its emancipatory aims to culminate in a reign of terror. Classic Enlightenment ideals—progress, universalism, and guaranteed freedoms once privileged at the end of the eighteenth century and throughout the nineteenth—now appear in contemporary culture as suspicious ideologies, masking special privileges and selfish materialism. For example, scientific medicine, long viewed as the paradigmatic expression of Enlightenment reason, has been lax regarding health matters (and this is further complicated by the race, gender, and class inequalities in health care) as well as for its acquisitive nature, which seems endless. Likewise, technical industrialism—in expanding as the Enlightenment had hoped—continues to pose a threat to myriad natural systems and to use up crucial natural resources on which we all depend. The African philosopher Emmanuel Eze has demonstrated that Enlightenment reasoning provided an ideal intellectual and epistemological basis for establishing and legitimating the hegemony of the West over the other traditions Europeans were beginning to encounter in Asia, Africa, and the Americas.[9] Those traditions, values, and folk practices not compatible with or radically different from the Enlightenment ideal were often viewed suspiciously and located low on the hierarchy, beneath the required level of cognition or scientificity. Ironically enough, the liberal humanism that espoused the universal rights of humans spawned a distinct set of discursive formations and cultural practices that justified unjust capitalist social relations in the West and their extension, via colonialism and imperialism, to other societies.

Jean-Paul Sartre depicts the violence of liberal universal discourse when he noted in 1961 that abstract assumptions of universality really served as a cover for the more realistic practices of slavery, colonialism, racism, and barbarity.[10] Here, Sartre alludes to the development of the Euro-Anglo construction of "whiteness" as the normative identity for human subjectivity. This form of racism depended on a logic of racial difference whereby one group (certain Europeans) defined itself as white in such a way that it became the standpoint from which other races or groups were judged on the basis of the degree to which they are less white. Thus the white liberal subject often identified itself as full and others (those identified as nonwhites) as empty or existing in the condition of lack. In modernity, this condition of lack eventually took on group associations, which often led to the dichotomy of fullness and hunger having symbolic form in antiblack societies as lightness and darkness, whiteness and blackness, and—one of the most complete and extreme form of this racialized binary logic—the construction of the superiority of the white race over the black races.[11]

"WE, TOO, ARE HUMAN": EXPRESSIONS OF AFRICAN AMERICAN HUMANISM

Addressing the adverse effects of this racialized reason (or reasoned racism) has been one of the hallmarks of an evolving African American intellectual trajectory. This legacy of thought is replete with ethical, aesthetic, and ontological implications involved in its ongoing task of justifying and defending the fact that African Americans, too, are human. In the late nineteenth century, for example, with her feminist collection of essays and speeches, *A Voice from the South* (1892), Anna Julia Cooper challenged the different ways African Americans were systematically dehumanized and denaturalized as the other—or how they were categorized below the normatively European model of humanity. Cooper rejected the narrow focus of this atomistic outlook and envisioned an ideal view of American culture that inspired and enabled each person to attain fullness of being and to flourish as part of the whole. Her humanistic discourse astutely reminded her contemporaries of a crucial insight: "The philosophic mind sees that its own 'rights' are the rights of humanity."[12] Cooper envisioned the young America of her day as the "consummation of the ideal of human possibilities" and, consequently, the proper arena for the interplay of forces and conflict in which diversity becomes standard.[13] For Cooper, the nation's growth would be structured by inevitable conflict as it evolves toward sustaining a rich, civilized culture of celebrating differences; in this context, healthy, stimulating, and progressive conflict consists of "the co-existence of radically opposing or racially different elements," which can contribute to a peaceful, harmonious way of life only when it is conceived as constituted by differences and where the general ethos is "the determination to live and let live."[14]

In the same era, Du Bois's conceptualization of life behind the "veil" of race and the resulting "double-consciousness"—a sense of always looking at one's self through the eyes of others—conveyed his aspirations for African Americans to look anew at themselves and to reinvent themselves.[15] His critiques of American life focused on perceived "truths" about blacks' inferior humanity, which he questioned, disassembled, and, in some cases, even reconstructed in his published work. Du Bois encouraged his contemporaries to recognize the transitory, fallible nature of racial constructions embedded in essentialist notions of superiority and inferiority. Du Bois's very early conception of life within the veil was always accompanied by his imagining the possibilities of life beyond it. He inspired his contemporaries to imagine African Americans as centers of value whose self-generating genius and potency had become obfuscated by the veil. Not only this, Du Bois sketched out a comprehensive, multifaceted approach to the social ills facing his generation:

> Work, culture, liberty,—all these we need, not singly but together, not successively but together, each growing and aiding each; and all striving toward that vaster ideal that swims before the Negro people, the ideal of human brotherhood, gained through the unifying ideal of Race, the ideal of fostering and developing the traits of the Negro, not in opposition to or contempt for other races, but rather in large conformity to the greater ideals of the American Republic, in order that some day on American soil two world-races may give each to each those characteristics both so sadly lack.[16]

Such theoretical acuity was also reflected in the writings of James Baldwin, who also dreamed of brave new conceptions of humanity beyond the vexed "raced" configurations he both experienced and witnessed in the U.S. during the mid-twentieth century. During a critical period of the civil rights era, Baldwin ingeniously unmasked racial, heterosexual, and other problematic privileges lurking in dominant constructions of the human in North American culture and religion. He emphatically rejected the distorted ontological gravitas of modern humanistic discourse, replacing it with a fuller, more capacious model of humanity. His essays and works of nonfiction expressed the sheer ontological weightiness of having to reinvent (or to justify) one's humanity over and over again, in every generation.[17] Baldwin's race-conscious rhetoric often focused on the hypocrisy of white Christians who, in an attempt to maintain their social order identity as white people, willfully denied their own moral connection to (and biological kinship with) those of African descent, whom they continued to exploit. Part of black Americans' exilic experience, Baldwin suggests, is owing to white Americans' denial of literal familial kinship with blacks.[18] Baldwin's insistence that whites see themselves in blacks (and as black) emphasizes the deep genetic homology structuring all life forms—what I will later describe as humans' interconnectedness with each other and with all natural organisms.

In select essays, Baldwin associates the term "love" with a critical awareness of our common humanity. As he eloquently asserts in *The Fire Next Time*, "We, with love, shall force our [white] brothers to see themselves as they are, to cease fleeing from reality, and begin to change it. . . . We can make America what America must become."[19] One thus finds in Baldwin a form of communal ontology that recognizes a humanity constitutive of our biotic materiality on which various identity markers are attached. As he emphatically states at one point: "It is so simple a fact and one that is so hard, apparently, to grasp:

Whoever debases others is debasing himself."[20] For Baldwin, what humans can become, and what we wish to be, depends on how we act in the here and now. In the most immediate sense, this construction of humanity is dependent on radical acts of love—of embracing otherness within oneself and as oneself.

At this juncture, it is important to note that in their brave attempts to formulate expanded views of humanity, Cooper, Du Bois, and Baldwin nonetheless operated within a paradigm of thought that retained an exclusive view of humanity. This tendency is not surprising given their historical locations during the first and middle parts of the twentieth century and their basic concern to address various forms of racist ideology with contemporaneous forms of knowledge. In the next section, I build on the intellectual legacy these historical figures helped to shape in order to explore an expansive view of black humanity free of this problematic anthropocentricism with the tenets of religious naturalism. Going beyond the level of human exclusivity in this context marks the emergence of what I deem an African American religious naturalism.

SACRED HUMANITY: STUBBORN, ENTANGLED MATERIALITY

Cooper, Du Bois, Baldwin, and a number of other iconic African American figures helped establish a cultural legacy that has confronted the persuasive force of a dualistic worldview, showing its far-reaching effects, including a dominant and impoverished view of black bodies as the immoral, devalued other. These authors promoted the humanity of African Americans at historical junctures when it was questioned and denied; more important, their brave efforts show that when blackness is defined in a narrow sense, or negatively marked as different, the more capacious visions of our entangled humanity become marred and distorted. Their collective work thus provides the impetus and vision for my formulation of "sacred humanity" in the twenty-first century. Grounded in the tenets of religious naturalism, the concept of sacred humanity evokes our essential entangledness with each other and with other natural organisms.

At the heart of religious naturalism in all its variants is a basic conviction: Any truths we are ever going to discover and any meaning in life we should uncover are revealed to us through the natural order.[21] In this essay, I embrace a contemporary strain of religious naturalism within the science and religion paradigm that is best associated with the works of Ursula Goodenough, Donald Crosby, and Loyal Rue—all of whom have been influential in my development as a religious naturalist.[22] This vein is particularly appealing because it focuses on the materiality of existence and includes human nature and human culture in its grasp of naturalism, thereby challenging some widely held paradigms about the nature of "nature." Another important feature of this strain of religious naturalism is its emphasis on emergence as an important new concept for thinking about biological and cosmic evolution. Consider, for example, that emergent properties arise as a consequence of relationships—for example, the relationships between water molecules that generate a snowflake, or the relationships between neurons that generate a memory. Emergent properties also give rise to yet more emergent properties, generating the vast complexity of our present-day cosmic, biological, ecological, and cultural contexts.[23]

These insights compel us to reflect meaningfully on the emergence of matter (and especially life) from the Big Bang forward, promoting an understanding of myriad nature as complex processes of becoming. The general view of humanity I hold, on which I build my concept of sacred humanity, arises from this context. With other religious naturalists, I believe that understanding the deep history of the cosmos is profoundly important for any basic understanding of the materiality of being human, of being alive in the manner we currently find ourselves. Humans are highly complex organisms, owing the lives we have to the emergence of hierarchies of natural systems. Expressed succinctly, humans are "ultimately the manifestations of many interlocking systems—atomic, molecular, biochemical, anatomical, ecological—apart from which human existence is incomprehensible."[24]

Human life is also part of an evolutionary history showing directionality or a trend toward greater complexity and consciousness. As Stephen J. Gould and other scientists have noted, there has been an increase in the genetic information in DNA and a steady advance in the ability of organisms to gather and process information about the environment and respond to it.[25] Goodenough has provided a persuasive view of evolutionary theory that celebrates such directionality; her descriptive account of how life works in terms of biophysics and biochemistry is useful here: "Life is getting something to happen against the odds and remembering how to do it. The something that happens is biochemistry and biophysics, the odds are beat by intricate concatenations of shape fits and shape changes, and the memory is encoded in genes and their promoters."[26] Biophysics is concerned with "electrochemical gradients" and the physics through which "channels and pumps" work to "span the cell membrane" and thus allow the chemical processes of the cell to work.[27] Basic biochemistry has to do with the shapes of proteins, particularly enzymes, and the sequences of shape changes or cascades, that is, those processes through which a cell perceives or interacts with that which is external. The "cell is set up to optimize the flowing of cascades." In other words, proteins that will interact with one another have "domains, called addresses, that target the proteins to the same cellular location," and each "destination proves to be optimal for particular biochemical reactions."[28] This means that a cell is like a community:

Its inner workings are segregated into interacting compartments, whereas its outer membrane defines its interactions with the rest of the world.

Biochemistry and biophysics generate the patterns that constitute pulsating organisms, including the development of multicellular organisms such as humans. To understand this idea fully, it is important to grasp the concept of natural selection and its role in our evolutionary heritage as human animals. As Goodenough points out, evolution can be viewed minimally as changes in the frequencies of different sets of instructions for making organisms. To understand evolution, then, one must grasp how the instructions become different (mutation) and then how the frequencies of those instructions are changed (natural selection). Within the context of describing mutation as simply a change in the sequence of nucleotides in a genome, the concept of natural selection raises two important questions: Does the new protein or promoter work better, worse, or the same as the old one? How important is this difference to the organism?[29] As a set of dynamic processes, evolution occurs through the chains of modification—a protein gets slightly modified by a mutation; this modification gets further modified or added to, and eventually over time, complex new systems are created. For example, bacteria flagella started out with a protein to improve acid transport that happened to also rotate; even though the rotation was originally an irrelevant consequence, this feature is eventually built onto as other proteins attach to it, and so on. In this context, for example, the term "bricolage" (the construction of things using what is at hand, similar to a patchwork quilt) is a colorful and helpful metaphor to help us grasp the complex process.[30]

Following Goodenough's lead, I consider the metaphor of music useful for understanding the complexities of evolution. Consider, for example, how a good musicologist can often detect older forms or variations (a fugal texture, a specific cadence, or even a melodic strain) in new compositions. This music analogy helps describe the complexity of evolutionary life, too, as a fascinating organic composition of intricate movements in which the old is woven with the new to generate something more:

> A good biochemical idea—a protein domain that binds well to a promoter, a channel that's just the right size for a calcium ion—gets carried along through time, tweaked and modulated to best serve the needs of the current composition/organism but recognizable through evolutionary history. These conserved ideas combine with novelty to generate new directions, new ways of negotiating new environmental circumstances.[31]

This technical, scientific discussion inspires a rich, even poetic, rendering of human life as one distinct biotic form emerging from and participating in a series of evolutionary processes that constitute the diversity of life. In this essay the scientific epic thus becomes the starting point for positing an African American religious humanism constituted by a central tenet: Humans are relational processes of nature; in short, as stated above, we are nature made aware of itself. In declaring such, I contend that our humanity is not a given, but rather an achievement. Consider that from a strictly biological perspective, humans are organisms that have slowly evolved by a process of natural selection from earlier primates. From one generation to another, the species that is alive now has gradually adapted to changing environments so that it could continue to survive. Our animality, from this perspective, is living under the influence of genes, instincts, and emotions, with the prime directive to survive and procreate. Yet this minimalist approach fails to consider what a few cognitive scientists, and most philosophers, humanists, and religionists tend to accentuate: our own personal experience of what it is like to be an experiencing human being. As I noted elsewhere, "Becoming human, or actualizing ourselves as human beings, in this sense, emerges out of an awareness and desire to be more than a conglomeration of pulsating cells. It is suggesting that our humanity is not reducible to organizational patterns or processes dominated by brain structures; nor do DNA, diet, behavior, and the environment solely structure it."[32] In positing fundamental questions of value, meaning, and purpose to our existence, human animals become human destinies. Our coming to be human destinies is structured by a crucial question: How do we come to terms with life?

With other religious naturalists, I share the sentiment that reveling in a sense of our connectedness with other living beings can only be described as sacred. On the molecular level, there is evidence to support the loftier (or religious) idea that in the very nature of life itself there is some essential joining force. This orientation toward joining with others, in establishing our common humanity, is what I imagine when using the term "sacred humanity." Humans are, by our very constitution, relational, and our wholeness occurs within a matrix of complex interconnectedness—put another way, ways of conjoining with others that transform us.[33] My usage of the term "sacred" is thus imbued with the quality of ultimacy within the confines of what religious naturalism affirms: Nature as the realm in which we move, live, and have our being. Sacrality is a specific affirmation and appreciation of that which is fundamentally important in life, or that which is ultimately valued: relational nature. Humans are interconnected parts of nature, and our sacrality is a given part of nature's richness, spectacular complexity, and beauty.

Notwithstanding the diverse cultural and individual options of articulating this truth, there is, quite simply, the sacrality of humanity's inextricable entanglement with all that is. Finding meaning and value in our lives within the natural order presupposes this fundamental interconnectedness. We claim and become our humanity in seeking and finding community with others—and with otherness. This is a simple value that religious discourse has reiterated again and again. Goodenough states:

> We have throughout the ages sought connection with higher powers in the sky or beneath the earth, or with ancestors in some other realm. We have also sought, and found, religious fellowship with one another. And now we realize that we are connected to all creatures. Not just in food chains or ecological equilibria. We share a common ancestor. We share genes for receptors and cell cycles and signal-transduction cascades. We share evolutionary constraints and possibilities. We are connected all the way down.[34]

My notion of sacred humanity within African American religious culture presupposes these truth claims, marking the emergence of an African American religious naturalism in the twenty-first century. This viewpoint's distinctive voice stands in contrast to other contemporary ones in positing and celebrating blacks as an emergent, interconnected life form amid spectacular biotic diversity. With the concept of sacred humanity, this model of religious naturalism declares that blacks essentially value our deep homology with other sentient beings. As such, the term can be used to challenge the most viral constructions of "isms" rooted in problematic and alienating self-other differentiations, especially racially constructed ones the enduring legacy of African American religiosity has targeted.

SACRED HUMANITY AND MEANING

I am aware that an attempt to derive an African American religiosity based on an understanding of humans as natural processes may seem troublesome to some. Recall, for example, that nineteenth-century scientific perspectives on natural processes within later evolutionary thought (the new "science of man") promoted Enlightenment racism, in which notions of racial differences often presented the social inequalities between various cultural groups as reflecting the prescripts of nature.[35] However, these ideological extensions are part of a discourse on nature that some contemporary feminists, religious scholars, and postcolonial critics are currently addressing. In resisting many of these problematic tendencies, the notion of sacred humanity that I ground in the language of religious naturalism also leads us to other possibilities, visions, and assumptions. Evoking our sacred humanity presupposes an evolutionary narrative that has propelled humans' efforts to create meaning and purpose. This conviction converges with Rue's descriptive account of human beings as star-born, earth-formed creatures endowed by evolutionary processes to seek reproductive fitness under the guidance of biological, psychological, and cultural systems that have been selected for their utility in mediating adaptive behaviors. Humans maximize their chances for reproductive fitness by managing the complexity of these systems in ways that are conducive to the simultaneous achievement of personal wholeness and social coherence.[36] The notion of sacred humanity thus emphasizes the social character of cognition in human animals and enriches our understanding of humanity as symbol makers, creators of a world imbued with value, and as social organisms.[37] Studies of the brain suggest that the human limbic system, which we share with mammals, is the center of emotions that mobilize action and makes possible richer forms of relationship that involve empathy and caring for the young. These factors, in turn, lead us to recognize emotion, social relationships, and values often associated with traditional religious symbols—all as part of human reality.

This brings to mind Terrence Deacon's exploration of the intricate connection between the evolution of human language and our brains, or what he calls their coevolution. He argues, for example, that language itself is part of the process that has been responsible for the evolution of the brain.[38] Language has changed the environments in which brains have evolved. Humans are a species that in part has been shaped by symbols, in part shaped by what we do. According to Deacon, ritual, mythology, and so on—simply ways of doing things that are organized conventionally, symbolically—have become a hallmark of our species. Humans have transformed and even reinterpreted much of our biology through this symbolic system: So much of what we do—for example, marriage and conflict, as in warfare—has been transformed by this linguistic tool; it has, in a sense, taken over and biased all of our interactions with the world. Expressed succinctly, our brain has evolved very differently in some regards than other species' brains in ways that are uniquely human.[39] Based on these insights, I share the conviction that humans seek meaning by viewing their lives in a cosmic and religious framework that is itself a human symbolic construct—the brain is part of the cosmos and a product of the cosmos. Its structures reflect the nature of the cosmos and whatever ordering and meaning-giving forces are expressed in its history.[40] In short, processes of "religious" valuing within humans are inextricably connected to the fact that we are organisms with built-in values.

Furthermore, evolutionary biologists, sociobiologists, evolutionary psychologists, and philosophers are currently debating the extent to which one can argue that humans are value-driven decision systems with primary values built into us. A consistent scientific view is that a successful life outcome consists of promoting the transmission of information conducive to maintaining the emergent dynamic logic that gives it its meaning—that is, promoting the production of emergent outcomes (called traits in biology) that collectively make their own continuation more likely. Deacon and Goodenough contend,

> Traits common to all organisms include such non-depressing and religiously fertile capacities as end-directedness and identity maintenance; traits common to all animals include awareness and the capacity for pleasure and suffering; traits common to social beings include co-operation and meaning making; traits common to birds and mammals include bonding and nurturance; traits common to humans include language and its capacity to share subjective experiences, and thus to know love.[41]

Here, the human self emerges as a biological process that is not only affected by genes but also by many other factors at higher levels. In human development, as in evolutionary history, selfhood is always social, a product of language, culture, and interpersonal interaction, as well as genetic expression. Additionally, as various scientific views make clear: Naturalistic views of the human indicate a complex, social organism that can connect deeply with others, symbolize its environment (or engage in world formation) through values and language, and express love.

SACRED HUMANITY AND RELIGIOUS VALUING

The concept of sacred humanity I advance here is a particular configuration of our humanity as finite organisms; this conceptual view inevitably raises important issues of how humans come to terms with the facticity of nonexistence, or, more popularly expressed, with the inevitability of death. It is this angle, I believe, that generates the fullest philosophic and religious connotations of the sacred humanity concept. It is also what distinguishes my model of religious naturalism from strict reductionist forms of naturalism. Religious naturalism posits a view of the human organism as a valuing entity; in this context, I suggest that affirming death also becomes the proper point of departure for appreciating the value and meaningfulness of human life. In advancing this perspective, I follow the general ideas of Konstantin Kolenda in his overlooked text, *Religion without God*.[42] In this work, Kolenda transforms the life-death dialectic within the framework of recognizing death as a necessary condition of humans having a destiny. For example, rather than posit a life-death contrast, one may think of these two contrasting terms as necessary conditions for the emergence and development of individual destinies. Hence, for Kolenda, the idea of destiny carries with it the notion of wholeness. And rather than assume, as many still do, that the ultimate destiny of every human is to die, Kolenda refocuses attention on the fact that the destiny of a human also includes all that an individual has experienced and will experience in life.[43] This insight encourages humans to immerse ourselves in our finite naturalness, and to see that death is not an experience or phenomenon of oppositional otherness.

Human life as a totality is captured more correctly when one understands death as a necessary condition of having a destiny, or of being a whole. Another aspect of this theory is that the universe acquires distinct meaning through human destinies. There is a sense of the universe coming to be in a different way with the entrance of human destinies, for "every upsurge of consciousness through the birth of a person is a triumph [for the universe] and a privilege [for the emerging individual]."[44] With Kolenda, I affirm each human birth as a glorious event and the starting point of yet another spectacular phenomenon that helps transform the complex cosmos into an even more dynamic, dramatic world. Consequently, when we value human life and find it interesting or even precious, we are aware of the countless opportunities to light up the cosmos with the varied and seemingly inexhaustible projects of human individuals and cultures.[45] Granted, I am not suggesting that the knowable universe is enlivened *only* through human activity, as that idea retains too much hubris of traditional humanism, devaluing the emergence of other forms of animal and plant life, as well as their concomitant levels of sentience, conscious awareness, and valuing. Rather, I am much more compelled by the subtler notion that humans are individual and collective destinies engaging an appreciable world. These ideas help us view humans as enmeshed natural organisms that add a particular dimension of value to an already vital, expansive, and valuable universe.

With each human being viewed as a unique destiny, I impress on myself the notion that when I meet a person, I am encountering another center of value. Moreover, assuming that the values realized in human lives are the highest values we know of, then even seeing a stranger on the street puts me face to face with a manifestation of myriad cosmic meanings—both potential and actual. Personal interaction with another becomes an intersection of entangled worlds, as well as our participation in the drama of life in its evolving transformation and expansion. As each human destiny is a recipient of modes of acting, speaking, and thinking that reach deep into the past, one may become aware of concrete links with other members of the human race who have prepared the emergence of our destinies by living out theirs. Seen from this perspective, my individual experience is a continuation, a development of a larger project, and I can view myself as its partial actualization. More eloquently, I am an important aspect of the evolving, unfolding, entangled stream of life.

SACRED HUMANITY, HUMAN DESTINIES, AND AFRICAN AMERICAN RELIGIOSITY

Both within the specifics of African American historical and cultural realities and on a more general level, the sacred humanity I have described in this essay accentuates the liberationist theme in African American religiosity as the uncontestable recognition of our longing to take our highest ideals or values seriously: the irrefutable, essential value of our humanity. Additionally, the theory of human destinies outlined above lends itself to a deepening understanding of this religiosity grounded in an expanded view of our human nature. As human destinies, sacred humans seek compensation, rounding out the actual with the ideal, according to our many talents, insights, and abilities. This experience of religiosity (both individually and culturally) is inextricably connected to how African Americans *become* our humanity—how we literally become human from the perspective of a dominant culture that has denied us that right. The very presence of this longing in African American culture attests to an awareness or sense of distance between what we are and what white supremacist racial views suggest we are, or between the world as it is and as it could be.

As I interpret these ideas within the context of African American religious life and culture, I consider such awareness and desire as instances of existential courage. African Americans have been compelled to seek realms of possibility. This ongoing task—of necessarily positing and celebrating African Americans as humans—has been one way, among many, for African American visionaries to contest impoverished models of black animality that justified slavery and helped fabricate the notion of white supremacy in the United States. Within the context of blacks' lived experiences in this nation, these insights reveal the significance of the liberationist theme in black religiosity: blacks' tenacious refusal to reduce our ennobled humanity to mere brute existence (or rather to determinist forces and mechanistic explanations of cause/effect). Both within and beyond the specifics of black culture, the task of becoming our humanity involves the recognition that participation in human affairs helps transform isolated individual destinies—we become enriched by diverse allegiances, identifications, and loyalties.

Within the context of U.S. racial discourse that has persistently placed blacks and other subjectivities outside the circle of humanity, the notion of *sacred* humanity symbolizes our rejection of conceiving our humanity solely as an individualistic phenomenon—a rigorously communal ontology is implied. As a materialist critique suggests, our humanity is inescapably entangled in other natural processes of becoming, such that in embracing our sacred humanity we are acknowledging that the "more-than-human" constitutes the human as such. Moreover, any inkling of white supremacy, or sense of cultural superiority of any kind, is antithetical to this naturalistic view; these eschewed cultural constructions are forced impositions on the wholeness of natural interrelatedness and deep homology that evolution has wrought. Sacred humanity conveys connectivity with all that is—with oneself, one's family, the larger human community, local and global ecosystems, and the universe. Religiously, this implies love, and love implies concern for the well-being of the beloved. The concept of sacred humanity also reinforces perennial, expansive perspectives from the wisdom traditions that adamantly promote kindness, empathy, and compassion for all natural processes, including human ones. With the capacity to influence each other and other natural processes, humans have a responsibility to act in ways that promote the flourishing of all life, and to urge other humans who may be less inclined to value our interconnectedness to do the same.

A major value of this model of African American religious naturalism is its emphasizing humans as relational natural organisms—we are part of the evolving universe. Those values that we confer on events are more or less the universe appreciating dimensions of itself. Granted, this process is not merely narcissistic introspection, isolating any one of us from the contexts in which we live our destinies. All manifestations of truth and any sense of the real emerge from the concrete actions of entangled, aspiring, stubborn individuals: we, right now, where we are. It is akin to discovering worlds of possibility beyond the sterile fear of nonexistence, beyond enforced solitariness founded on illusions of separateness and universal abstractions. In *Enfleshing Freedom: Body, Race, and Being*, M. Shawn Copeland suggests that being human is neither hypothetical nor abstract. Rather, while advancing a theological anthropology based on critical analysis of the body—physical and social, enfleshed, historical, and concrete—and in particular the bodies of black women, Copeland points to an expanded view of selfhood that is concrete, visceral, and embodied in everyday experience and relationships—all determinants of who we are.[46] Ascertaining this fuller sense of humanity evokes the speech of Miranda in *The Tempest* as she envisioned an enchanted new world of possibility: "O wonder! How many goodly creatures are there here! How beauteous mankind is! O brave new world! That has such people in it!"[47] In light of the emerging theory I advance here, we might now say, "Oh what wondrous worlds that honor such entangled creatureliness—our becoming human!" As I mentioned earlier, human life as a totality is captured more correctly when one understands death as a necessary condition of having a destiny, or of being a whole. Thus human sacrality is inescapably tied to our sense of finitude, which is only possible and truthful when we rest confidently in our natural, material bodies. The concept of sacred humanity and the African American religious naturalism from which it emerges honor this truth. They remind us that the holiness of life is recognizing that each moment of existence is an opportunity to honor our stubborn materiality and entangled finitude.

NOTES

1. See, for example, Karen Barad, *Meeting the Universe Halfway: Quantum Physics and the Entanglement of Matter and Meaning* (Durham, N.C.: Duke University Press, 2007); Jane Bennett, *Vibrant Matter: A Political Ecology of Things* (Durham, N.C.: Duke University Press, 2010); Diana Code, *New Materialisms: Ontology, Agency, and Politics* (Durham, N.C.: Duke University Press, 2010); Rick Dolphijn and Iris van der Tuin, *New Materialism: Interviews & Cartographies* (Ann Arbor: MPublishing—University of Michigan Library, 2012); Rosi Braidotti, *Metamorphoses: Towards a Materialist Theory of Becoming* (Cambridge: Polity Press 2002); Elizabeth Wilson, *Psychosomatic: Feminism and the Neurological Body* (Durham, N.C.: Duke University Press, 2004); Elizabeth Wilson, "Gut Feminism," in *Differences: A Journal of Feminist Cultural Studies*, 15, no. 3 (2004): 66–94; Cary Wolf, ed., *Zoontologies: The Question of the Animal* (Minneapolis: University of Minnesota Press, 2003); Cary Wolf, *Animal Rites: American Culture, the Discourse of Species, and Posthumanist Theory* (Chicago: University of Chicago Press, 2003).
2. Carol Wayne White, *The Legacy of Anne Conway: Reverberations of a Mystical Naturalism* (Albany: State University of New York Press, 2009).
3. Elements of the concept of sacred humanity in this essay draw on my fuller discussion found in *Black Lives and Sacred Humanity: Toward an African American Religious Naturalism* (New York: Fordham University Press, 2016), 28–46.
4. Religious naturalism offers a variety of perspectives and ideas that often depart from traditional forms of Western religion, specifically in rejecting, reinterpreting, or reconceptualizing traditional concepts of God (or supernatural theism), and in using current developments in science to primarily conceptualize humanity and our ethical orientations, aesthetic appreciations, and religious values. Representative works that have shaped my thinking include Loyal Rue, *Religion Is Not about God* (New Brunswick, N.J.: Rutgers University Press, 2006); Donald Crosby, *Living with Ambiguity: Religious Naturalism and the Menace of Evil* (Albany: State University of New York, 2009); Ursula Goodenough, *The Sacred Depths of Nature* (New York: Oxford University Press, 2000); Mordecai Kaplan, *The Meaning of God in Modern Jewish Religion* (Detroit: Wayne State University Press, 1995); Chet Raymo, *When God Is Gone, Everything Is Holy* (Notre Dame, Ind.: Sorin Books, 2008).
5. Religious naturalism is a way of describing a worldview that is scientifically credible and emotionally satisfying. As such, it is not common naturalism if that is understood as cold, heartless reductionism. Some religious naturalists have targeted a type of reductionistic naturalism sometimes expressed by Richard Dawkins and Peter Atkins. Accordingly, these figures assume reality to be self-explanatory, and no further explanation is needed. For religious naturalists, however, the universe is not reducible to the categories of analysis used to explain it. The natural explanations within the framework of science are not decisive with respect to the

explanation of that framework. Nor is my approach another form of natural religion, or a form of apologetics using scientific and natural investigations to support religious claims or doctrines in traditional theology. In this essay, I follow the trajectory of religious naturalism within the science and religion paradigm that is best associated with the works of Ursula Goodenough, Donald Crosby, and Loyal Rue. I discuss this viewpoint in much more detail later in the essay.

6. Mark C. Taylor, *Erring/A/Theology* (Chicago: University of Chicago Press, 1986), 4.

7. This list is not fully representative of the complexity and diversity of liberal theology; it does, however, list some of the major post-Enlightenment Protestant scholars who were concerned with the vitality and necessity of religious interpretations in an increasingly secularized society. For representative works, see Paul Tillich, *Systematic Theology*, vol. 1 (Chicago: University of Chicago Press, 1967); Paul Tillich, *Dynamics of Faith* (New York: Harper & Brothers, 1958); Emil Brunner, *Man in Revolt* (Louisville, Ky.: Westminster John Knox Press, 1979); Reinhold Niebuhr, *Moral Man and Immoral Society* (Louisville, Ky.: Westminster John Knox Press, 2013); John Macquarrie, *In Search of Humanity: A Theological and Philosophical Approach* (New York: Crossroad, 1983).

8. With this insight one is reminded of the rise of feminist and liberation theologies or religious critiques (African American, Asian, Latina/o, African, Queer, etc.) where marginalized histories are taken into account. For further discussion, see Claudia Schippert, "Implications of Queer Theory for the Study of Religion and Gender: Entering the Third Decade," in *Religion and Gender* 1, no. 1 (2011): 66–84. Another helpful, albeit earlier, articulation is given by Gordon Kaufman, who has expressed the necessity of acknowledging pluralism and dialogue in critical theology in "Critical Theology as a University Discipline," in *Theology and the University: Essays in Honor of John B. Cobb Jr.*, ed. David Ray Griffin and Joseph C. Hough Jr. (Albany: State University of New York Press, 1991), 35–50.

9. Emmanuel Eze, ed., *Race and the Enlightenment* (Malden, Mass.: Blackwell, 1996), 1–9. See also David Theo Goldberg, *Racist Culture* (Cambridge, Mass.: Blackwell, 1993). For more information on the notion of certain "subjugated knowledges" becoming lost amid the triumphant march of European reasoning and progress, see Michel Foucault, *Power/Knowledge: Selected Interviews and Other Writings, 1972–77* (New York: Pantheon Books, 1980), 81–82.

10. Jean-Paul Sartre, Preface to Frantz Fanon, *Wretched of the Earth* (New York: Grove Press, 1961), vii–lxii. Insights emerging from such developments as cultural studies, feminism, and postcolonial thought show that discussions of liberal (and influential) political philosophies, such as those of Jefferson, Hume, or Locke, must also take into account the extent to which their doctrines were implicated in acts of racial supremacy, sexism, exploitation of other nationalities, colonialism, and slavery. For further readings, see Sandra Harding, ed., *The "Racial" Economy of Science* (Bloomington: Indiana University Press, 1993); William Tucker, *The Science and Politics of Racial Research* (Urbana: University of Illinois Press, 1994); Elazar Parkan, *The Retreat of Scientific Racism: Changing Concepts of Race in Britain and the United States between the World Wars* (Cambridge: Cambridge University Press, 1992); Wil-

liam B. Cohen, *The French Encounter with Africans: White Response to Blacks, 1530–1880* (Bloomington: Indiana University Press, 1980); Katherine Faull, *Anthropology and the German Enlightenment: Perspectives on Humanity* (Lewisburg: Bucknell University Press; London: Associated University Presses, 1995); Henry Louis Gates Jr. and Kwame Anthony Appiah, eds., *"Race," Writing and Difference* (Chicago: University of Chicago Press, 1985); George Fredrickson, *The Black Image in the White Mind: The Debate on Afro-America Character and Destiny, 1817–1914* (Middletown, Conn.: Wesleyan University Press, 1987); Stephen J. Gould, *The Mismeasure of Man* (New York: W. W. Norton, 1981); Nancy Stepan, *The Idea of Race in Science: Great Britain, 1800–1960* (London: Macmillan, 1982); Leon Poliakov, *The Aryan Myth: A History of Racist and Nationalist Ideas in Europe* (New York: Basic Books, 1974).

11. Recent studies in race theory show that even though such terms as blackness and whiteness are constructions that are projected, they take on certain meanings that apply to certain groups of people in such a way that makes it difficult not to think of those people without certain affectively charged associations. Thus the blackness and whiteness of individuals and groups become regarded by a racist culture, which takes its associations too seriously, as their essential features—as, in fact, material features of their being. For further readings, see Kwame Anthony Appiah, "Racisms," in *Anatomy of Racism*, ed. David Theo Goldberg (Minneapolis: University of Minnesota Press, 1990), 3–17. For further readings in critical race theory, see Richard Delgado and Jean Stefancic, *Critical Race Theory: An Introduction* (New York: New York University Press, 2011); Kimberlé Crenshaw and Garry Peller, *Critical Race Theory: The Key Writings That Formed the Movement* (New York: New Press, 1995).

12. Anna Julia Cooper, *A Voice from the South* (New York: Oxford University Press, 1988), 118.

13. Ibid., 164.

14. Ibid., 151; 149.

15. W. E. B. Du Bois, *The Souls of Black Folk*, ed. Henry Louis Gates Jr. and Teri Hume Oliver, Norton Critical Edition (New York: W. W. Norton, 1999).

16. Du Bois, "Of Our Spiritual Strivings," in *The Souls of Black Folk*, 370.

17. James Baldwin, *The Cross of Redemption: Uncollected Writings* (New York: Vintage, 2010); *Baldwin: Collected Essays* (New York: Library of America, 1998). Hereafter cited as *BCE*.

18. See Will Walker, "After *The Fire Next Time*: James Baldwin's Postconsensus Double Bind," in *Is It Nation Time? Contemporary Essays in Black Power and Black Nationalism*, ed. Eddie S. Glaude (Chicago: University of Chicago Press, 2002), 229.

19. Baldwin, "The Fire Next Time," *BCE*, 294.

20. Ibid., 334.

21. See Jerome Stone, *Religious Naturalism Today: The Rebirth of a Forgotten Alternative* (Albany: State University of New York Press, 2008), chap. 1.

22. See, for example, Goodenough, *Sacred Depths of Nature*; Rue, *Religion Is Not about God*; Crosby, *Living with Ambiguity*.

23. Ursula Goodenough and Terrence Deacon, "From Biology to Consciousness to Morality," *Zygon: Journal of Religion and Science* 38, no. 4 (2003): 801–19.

24. Rue, *Religion Is Not about God*, 25.
25. Stephen J. Gould, *Wonderful Life: The Burgess Shale and the Nature of History* (London: Hutchinson Radious, 1989); see also Terrence Deacon, "The Hierarchic Logic of Emergence: Untangling the Interdependence of Evolution and Self-Organization," in *Evolution and Learning: The Baldwin Effect Reconsidered*, ed. B. Weber and D. Depen (Cambridge, Mass.: MIT Press, 2003), 273–308; Terrence Deacon, "Emergence: The Hole at the Wheel's Hub," in *Re-Emergence of Emergence*, ed. Philip Clayton and Paul Davies (Oxford: Oxford University Press, 2006).
26. Goodenough, *Sacred Depths of Nature*, 63.
27. Ibid., 40.
28. Ibid., 42, 44.
29. Ibid., 66.
30. Ibid., 71.
31. Ibid., 64.
32. White, *Black Lives and Sacred Humanity*, 33.
33. Granted, this is not your typical approach to the sacred, which admittedly is a complex word that has been used for a wide range of phenomena: places, times, persons, events, and deities. Traditionally, when people designate something as sacred, they view the thing in question as "other than ordinary." Thus, in the broadest sense of the term, the sacred has been used by scholars, especially those sympathetic to the work of Mircea Eliade, to convey the "extraordinary." See Mircea Eliade, *The Sacred and the Profane* (New York: Harcourt Brace Jovanovich, 1987).
34. Goodenough, *Sacred Depths of Nature*, 73.
35. The development of various theories of evolution from Lamarck to Darwin was crucial to the creation of racism in the nineteenth century. For further reading, see Peter Reill, "Anti-Mechanism, Vitalism, and Their Political Implications in Late Enlightened Scientific Thought," *Francia* 16, no. 2 (1989): 195–212.
36. Rue, *Religion Is Not about God*, 75. Rue also observes: "The meaning of human life should be expressed in terms of how our particular species pursues the ultimate telos of reproductive fitness. Like every other species, we seek the ultimate biological goal according to our peculiar nature. That is, by pursuing the many teloi that are internal to our behavior mediation systems, whether these teloi are built into the system by genetic means or incorporated into them by symbolic means. For humans there are many immediate teloi, including the biological goals inherent in drive systems, the psychological goals implicit in our emotional and cognitive system, and the social goals we imbibe through our symbolic systems. Human life is about whatever these goals are about" (75).
37. Carol Albright and James Ashbrooke, *The Humanizing Brain* (Cleveland: Pilgrim Press, 1997), 34.
38. Terrence W. Deacon, *The Symbolic Species: The Co-Evolution of Language and the Brain* (New York: W. W. Norton, 1997).
39. Ibid., 36; 43–46.
40. Michael Arbib, *The Metaphorical Brain 2; Neural Networks and Beyond* (New York: John Wiley, 1989); Joseph LeDoux, *The Emotional Brain: The Mysterious Underpin-

nings of Emotional Life (New York: Simon and Schuster, 1996); Leslie A. Brothers, *Friday's Footprint: How Society Shapes the Human Mind* (New York: Oxford University Press, 1997).
41. Terrence Deacon and Ursula Goodenough, "The Sacred Emergence of Nature," in *The Oxford Handbook of Religion and Science*, ed. Philip Clayton (New York: Oxford University Press, 2008), 860.
42. Konstantin Kolenda, *Religion without God* (Buffalo, N.Y.: Prometheus Books, 1976). Kolenda rejects the life-death dialectic that has become an influential conception of human reality for many Westerners. In this sense, his perspective is not unlike that of Heidegger, Camus, Sartre, and other major twentieth-century existential figures, whose philosophical and phenomenological reasoning offer important and serious appraisals of death. I build on his ideas in this chapter.
43. Ibid., 21.
44. Ibid., 26.
45. Ibid., 26–27.
46. Shawn Copeland, *Enfleshing Freedom: Body, Race, and Being* (New York: Fortress Press, 2009).
47. William Shakespeare, *The Tempest* (New Haven, Conn.: Yale University Press, 2006), 125.

ꙮ Grace in Intra-action: Complementarity and the Noncircular Gift

TERRA S. ROWE

And we don't have to do anything to bring about this future. All we have to do is nothing. . . . All we have to do is not react as if this is a full-blown crisis.
—NAOMI KLEIN, This Changes Everything

We are given to understand . . . that we can do nothing of ourselves. . . . [This is] the grace of God.
—MARTIN LUTHER, On the Bondage of the Will

In August 2014 an op-ed on faith and climate change appeared in the *New York Times*. Frustrated with the prevalent antagonism between science and faith in climate change literature, Kristin Dombek describes a bridge she has constructed between her old Christian beliefs and her current climate concerns. Christianity—specifically the Lutheran-Reformed Heidelberg Catechism—taught her, she explains, how to "belong, body and soul," to something bigger than herself. She compares overlapping journeys—one, toward a new faith in Darwin, the other, toward a way to cope with paralyzing, climate change–induced anxiety—to learning to float in the ocean. Just as one learns buoyancy, she had to learn to trust that her life was "linked to something bigger: that [she] belong[ed], body and soul, to a larger story for which [she was] responsible."[1] Shifting from the poetic to the practical, she concludes that to "stay awake, active, useful is a matter of feeling as much as knowing." In other words, we must learn to feel in our bodies the world's "vast and humbling contingency."[2]

Dombek points out the need for major shifts in our cosmologies and materialisms. The inconvenient truth is that climate change only lends further evidence to the theory that we live in what Karen Barad calls an "intra-active" universe. For better or worse, we humans cannot extract ourselves from the natural world we seek to know and, too often, control or use—climate change presses this reality on us. Where, at least since Descartes and Newton, science, religion, and philosophy have conspired to produce an illusion of human separability from the other-than-human world, the "new materialisms" emphasize human inseparability from worldly entanglements. Even in the name of gaining scientifically reliable, objective, and verifiable knowledge of the world we cannot extricate ourselves from these ties. Karen Barad, for example, insists that "'we' are not outside observers of the world. Nor are we simply located at particular places *in* the world; rather, we are part *of* the world in its ongoing intra-activity."[3] If we are indeed "part of the nature we seek to understand,"[4] with no radical distinction between humanity and the other-than-human world, then what we humans do, the economic and cultural choices we make, cannot remain inconsequential for the larger systems and ecologies of the world. Such influence, furthermore, is not one-directional. The other-than-human world retains what Jane Bennett calls *"Thing Power"*: Materialities do not remain purely passive recipients of human agency but talk back in the form of pollution and climate change.[5]

I agree with Dombek that such shifts in our cosmologies and materialisms are a necessary part of feeling our way into a new mode of being with the world. More than a necessary coherence with our current best understandings of the world, Dombek insightfully suggests that such shifts will also motivate and sustain us as we face the challenges associated with mitigating and adapting to climate change. In this light, her reference to the Protestant Heidelberg Catechism is startling—mainly because important aspects of Reformation theology have generally been regarded, for good reason, as anything but motivational for combating climate change.[6]

ALL WE HAVE TO DO IS NOTHING: GRACE AND CLIMATE CHANGE

> The Puritans wanted to be men of the calling—we, on the other hand, must be. For when asceticism moved out of the monastic cells and into working life, and began to dominate innerworldly morality, it helped to build that mighty cosmos of the modern economic order (which is bound to the technical and economic conditions of mechanical and machine production). Today this mighty cosmos determines with overwhelming coercion, the style of life not only of those directly involved in business but of every individual who is born into this mechanism, and may well continue to do so until the day that the last ton of fossil fuel has been consumed.
>
> MAX WEBER, *The Protestant Ethic and the "Spirit" of Capitalism*

Max Weber famously identified a unique set of conditions within the Protestant tradition that contributed to a personally and socially brutal production drive. Since the publication of *The Protestant Ethic*[7] at the opening of the twentieth century, however, concerns about the unaddressed ecological implications of this drive have increased exponentially. Engagement with twentieth- and twenty-first-century gift theory demonstrates that Weber's thesis identifies only the tip of a religio-economic iceberg. Gift theory takes the discussion of a link between Protestantism and capitalism straight to the heart of the religious tradition—to the theology of grace.

The Reformation arguably took unprecedented steps to align divine grace with God's action over and against human passive receptivity.[8] Such a clear active/passive binary has been a continued locus of debate among Christian traditions. Starting in the late twentieth century, Protestant grace famously became a particular locus of feminist theological critique, which targeted the construction of purely receptive, feminine selves opposite an active, masculine deity.[9] The social and economic relevance of this theological debate becomes clear through the critical lens of gift theory.

As it turns out, grace in twenty-first-century U.S. economic practices and values is not an exclusively religious issue. The Protestant understanding of grace as the ideal kind of gift has so saturated U.S. society that an understanding of U.S. ecological and economic values remains incomplete without it. Even after all religious notations have been suppressed or forgotten, the fundamental logic of the gift remains distinctively Protestant. This has been the argument, anyway, of the Anglo-Catholic John Milbank. He contends that the Protestant Reformation introduced an entirely new gift ideal based on an unprecedented dualism: gift versus exchange. This dualism resulted in the prevailing assumption that the ideal form of the gift, as modeled after the divine gift, occurs free of exchange, return, reciprocity, debt, or obligation. Defining the gift as fundamentally "free" shifted it away from medieval and ancient ideals that were fully embedded in notions of exchange.[10]

Those both inside and outside the Protestant tradition (whether expressly or implicitly) describe grace as a pure, blissful receptivity of divine or creaturely gifts. As such, grace is conceptually opposed to the active reciprocity exchange. As evidence of the pervasive lure of the gift/exchange binary, Milbank points to the distinction Jacques Derrida famously insisted on, even as he deconstructed socio-anthropologist Marcel Mauss's theory of the gift. According to Derrida, Mauss never approaches the gift. He describes only exchange, since "for there to be a gift, there must be no reciprocity, return, exchange, countergift, or debt."[11] Such phrases make clear the permeating influence of the perception of "gift" as free and thus outside exchange. After all, Derrida, the Jewish Algerian–born, "rightly passing as an atheist,"[12] philosopher certainly was not intending to recapitulate a Protestant theological doctrine. However, Milbank suggests that this particular theological definition has so saturated our cultural and even philosophical definitions of gift that Derrida ends up merely repeating the Reformation era binary.[13]

The grammar checker on my word processor provides further evidence of a ubiquitous gift/exchange dualism: It gets huffy every time I type "free gift." The objection? My computer insists the phrase is a "redundant expression." The widespread influence of a Protestant theological doctrine might be rather quaint if it were not also tied to such damning consequences. The problem with this definition of gift—particularly from an ecological and climate-concerned perspective—is that, when defined as an exclusion of exchange or reciprocity, the "free" gift is not only free of obligation, but also of the intra-dependent relations that we now understand to make up the fabric of every level of existence. If the earth gives freely and unconditionally, nothing ties us to the world. We are free to do what we please, free of demands, free of responsibility, and free to separate ourselves and see the earth as something "other"—an expendable, usable "resource." From this perspective, it seems that grace is, like the modern separative subject, merely a convincing set of blinders allowing us to believe falsely that we exist essentially as sovereign islands. Accordingly, Milbank and others have persuasively demonstrated that the kind of gift prioritized marks the difference between individualist ontologies, on the one hand, and interdependent, relational, or "exchangist" ontologies, on the other.[14]

As images and ideals of substance, solidity, and the isolated self lose credibility, the complexity of a universe unfolding through a continual flow of relational exchanges captures the imagination. However, while I join the challenge of the free gift's tendencies toward separative individualism and commoditization, I also argue that we must reserve space for uncapitalized returns and even losses in our gift-investments. This is something that grace as a *noncircular* gift can resist.[15] Consequently, I argue that what are commonly held to be competing and incompatible concepts—unconditional grace and the exchangist (or intra-active) universe—are actually complementary disruptions of key economic practices contributing to climate change. This will, however, also initiate a transformation of grace so that it refuses a merely disruptive relationship to life-sustaining exchanges of the world, animating the motivation and courage needed to face the challenges of climate change.

GIFTING MATTER(S)

"Agential" (Barad) or "vibrant" (Bennett) materialisms assume a model of reality as constructed of or through various material agents. These materialisms consciously offer alternatives to perspectives that assume a radical distinction between human agents and nonhuman passive objects. Jane Bennett explains the leveling effect of this shift on our worldviews:

> I am a material configuration, the pigeons in the park are material compositions, the viruses, parasites, and heavy metals in my flesh and in pigeon flesh are materialities, as are neurochemicals, hurricane winds, E. coli, and the dust on the floor. Materiality is a rubric that tends to horizontalize the relations between humans, biota, and abiota. It draws attention sideways, away from an ontologically ranked Great Chain of Being and toward a greater appreciation of the complex entanglements of humans and nonhumans.[16]

When we see reality as constructed through the engagements of a variety of material agents, gift theory becomes an insightful means of drawing connections among economic, ecological, and religious practices.

For example, "The Preoriginal Gift—and Our Response to It," an essay by the ecotheologian Anne Primavesi, coaxes gift theory from its pervasive anthropocentrism toward an ecological gift.[17] Here, life itself—not just social relations and their gift "things"—emerges through a continual constitutive flow of gift exchanges. For example, Primavesi describes the "essential contributions to present gift events made by 'more than' those participating in them now. They include antecedent generations of living beings: all those who, by their lives, their labor, their deaths, their vision, and their patient endeavors have made such events presently possible."[18] Furthermore, when the relation between materialities is constituted by the exchange of "gifts," an absolute distinction between economy and ecology can no longer be drawn along the same boundaries that would distinguish humanity from the other-than-human, or culture from nature.[19]

Primavesi's eco-gift intuits something of Barad's intra-action; the fact that matter doesn't just interact at a distance might be interpreted as a kind of gift exchange. Employing the terminology without citing the (gift) theory Barad writes, "'Individuals' are infinitely indebted to all others, where indebtedness is about not a debt that follows or results from a transaction but, rather, a debt that is the condition of possibility of giving/receiving."[20] Just as Primavesi broadens the framework of gift theory beyond the social realm to all life, intra-active gift exchange would again expand to material reality. Barad's distinction between *inter*action and *intra*-action marks a shift from Cartesian-Newtonian assumptions of human alienability from the world to the interdependence and intra-constitution of all things. In terms of gift theory, we might describe intra-action as a process of continual reciprocal relations of exchange with the other-than-human world, emphasizing that we all fully rely on an exchange of gifts with others. No original starting place of pure human selfhood or materiality remains outside of, or prior to, these exchanges; they are constitutive of human selves, agency, cognition, language—everything traditionally thought to distinguish us from the nonhuman world.

Interconnectedness, or rather intra-connectedness, is also a starting place that has made us vulnerable to the emergence of realities such as climate change. Our inability to avoid consequences that are not merely of our own making presses the reality of an intra-active universe upon us. We have all participated in a climate-poisoning economy (of course, some with more responsibility and knowledge than others) and because we intra-act, exchanging gifts from within nature and not some transcendent exterior, we now face this crisis together. If we see reality as constituted through continual exchanges, then an engagement with gift theory should highlight profound eco/nomic implications of such new materialisms while initiating significant reverberations within Christian theologies of grace.

AN "EXCHANGIST" UNIVERSE

Just as Anne Primavesi's eco-gift stands out for its explicit defiance of anthropocentric gift exchange, John Milbank's gift-exchange theology stands out for its interrelated condemnation of capitalism and Protestant grace. Like many of the Reformers, Milbank places himself within the Augustinian tradition. However, in his interpretation of Augustine he emphasizes what Protestant theology arguably left behind: a Trinitarian relational gift exchange as a basis of and necessary precursor to any divine unilateral gift. "For Augustine," Milbank explains, "the *donum* that is the Holy Spirit is not only a free one-way gift (though it is also that), but in addition the realization of a perpetual exchange between the Father and the Son."[21] Creation participates in this relational exchange, which becomes the basis for Milbank's ontology of gift exchange. The political and economic stakes for such a relationship between gift and exchange soon become clear. Milbank insists that this Trinitarian exchangist reality fundamentally contradicts capitalism's foundation in individualism and commodification. Since these exchanges are also participation in Trinitarian gift exchange, capitalism is, furthermore, "heresy."[22]

Milbank does not deny that the possibility for an "offering without return" always remains, but he insists that such a gift is always supplemental to a continual gift exchange.[23] Christianity, he argues, has successfully integrated the unilateral (or free) and the reciprocal gifts. Milbank therefore locates his concern with redefining the exemplary divine gift. He argues that the epitome of the good (divine) gift is reciprocity, since the gift is never good "outside the hope for a redemptive return to the self."[24] When we do give sacrificially (that is, without a return), as is sometimes necessary in the fallen world, a truly

Christian theology must insist that "we are always receiving back as ever different a true, abundant life (this is the Gospel)."[25]

In reemphasizing gift exchange rather than the "free gift," Milbank relies on Marcel Mauss's 1924 work, *The Gift: The Form and Reason for Exchange in Archaic Societies*. Since this work established Mauss's place as a pioneer of socio-anthropology, its influence remains difficult to overstate. In current debates about economic, social, and ecological issues, many theorists and theologians (including Milbank) hold up Mauss's gift economy as an exemplary alternative to capitalist commodity economies. *The Gift* proves invaluable for Milbank as Mauss uncovers a critical link between ideal models of the gift, separative individualism, and the utilitarianism that drives human lives and decisions in capitalist society.[26]

According to Mauss, the idealized version of the gift as "free" has not always been the case. Contrary to common belief at his time, Mauss's research shows that economies in "archaic" or "primitive" societies were not characterized by barter, but by gift exchange. Where barter is understood "as a highly rational, purely material, and economic process . . . which smoothly gives way to the higher convenience of money," Mauss demonstrates that the impetus of exchange in these economies was more than utilitarian.[27] Consequently, he concludes, forms and reasons for exchange other than capitalist utilitarianism are possible. Furthermore, since such economies are more ancient, they remain more "natural" than capitalism.[28]

Mauss demonstrates that in gift-giving societies, exchange was not simply a transfer of commodities from one person to another.[29] Gift exchange connected people—even over generations—by binding them to each other through the moral principle that a gift given demands the return of a gift after a certain amount of time.[30] More than mere objects, property, or goods, these societies exchanged "acts of politeness," giving and returning "respects" and "courtesies" in such a way that gift exchange was not merely a guiding economic principle but constituted *the society* itself.[31] "By giving," Mauss explains, "one is giving oneself, and if one gives oneself, it is because one 'owes' oneself—one's person and one's goods—to others."[32] Such a society consists of lives "mingled together," since this is, he says, "precisely what contract and exchange are."[33] Inasmuch as a gift is "free" of exchange and "mutual ties," it therefore "does nothing to enhance solidarity," and it is a "contradiction."[34]

Milbank explains that in demonstrating such "intermingling of subject and object" (whether between person and neighbor, person and thing, or giver and gift and so on), Mauss essentially "wrote a meditation against Descartes."[35] Rather than separable individuals or an atomistic ontology, premodern views envisioned society as an organism "on which all depend, whose destiny is tied to the exchange of these particular sacred things."[36] Extending Mauss's argument that there is no such thing as a "free" gift, Milbank demonstrates that without a primary and underlying assumption of continual relational exchange, anything that we might call a "gift" turns out to be anything but. Inviting a consideration of the relational implications of the free gift, Milbank notes that "it seems not insignificant that within romantic love an asymmetry of giving, where only one partner gives presents and favors, suggests not at all freedom and gratuitousness, but rather an obsessive admiration that subsists only at a willfully melancholic distance, or still worse a purchase of sexual satisfaction, and in either case the slide of desire towards one-sided private possession."[37] When reality can be framed through a subject/object dualism, then corresponding gifting practices emerge, resulting in relationships of control, possession, and utilitarianism.

Milbank portrays the modern gift as "strictly formalist and unilateral . . . as non-compulsory for donor or donee, as not expecting a return, and as indifferent to its own content."[38] Beyond the Protestant theological tradition, he also sees this model of gift working in modern ethics. By idealizing altruism and self-sacrifice, modern ethics assumes that a gift (in the form of an ethical gesture) is best given without expectation of a return. Milbank's condemnation of such "other-oriented ethics" aligns with previously mentioned feminist critiques of the Protestant theological tradition as valorizing self-emptying. Furthermore, he positions Derrida's ethics alongside the Protestant gift since Derrida too refuses to expect a return and thus can only envision the ethical act as self-annihilating.[39] Despite Derrida's moves against the Cartesian selfsame, self-aware subject, Milbank argues that his ethic remains fundamentally separative and individualistic since it remains tied to the alienable gift.

Beyond the personal individualism of the free gift, Mauss's work also demonstrates its effects on human perception of other-than-human "things." The gift/exchange dualism produces an active/passive binary that enables the commodification of objects. A capitalist economy trades in commodities—objects with utilitarian value only—but a gift economy does not see gifts as merely passive objects. Even as they are given, gifts remain inalienable from the giver because they are (in what Mauss calls a spiritual or magical way) imbued with the personhood of the giver. A gift given cannot be conceptualized as a mere object because a part of the giver is given with the gift. The gift remains "inalienable" since it must return to the giver in the form of a counter-gift. Such an exchange creates a social bond because it drives the principle that once given, a gift must be returned; it guarantees not merely the return of an object but a return of relation. As a consequence of the shift from gift economy to commodity economy, what were once gifts imbued with the personality of the giver are now just passive objects awaiting human inscription of meaning.

When we assume that a gift, by definition, must be free of the obligation to reciprocate, we also assume an ontology of separability among exchange partners and their gifts. The free, alienable, passive object shuttles between separative parties, leaving no connecting ties of relationship over time. Such alienable gifts make capitalist purchasing practices possible. I can go to a grocery store for milk with no prior connection to the farmers, their cows, or the grocers. In fact, there is very little chance that I will ever even have a face-to-face encounter, let alone a conversation, with the owner of the store. This kind of separability of people from one another and their material "things" is the legacy of the alienable gift. Ultimately, Milbank desires an articulation of an ideal Christian gift that does not suppress but transforms our "given social nature which is exchangist."[40] In Milbank's and Mauss's exchangist universe, no legitimate space remains for a gift that does not circle back to the giver, since any such gift would assume and institute an ontology based on separation that results in individualism and commodification.

DERRIDA AND A PLACE FOR GRACE?

Considering a climate-concerned desire to move away from capitalist separative individualism and commodification of matter, the condemnation of the free gift becomes compelling. Yet I remain convinced of the importance of unconditioned grace within such an intra-active universe. Or in Derridean terms, I believe space must be reserved for the gift that disseminates rather than returning to its origin—the giver. Without such a space for grace, the intra-active universe would be in danger of falling into a different capitalistic pitfall. Where Protestantism without participation in an exchangist universe falls into separative individualism and commodification, the exchangist universe that cannot give without the certainty of return is liable to fall into growth-dependent economics.

While Mauss and Milbank emphasize the power of gift exchange to create ties that bind in relationship, Kathryn Tanner points out that these gifts are not as different from capitalizing investments as Milbank would lead us to believe.[41] With Tanner, I would argue that Milbank fails to take seriously the counter-capitalist potential of the unconditioned gift. I'd add that as Milbank rails against the heresy of capitalism and the "nihilism" of Derrida's other-oriented ethic, he fails to take seriously Derrida's countercapitalist gift argument. Rather than leading toward nihilistic individualism, Derrida's insights, I believe, may be an important complement to the strengths and vulnerabilities of an exchangist universe in the form of a disruptive responsibility from within.[42]

Mauss's insight that the basis of society emerges through gift exchange has opened new avenues of thought for generations of theorists and theologians after him. Mauss's work piques Derrida's deconstructive interest by means of its overlapping foundational themes: time, being, and gift. These three themes, Derrida notes, share a particular economy of symbolic exchanges. In Western metaphysics and in common sense, time, being, meaning, and economy are figured—symbolized—as a circle; their end goal predictably emerges as a return to their origin. Derrida describes this as an Odyssean economy beginning at home, journeying away from home, and struggling to return to one's home origin.[43]

From the perspective of Derrida's effort to unsettle every "restricted" economy, or circle of the same, Mauss's "gift" does not appear to be anything of the sort. Derrida laments that for a text titled *The Gift*, the eponymous event can be found nowhere; everything Mauss wrote concerned economy and exchange. Reciprocity, return, exchange, and debt all turn a gift into poison—the fitting translation of the German *Gift*. Mauss's gift exchange "puts the other in debt [appearing] to poison the relationship, so that 'giving amounts to hurting, to doing harm.'"[44] Even a verbal expression of gratitude turns a gift into exchange, since the recognition of a gift acknowledges a good deed and thus reciprocates the gift. Mauss insists on a definition of gift that can be given only with interest, with the expectation of a return, but to Derrida a gift that returns to its origin erases the gift.

As I've explained above, despite vastly divergent ethical, theoretical, and religious concerns, Derrida's gift is commonly aligned with the traditional Protestant conception of gift. Indeed, some of Derrida's language does seem perplexingly close to theological proponents of absolute transcendence. In *Given Time*, for example, he writes, "But is not the gift, if there is any, also that which interrupts economy? That which, in suspending economic calculation, no longer gives rise to exchange? That which opens the circle so as to defy reciprocity or symmetry, the common measure, and so as to turn aside the return in view of the non-return?"[45] If "economy" is implied not just in the exchange of finances or goods but in every exchange—even those necessary for the founding of relationships and our ecosystems—then what kind of interruption does Derrida desire? A welcome intrusion? If it is an intervention by something wholly other, does the gift, as wholly other, transcend the world and all exchanges that make life possible in the world? "Not that it remains foreign to the circle," Derrida writes, "but it must *keep* a relation of foreignness to the circle."[46] The gift is not foreign to the circle and all those economies that make up our human and other-than-human worlds. Even in the interruption of economy, we must "give economy its chance."[47]

Where Derrida may give comfort with one phrase, he seems to undermine it with the next. He continues by noting that the gift "must *keep* a relation of foreignness to the circle, a *relation without relation* of familiar foreignness.

It is perhaps in this sense that the gift is the impossible."[48] But what kind of bond is a relation without relation? If the gift is related without being related, must it enter the circle by force? For theologians in particular, Derrida's phrasing, while entirely coincidental, may carry extra baggage. The theologian Karl Barth's absolutely transcendent God of wholly otherness also retained some necessary relation to the world of economy and exchange: "In the resurrection the new world of the Holy Spirit touches the old world of the flesh. But it *touches* it as a tangent to a circle, which is, *without touching it*."[49] Analogously, does Derrida's gift touch the circle of worldly exchanges only as a tangent touches a circle: that is, without *really* touching it? How does a touch without touch or relation without relation *materialize*?

Barth's emphasis on God's absolute transcendence may have been a powerful critique of Hitler's seizure of Christian themes for ungodly aims, but the Reformed theologian is also not known for a particularly eco- or feminist-friendly theology. Given such seeming congruence between Derrida's gift of "relation without relation" and Barth's divine gift of "touch without touch," the ecofeminist issues at stake here are worth stating explicitly: For the sake of a desire for something wholly other, does Derrida himself succumb to a *phallogocentrism* wherein the gift inseminates the world of economic and ecological exchange from the outside? Gregory Walter explains the eco-concerns from another theologically weighted angle: "The gift that has no relationship to reciprocity or what is already given cannot interact with what comes before the gift, the field into which the gift is given, in any other way than to erase it, to trump it, to completely overcome and end that economy."[50] This, Walter adds, is exactly the traditional problem Protestant theologies such as Barth's have had with nature and grace. Is it also the case with Derrida's gift?

Further into *Given Time* Derrida clarifies: The gift is not a transcendent exteriority. The gift (*if* there is any, he often adds) "does not lead to a simple, ineffable exteriority that would be transcendent and without relation."[51] How is it that the gift, if there is any, might interrupt economy while not remaining exterior to it? Derrida continually returns to Baudelaire's short story "Counterfeit Money" as he engages the question of the gift and economy. In Baudelaire's story, the narrator's friend makes a gift of a counterfeit coin to a beggar. Derrida notes the true ambiguity of this gift of alms. On the one hand, it may have the effect of simply circulating as a "true" coin, allowing the beggar relief from his poverty. In this case, since the counterfeit coin would have the same effect as a true coin, the difference between them would effectually be none. This result, however, is not guaranteed, for, on the other hand, its use could result in the unsuspecting beggar's arrest. In effect, the narrator's friend has interrupted or tainted a closed and trustworthy economy (where every symbol hits its mark) with a measure of indeterminacy, which is the mark of the gift itself. Alongside authentic money and inside this economy, the gifted counterfeit coin circulates, all the while creating the possibility for something new, unknown, and disruptive from within the circle of the same.

Counterfeit money masquerades as honest currency and thus participates in the economy while also disrupting it. Recall the above quotation, now read in a different light: "It must *keep* a relation of foreignness to the circle, a relation without relation of *familiar foreignness*." Retaining a relation of familiar foreignness, the counterfeit coin interrupts, though not from a purely exterior location. The gift of a counterfeit coin introduces some wiggle room, some play, into the economy of the same. This play leaves the circle just loose enough to allow for the possibility of something new and different disrupting the circle of the same. *If* the counterfeit coin does interrupt the circulation, it will not be from outside the economy. The effects of the counterfeit coin come forth from *within* the depths of that otherwise circumscribed circulation.[52] As in a 1960s segregated soda shop counter sit-in strategy, when protesters—like the counterfeit coin—inserted themselves into an oppressive system and with an active passivity waited for the system to collapse from its own vulnerabilities, such a *contre / faire* (against / doing) resists the drilling, pumping drive[53] of *laissez-faire* (let, leave to do) capitalization and production not over and against, but from within.[54]

Such counter-economic implications emerge even more clearly when the above texts are read alongside Derrida's 1978 essay "From Restricted to General Economy: A Hegelianism without Reserve," on economy and Hegel's system of absolute knowledge.[55] In this essay, too, Derrida's general economy can be seen as a strategy to create some space, some wiggle room for the play of difference, uncertainty, and even loss in a restricted economy. Derrida aligns Hegel's closed system/circle of absolute knowledge with the capitalist desire for efficiency and production. Mark C. Taylor writes that "Derrida, following Bataille, sees in Hegelianism a transparent translation of the foundational principles of a capitalistic market economy."[56] Similarly, Arkady Plotnitsky demonstrates the influence of capitalist economy on Hegel, stating that "Adam Smith's political economy was a major influence on Hegel during his work on *The Phenomenology of the Spirit*. . . . From *The Phenomenology* on, economic thematics never left the horizon of Hegel's thought, the emergence of which also coincides with the rise of economics as a science, which conjunction is, of course, hardly a coincidence."[57] Hegel overlays Smithian economics with a resurrection narrative—every loss is preserved. Similarly, capitalistic economy emerges as the ultimate Christian economy, working by a principle of resurrection. In short, Hegelian, material, and religious economies work "by securing a return on every investment."[58] Although Hegel claims to preserve the negative through double negation, Derrida contests that the negative is just what is lost. Nothing can be lost because no space remains outside of sublation/capitalization/resurrection.

Bataille labels Hegel's economy "restricted," since loss is impossible; the (circular) return is inescapable. Such a system cannot abide difference, uncertainty, or anything like "the impossible" that resists calculation.[59] The restricted economy is a swirling vortex, sucking everything into its singular aim; no room remains for loss or the unredeemed. Ideally, everything finds its meaning; the symbol always hits its mark; and every investment gains a return. Where no space remains for loss in this Odyssean economy, neither can there be room for an Abrahamic economy of traveling to new lands with no hope for a return home. Derrida the Jew (who can pass for an atheist) has no space in Hegel's capitalizing/resurrection economy. So he (with Bataille) makes room for some things to exceed or fall outside this restricted economy. In this essay, Derrida resists this particular kind of economy with a different (Abrahamic, "general") economy, not the anti-economic gift.[60] General economy emerges as the space for difference at the margins of a restricted economy.

By proposing a different kind of economy with Bataille, one which abides difference, loss, and the unexpected or the surprise of the impossible, Derrida is clearly not opposing the gift to all forms of exchange—just those modes of exchange that have no room for him, difference, or investments of uncertain returns. Where Derrida claims to desire the gift as if it were "outside" economy, it is primarily the symbolic, commercial, Hegelian, restricted economy he seems to have in mind. For Derrida, the gift, exemplified by the counterfeit coin, makes space within the constant flux of exchanges that constitute the world for the *différance* of the gift. Instead of a new binary between gift and economy, Derrida's gift retains a *supplementary* relation to economy. Rather than a fulfillment or addition to something more primary or original, the gift that seems to be secondary or an addition actually remains a vital part of any economy in that it opens the possibility of and desire for economy. For Derrida the possibility of desire depends on the play or supplementarity of signifiers. We do not desire what is fully present to us—that which is fully within our grasp. For desire to function, a certain play between absence and presence must be available to us.[61] Similarly, with the gift and economy, the economic circle is set in motion by the gift. "For finally," Derrida elucidates, "the overrunning of the circle by the gift, if there is any, does not lead to a simple, ineffable exteriority that would be transcendent and without relation. It is this exteriority that sets the circle going, it is this exteriority that puts the economy in motion. It is this exteriority that *engages* in the circle and makes it turn."[62] Desire for economy and reciprocity depends on a certain play or supplementarity between gift and economy.

The supplement provides an alternative relationality between the seemingly mutually exclusive noncircular gift and exchange. Milbank interprets the Reformation tradition as displacing a primary exchange with an originary free gift. He reverses this theological shift by again reasserting (Trinitarian) exchange as primary. A relation of Derridean supplementarity, however, recognizes and maintains the difference of these gifting modes while insisting on their interdependence or entangled intra-action. Both retain important challenges and counterpoints to ecologically degrading capitalistic practices.

Refusing to recognize the counter-capitalist potential of Derrida's gift proves unfortunate for Milbank. In ignoring the connection Derrida draws between capitalism and resurrection, Milbank creates his own economy of exclusion and capitalized gains. As Derrida strives to make room for uncertainty and difference, we might also add that he makes space for grace as a gift that might not be a good investment. With Milbank's ethic, Derrida's critique of the necessary capitalization of every loss hits its mark. "So long as there is loss," Milbank argues, "there cannot be any ethical, not even in any degree.... *To be ethical therefore is to believe in the Resurrection*, and somehow to participate in it. *And outside this belief and participation there is, quite simply, no 'ethical' whatsoever.*"[63] Milbank's ethic *requires* the confession of Christian orthodoxy. Outside this confession no hope remains for an ethical gesture. Such exclusivism goes to the root of his concept of gift. Again, for Milbank, the gift is never good "outside the hope for a redemptive return to the self."[64] For him this is the definition of the Gospel—that we *always* receive back. Milbank draws his circle of gift exchange tight so that his ethical act must always be a good investment, continually returning to shore up the security of an orthodox self. Where Derrida desires some wiggle room within economy, Milbank is not willing to grant him that hospitality.

In addition to Milbank's failure to recognize the counter-capitalist potential of Derrida's gift, Catherine Keller also argues that Milbank does not recognize his own vulnerabilities to separative individualism. Keller affirms the move toward an ontological reciprocity in her essay "Is That All?: Gift and Reciprocity in Milbank's *Being Reconciled*," yet offers a caveat regarding Milbank's particular gift exchange: "Whether Milbank himself breaks out of the trap of the substantial subject remains to be seen."[65] For all of Milbank's emphasis on participation and reciprocity, in the end his doctrine of God cannot abide the flux and uncertainties of life in and with the world. For him, the relational, reciprocal flow between God and creation only goes so far because of his commitment to a traditional orthodox doctrine of God. This would account for his careful avoidance of any language of "interdependence," suggesting that Milbank is not prepared to think seriously about reciprocity between God and the world. Milbank's ontology of participation, Keller suggests, reaches into the world only "as a supernatural donation, from the transcendent outside, beyond, after all."[66] With an anthropology and ontology of participation in God from the transcendent outside, material relations must follow suit. By contrast, Keller argues that although grace would be meaningless apart from its unconditionality, we may yet affirm a graceful reciprocity that emerges as asymmetrical but interdependent.

ON THE COMPLEMENTARITY OF THE NONCIRCULAR GIFT AND INTRA-ACTION

In order to interrupt the economic ideologies fueling climate change, we need both an other-oriented, unconditioned gift *and* a relational ontology of continual gift exchange. In Milbank's narrative of gift theory the two are mutually exclusive: The unconditioned gift lacks exchange, whereas a relational ontology is characterized by a continuous exchange. So what kind of comprehensive framework will simultaneously allow for their distinct characteristics and strategic roles?

Niels Bohr introduced the theory of complementarity to a scientific community stuck in confounding binary oppositions. Physicists had, for some time, debated whether the essential property of light could be characterized as a wave or particle. By Bohr's time startling evidence demonstrated that in some experimental arrangements light demonstrated characteristics of waves and in others particles. The scientific community became further divided when not just light but matter itself demonstrated this baffling indeterminacy.[67] Bohr introduced his theory of complementarity in response to this seeming aporia. His theory of complementarity accounts for the indeterminacy of quantum phenomena, explaining that some phenomena such as waves and particles cannot be measured simultaneously—not because they are *essentially* exclusive but because the measurement (in Barad's terminology) is the result of an intra-action between the measuring apparatus and the object of observation (light beam or atom) resulting in temporary, not absolute, determinacy.

In a distinctive analysis of the complementarity of Derrida's and Bohr's work Arkady Plotnitsky describes Bohr's complementarity as "a very broadly conceived interconnectivity, except that it equally implies under certain conditions, the possibility of mutual exclusivity, conflictuality and other forms of discontinuity."[68] In other words, "at times . . . [complementary] constituents act jointly, at times complementarily, at times conflicting with or inhibiting each other and at times mutually exclusive, but never allowing for a full synthesis."[69] Plotnitsky broadly describes complementary elements as "interactively heterogeneous and heterogeneously interactive."[70] He describes how Bohr maintains difference by avoiding a synthesis without falling into an absolute dualism. Consequently, Plotnitsky argues that the economy of complementarity is a general economy.[71] In this framework he also links complementarity and supplementarity: "The possibility of these [supplemental] effects will always produce what it is (classically) supposed to be added on—the form or the very possibility of any representation, or unrepresentability, of this efficacy. The same economy is at work in quantum mechanics. Complementarity is concerned with a supplementary efficacy, rather than a causal efficacy. It is supplementary."[72]

Reciprocally, it turns out that Derrida's supplementarity has emerged in close relation to Bohr's theory of complementarity. Plotnitsky notes the influence of quantum theory on Bataille and thus describes the historically and theoretically intra-active[73] relation between general economy and complementarity: "One can ascertain not only the general economic character of quantum mechanics, particularly Bohr's complementarity, but also a kind of 'quantum mechanical' and complementary character of general economy. Genealogies of both ideas overlap."[74] Complementary to Derrida's supplement, Bohr describes the relation between the particle and the wave as neither mutually exclusive nor hierarchical, with one term primary and the other secondary. In Plotnitsky's reading, Bohr's complementarity retains difference within interrelation. By viewing the relation between the noncircular gift and intra-action as complementary, we begin to see that the kind of buoyancy Dombek describes has everything to do with grace.

ALIEN RIGHTEOUSNESS? GRACE IN AN INTRA-ACTIVE UNIVERSE

I have shown how the noncircular gift that is not guaranteed to circle back to the giver keeps exchange from remaining stuck in a growth-addicted economy. However, I have also shown that, if unaffected by an intra-active universe, the unilateral gift merely perseveres in a destructive separative individualism. In this final section, I explore an understanding of grace that, as Keller suggests, remains differently asymmetrical just as it complements and even deepens the primary exchanges that constitute reality and sustain life.

Themes of self and interest have threaded their way through this entire analysis of grace, capitalism, and an intra-active universe. For the Reformers, grace was the salvific remedy to the human condition of being turned in on oneself (*incurvatus in se*) in a circular self-interest. Grace, in the form of Christ's "alien righteous," came to humans as pure exteriority, turning them out of themselves and toward their neighbors for acts of service and love.[75] After the Reformers designated sin as a kind of self-interest wreaking havoc on our relationships with God and created others, Adam Smith famously suggested that this same human tendency drives capitalistic exchange. Through self-interest, the "invisible hand" of capitalism will work out the greatest good for society: "By pursuing his own interest he frequently promotes that of the society more effectually than when he really intends to promote it."[76]

Sidestepping the polarized positions of the Reformation demonization of self-interest and the capitalistic sanction of it, Jane Bennett turns the whole concept inside out and outside in. In the conclusion to *Vibrant Matter*, Bennett argues that the very recognition of our entanglements transforms our understanding of both "self" and "interest." She renders this hackneyed theme pleasingly odd by disconnecting it from a circle of the same: Self-interest is no longer same-interest. Bennett marks this disconnection with a more complex inside/outside dynamic, emphasizing the many bodies that make up a self:

> Vital materialism better captures an "alien" quality of our own flesh. . . . My "own" body is material, and yet this vital materiality is not fully or exclusively human. My flesh is populated and constituted by different swarms of foreigners. The crook of my elbow, for example, is "a special ecosystem, a bountiful home to no fewer than six tribes of bacteria. . . . The bacteria in the human microbiome collectively possess at least 100 times as many genes as the mere 20,000 or so in the human genome." The *its* outnumber the *mes*.[77]

When we consider the "alien quality of our own flesh," self can no longer be aligned with "same," and "interest" becomes something other than gains from gift-investments. I associate such a recognition of our entangled selves with a moment of grace, turning us from separative individualist selves (*incurvatus in se*) to selves always already turned inside out and outside in by a multitude of divine and creaturely others. As Keller concludes in her essay on Milbank's gift exchange, God's life-giving energy comes to us not from outside creation but from outside "our entrapped, self-possessive subjects."[78]

Emphasizing the alien quality of Christ's righteousness given *extra nos* (outside of ourselves), the Reformers insisted that righteous being and action cannot come from a purely interior or self-possessed place. Either in doing good or ill, our wills are not entirely our own in the modern, autonomous sense. Even still, the Reformers insisted on a binding call to responsibility through gifts that are not purely our own. This soteriological drama depended on a stringent inside/outside duality. Maintaining it today in our concepts of gift exchange would only preserve an illusion that continues to conform to capitalist separative individualism.

Where grace was the remedy for the self turned in on itself, the associated spirituality desired the no-self, poured-out-self, or annihilated self. As we have seen, Milbank finds this logic still at work in postmodern, other-oriented ethics. However, I wonder: Might this tradition also resonate with, and potentially inspire, a spirituality that shifts away from a self/no self dualism toward a multiplied self embedded in a web of a host of relations, riddled with familiar foreignness and alien insiders? One would need to emphasize then that the opening of grace, classically signaled by reversing the human condition of sin as self-interest, not only opens in order to pour out the self. It also opens to receive itself from both divine and creaturely alien insiders always anew.

Milbank criticizes other-oriented grace as self-negating, but his negative appraisal also assumes a separative self. Other-orientation can manifest itself as self-negation only if the self can be secured as an isolated, self-same island in the first place. For the self to be negated there must be a self-contained, isolated bottom to approach. However, if the self is always already embedded in an ocean of otherness, no self-same, self-contained bottom remains. Only an isolated, self-contained individual is in danger of exhausting itself—of pouring the whole self out. Reciprocally, from the side of traditional grace, we must insist that grace is never an opening just for pouring out, but for letting in—for living from our interdependent embeddedness in an ocean of otherness never purely exterior or interior.[79]

When grace does not necessarily circle back to the giver, it may emerge as an opening event, exposing the self to a multitude of others that always already make it up. Never outside of entangled relations, but within them, grace gives space for others to be more than extensions of my own capitalizing interests. Just as Derrida's gift of counterfeit money makes room within economy (which is always vulnerable to sliding into a circle of the same in the form of self-interest, capitalizing on loss, etc.) for the gift that might not return to its origin, grace is the occasion that allows for a different kind of encounter, a kind of love given with the understanding that it may not be a good investment returning gains to the giver.

The return on our gift-investments may not be certain, but our debts remain constant. As we've seen, such debts are not the consequences of a transaction but open the possibility of giving and receiving in the first place. Meanwhile, the possibility for responsibility blooms. Being "part of the nature we seek to understand," use, or control is commitment as much as fact. Responsibility, like knowledge, comes through entangled relations entailing a response to the other that acknowledges interconnection. Far from shackling ethical agency, such entanglements emerge as bonds of commitment through "irreducible relations of" "response/ability" where response is a participation in the world's becoming by having a say in what comes to *matter*.[80] As Donna Haraway insists, "Response and respect are possible only in those knots"[81] between self and other, human and other-than human and, we might add, divine and creature.

Grace is not the same as interrelationship. However, we may yet retain their difference while avoiding a contradictory relationship and thus the gift/exchange binary. In a complementary or supplementary relationship grace elicits our desire for intra-connection. Interrelation may just as easily be our undoing as our salvation. Yet more than a unilateral divine gift from an absolute transcendent, a supplementary play between grace and interdependence might fuel the motivation and courage necessary to continue work toward climate change mitigation and adaptation.

Grace emerges as a possibility of familiarly foreign alien righteousness within an entangled relationality when we are turned with and through others to see ourselves as a nexus of gift exchange. This might be just the kind of buoyant "feeling" Dombek conjures and the "sensibility" Barad calls for in which "we are exposed to the outside, to the world's being, in such a way that we are bound to answer for it."[82] Alien righteousness indeed. In this light grace may emerge, after all, as a revelatory gift as we summon the courage and inspiration to face the challenges of adaptation and strive to mitigate climate change. This will, however, be the case only where grace deepens a sensibility of the outside in and inside out, calling for and calling forth our response/ability.

NOTES

1. Kristin Dombek, "Swimming against the Rising Tide: Secular Climate-Change Activists Can Learn from Evangelical Christians," *New York Times*, August 9, 2014.
2. Ibid.
3. Karen Barad, "Posthumanist Performativity: Toward an Understanding of How Matter Comes to Matter," *Signs: Journal of Women in Culture and Society* 28, no. 3 (2003): 828.
4. Ibid.
5. On "Thing Power" see, for example, Jane Bennett, *Vibrant Matter: A Political Ecology of Things* (Durham, N.C.: Duke University Press, 2010), 13. Regarding the ability of these things to talk back, Bennett writes in the same text that she's exploring the "capacity of things . . . not only to impede or block the will and designs of humans but also to act as quasi agents or forces with trajectories, propensities, or tendencies of their own." She hopes that such a shift will have an effect on the way we consume: "How, for example, would patterns of consumption change if we faced not litter, rubbish, trash, or 'the recycling,' but an accumulating pile of likely and potentially dangerous matter?" (ibid., viii).
6. See Sallie McFague, *Life Abundant: Theology and Economy for a Planet in Peril* (Minneapolis: Fortress Press, 2001) for an example of a common ecotheological critique of the Protestant tradition, tying it to a shift to individualism, a radical separation of God from creation, and consequently a separation of humanity from creation.

   There are noteworthy exceptions to such Protestant trends. Already in the 1950s (before Lynn White's famous thesis tying the Christian tradition in general to ecological degradation) the Lutheran theologian Joseph Sittler offered a powerful internal critique and alternative vision of the tradition. His 1961 address to the World Council of Churches is the most famous (Joseph A. Sittler, "Called to Unity," *Ecumenical Review* 14 (January 1962): 177–87). Dieter Hessel and Rosemary Radford Ruether suggest that this speech was a key inspiration that led to the ecotheological movement (Dieter T. Hessel and Rosemary Radford Ruether, Introduction to *Christianity and Ecology: Seeking the Well-Being of Earth and Humans*, ed. Hessel and Ruether [Cambridge, Mass.: Harvard University Press, 2000], xxxiv).

In the late twentieth century a number of critiques of capitalist values and practices began to emerge from Protestant perspectives. The German theologian Ulrich Duchrow is a noteworthy example (see especially his *Alternatives to Global Capitalism: Drawn from Biblical History, Designed for Political Action*, trans. Elizabeth Hickes et al. [Utrecht, The Netherlands: International Books, 1995]).

7. Max Weber, *The Protestant Ethic and the 'Spirit' of Capitalism and Other Writings*, trans. and ed. Peter Baehr and Gordon C. Wells (New York: Penguin Books, 2002), 120–21. The quotation in the epigraph to this section continues with the famous line (previously translated as an "iron cage"): "In Baxter's view, concern for outward possessions should sit lightly on the shoulders of his saints 'like a thin cloak which can be thrown off at any time.' But fate decreed that the cloak should become a shell as hard as steel."

8. "For, as regards justification, faith is something merely passive, bringing nothing of ours to the recovering of God's favor but receiving from Christ that which we lack" (John Calvin, *Institutes of the Christian Religion*, ed. John McNeill, trans. Ford Lewis Battles [Philadelphia: Westminster, 1960], bk. 3, chap. 13.5, 768).

9. See, for example, Catherine Keller, *From a Broken Web: Separation, Sexism, and Self* (Beacon Press, 1988).

10. See John Milbank, *Being Reconciled: Ontology and Pardon* (New York: Routledge, 2003), especially the chapter "Forgiveness: The Double Waters," in which he contrasts medieval Thomistic forgiveness with Protestant "negative" forgiveness, which essentially does the work of "decreation" (44–60). Furthermore, see the chapter "Grace: The Midwinter Sacrifice" for his critique of Protestant grace alongside Derridean/Levinasian ethics (138–61). His reference to the Protestant nature of the pure gift is less explicit in "Can a Gift Be Given?: Prolegomena to a Future Trinitarian Metaphysic," *Modern Theology* 11, no. 1 (1995): 119–61. Here the Protestant theologian Anders Nygren's work becomes the locus of what Milbank later identifies more broadly with the Reformation: "I would suggest that modern purism about the gift, which renders it unilateral, is in part the child of *one* theological strand in thinking about *agape* which has sought to be over-rigorous in a self-defeating fashion" (132). An attached footnote cites Anders Nygren, *Agape and Eros*, trans. Philip S. Watson (London: SPCK, 1982).

11. Jacques Derrida, *Given Time: I. Counterfeit Money*, trans. Peggy Kamuf (Chicago: University of Chicago Press, 1994), 12.

12. Jacques Derrida, *Circumfession*, in *Jacques Derrida*, trans. Geoffrey Bennington (Chicago: University of Chicago Press, 1993), 155.

13. Milbank summarizes his argument with regard to the Protestant perception of gift, its necessarily self-sacrificial or "other-regarding" ethic and the ethical trajectory of Patocka, Levinas, and Derrida in the chapter "Grace: The Midwinter Sacrifice" in *Being Reconciled*: "In the course of this argument I shall try to show that these received ideas of the ethical, which may or may not permit some play to 'moral luck,' all subscribe to a 'sacrificial economy.' And that they do so in two different variants: either in terms of the giving up of the lesser for the greater, or else of a more radical notion of absolute sacrifice of self for the other, without any 'return' for, or of the self, in any guise whatsoever. The second variant, which would usually see itself as *escaping* the sacrificial economy of *do ut des* [giving with expectation of a return], but which I will argue is but the same economy taken to its logical extreme, has been recently espoused in different but profoundly analogous ways by Jan Patocka, Emmanuel Levinas and Jacques Derrida. Against this view, which now enjoys wide consensus, I shall argue that a self-sacrificial view of morality is first, immoral, second, impossible, and third, a deformation, not the fulfillment, as Patocka echoed by Derrida claims, of the Christian gospel" (139).
14. "Just as Christianity transforms but does not suppress our 'given' social nature which is exchangist, so also Christian theology transforms, utterly appropriates to itself the ontological task, but does not abandon it in suspension, by elevating itself above it . . . in the name of a purely unilateral (and univocal) gift prior to that circular reciprocity which is, indeed, consequent upon *esse*" (Milbank, "Can a Gift Be Given?" 131–32). Milbank's gift theory does not engage ecology, but remains firmly in the social and economic spheres. I am uncertain as to how he would respond to the shift I propose with Primavesi, as discussed later in this essay, beyond the human economic gift exchange to a gift exchange that constructs all material reality.
15. A distinction between the circular and circulating gift is important here. The circular gift returns to its origin and thus creates a closed or "restricted" circle (more in the section on Derrida). I differentiate that the circulating gift, though, does not necessarily return to its origin but circulates in the exchange economy in a disseminating mode.
16. Bennett, *Vibrant Matter*, 112.
17. Anne Primavesi, "The Preoriginal Gift—and Our Response to It," in *Ecospirit: Religions and Philosophies for the Earth*, ed. Laurel Kerns and Catherine Keller (New York: Fordham University Press, 2007). See also Anne Primavesi, *Gaia's Gift: Earth, Ourselves and God after Copernicus* (New York: Routledge, 2004).
18. Primavesi, "Preoriginal Gift," 218.
19. The boundary between ecology and economy seems generally to depend on a more basic assumption of the nature/culture dualism which depends on a view of human agents and passive non-human "objects."
20. Karen Barad, "On Touching—The Inhuman That Therefore I Am," *differences: A Journal of Feminist Cultural Studies* 23, no. 3 (2012): 7.
21. Milbank, *Being Reconciled*, x.

22. Fellow Radical Orthodoxy theologian Stephen Long describes Milbank's position succinctly: "Capitalism is a Christian heresy because of the loss of the orthodox doctrine of the Trinity according to which the world is created through, in, and for participation with God, who is not some bare divine unity defined in terms primarily of will, but who is a gift who can be given and yet never alienated in his givenness. Once the doctrine of the Trinity is reduced to bare divine simplicity, a new 'secular' politics emerges from within Christianity that makes capitalism possible" (*Divine Economy: Theology and the Market* [New York: Routledge, 2000], 259).
23. Ibid., xi.
24. Ibid., 155.
25. Ibid.
26. Mary Douglas explains in the foreword that Mauss's text belongs to a wider critique of utilitarianism—the essence of which was the rejection of individualism. Marcel Mauss, *The Gift: The Form and Reason for Exchange in Archaic Societies*, trans. W. D. Halls (New York: W. W. Norton & Company, 1990).
27. Milbank, "Can a Gift Be Given?" 126.
28. This assumption, that since these practices are more ancient they are more natural, reveals, I believe, an assumption of a nature/culture binary, which Milbank surprisingly also accepts when he indicates in "Can a Gift Be Given?" that an inability to distinguish gift conceptually and linguistically from exchange is evidence of primitive practices such as those Mauss points out. For him these practices serve as evidence of a "universal human condition," implying the common perception based on a culture/nature binary that as society has advanced culturally we have moved further and further from our pure "natural" roots. As we will see, Derrida offers an alternative to this jockeying for what is more original or natural.
29. "We live in societies that draw a strict distinction . . . between real rights and personal rights, things and persons. Such a separation is basic: it constitutes the essential condition for a part of our system of property, transfer, and exchange. Now, this is foreign to the system of law we have been studying" (Mauss, *Gift*, 47).
30. Milbank notes Bourdieu's insight here, that a gift must be returned differently and after a certain amount of time ("Can a Gift Be Given?" 125).
31. Mauss, *Gift*, 5 and 46, respectively.
32. Ibid., 46.
33. Ibid., 20.
34. Mary Douglas, foreword to Mauss, *Gift*, vii.
35. Milbank, "Can a Gift Be Given?" 133.
36. Ibid., 127–28.
37. Ibid., 124.
38. Ibid., 123.
39. See ibid., esp. 130–33.
40. Ibid., 131.
41. Kathryn Tanner, *Economy of Grace* (Minneapolis: Fortress Press, 2005), 49–51.

42. See also Marion Grau, *Of Divine Economy: Refinancing Redemption* (New York: T&T Clark, 2004). Grau more fully explores such a space for agency within an economic system that tends to drive toward binary positions.
43. See Derrida, *Given Time*, 7–8.
44. Ibid., 12.
45. Ibid., 7.
46. Ibid.
47. Ibid., 30.
48. Ibid., 7, emphasis added.
49. Karl Barth, *Der Römberbrief, Zweite Fassung 1922* (Zürich: Theologischer Verlag, 1989), 6, quoted in Gregory Walter, *Being Promised: Theology, Gift, and Practice* (Grand Rapids, Mich.: Wm. B. Eerdmans, 2013), 39, emphasis added.
50. Walter, *Being Promised*, 40.
51. Derrida, *Given Time*, 30.
52. In "Erasing 'Economy': Derrida and the Construction of Divine Economies," Marion Grau suggests that the differential economy of the early Derrida is an "'economy that is ambiguous enough to seem to integrate noneconomy'" (*Cross Currents* 52, no. 3 [2002]: 365–66, citing Derrida, *The Gift of Death*, trans. David Willis [Chicago: University of Chicago Press, 2008]), whereas the later Derrida (especially in *Given Time*) desires the gift that is outside of economy altogether. As I have shown above, there are indeed phrases in *Given Time* that seem to suggest he desires something that transcends economy. I have intended to demonstrate, though, that this is not where he ends up with his main trope of the counterfeit coin.
53. Bush's defiant assertion that "we will not do anything that harms our economy," defending his position on the Kyoto agreement, comes to mind here. Cited in Grau, "Erasing Economy," 360.
54. While the prefix *contre* can take the form of a binary opposition—"against"—Catherine Keller's counter-apocalypse and Marion Grau's counter-economy reveal the Derridean-influenced strategy: "To criticize without merely opposing; to appreciate in irony, not deprecate in purity," "it knowingly performs an analog to that which it challenges" (Catherine Keller, *Apocalypse Now and Then* [Minneapolis: Augsburg Fortress, 2006], 19 and 19–20 respectively).
55. Jacques Derrida, *Writing and Difference*, trans. Alan Bass (Chicago: University of Chicago Press, 1978).
56. Mark C. Taylor, "Capitalizing (on) Gifting," in *The Enigma of Gift and Sacrifice*, ed. Edith Wyschogrod, Jean-Joseph Goux, and Eric Boynton (New York: Fordham University Press, 2002), 53.
57. Arkady Plotnitsky, "Re-: Re-Flecting, Re-Membering, Re-Collecting, Re-Selecting, Re-Warding, Re-Wording, Re-Iterating, Re-et-Cetera-Ing . . . (in) Hegel," *Postmodern Culture* 5, no. 2 (1995): 1, cited in Grau, *Of Divine Economy*, 8.
58. Taylor, "Capitalizing (on) Gifting," 55.
59. On "the impossible" see Jacques Derrida, "A Certain Impossible Possibility of Saying the Event," trans. Gila Walker in *The Late Derrida*, ed. W. J. T. Mitchell and Arnold I. Davidson (Chicago: University of Chicago Press, 2007), 223–43.

60. See Grau's comparison of early and later Derrida on the gift in "Erasing 'Economy.'"
61. See Jacques Derrida, ". . . That Dangerous Supplement . . . ," in *Of Grammatology*, trans. Gayatri Chakravorty Spivak (Baltimore: Johns Hopkins University Press, 1997), 141–64.
62. Derrida, *Given Time*, 31.
63. Milbank, *Being Reconciled*, 148, emphasis added.
64. Ibid., 155.
65. Catherine Keller, "Is That All?: Gift and Reciprocity in Milbank's *Being Reconciled*," in *Interpreting the Postmodern: Responses to "Radical Orthodoxy,"* ed. Rosemary Radford Ruether and Marion Grau (New York: Bloomsbury T&T Clark, 2006), 21.
66. Ibid., 31.
67. See Karen Barad, *Meeting the Universe Halfway: Quantum Physics and the Entanglement of Matter and Meaning* (Durham, N.C.: Duke University Press, 2007), 100.
68. Arkady Plotnitsky, *Complementarity: Anti-Epistemology after Bohr and Derrida* (Durham, N.C.: Duke University Press, 1994), 75.
69. Ibid., 73.
70. Ibid., 12.
71. Plotnitsky cites Bataille's description of general economy, which shows a clear affinity to Bohr's economy of quantum mechanics: "The *general economy* . . . makes apparent that *excesses of energy are produced, which, by definition, cannot be utilized. The excessive energy can only be lost without the slightest aim, consequently without any meaning,*" (*Complementarity*, 2).
72. Ibid., 51.
73. This is, of course, not the term Plotnitsky uses, but Barad's neologism. It seems Barad's "intra-action" is here a more helpful term for Plotnitsky's description of complementarity as "interactively heterogeneous and heterogeneously interactive" (ibid., 12).
74. Ibid., 18. See also Barad's essays on Derrida: "On Touching—The Inhuman That Therefore I Am," and "Quantum Entanglements and Hauntological Relations of Inheritance: Dis/continuities, SpaceTime Enfoldings, and Justice-to-Come," *Derrida Today* 3, no 2 (2010): 240–68.
75. In his famous "Freedom of a Christian" essay Luther emphasized that through Christ's righteousness we are freed from the impossible task of doing good works for God, and thus our time and energies are freed up for service to our neighbors (*Martin Luther's Basic Theological Writings*, second ed., ed. Timothy Lull [Minneapolis: Fortress Press, 2005], 386–411). See the essay "Two Kinds of Righteousness" for his articulation of alien versus proper righteousness (ibid, 134–40).
76. Adam Smith, *The Wealth of Nations* (New York: Oxford University Press, 2008), 292.
77. Bennett, *Vibrant Matter*, 112–13. The quote continues: "In a world of vibrant matter, it is thus not enough to say that we are 'embodied.' We are, rather, an array of bodies, many different kinds of them in a nested set of microbiomes. If more people marked this fact more of the time, if we were more attentive to the indispensable

foreignness that we are, would we continue to produce and consume in the same violently reckless ways?"
78. Keller, "Is That All?" 31.
79. More space connections could, I believe, be drawn here to Barad's insight that for Bohr the "cut together" is simultaneously the "cut apart," or that differentiation is always simultaneously an entanglement, since difference, in Barad's view does not need to rely on distance. See Barad, "Quantum Entanglements and Hauntological Relations of Inheritance."
80. For Barad's use of "response/ability" see "On Touching," 215. Her essay "Posthumanist Performativity" is also illuminating in this regard: "Particular possibilities for acting exist at every moment, and these changing possibilities entail a responsibility to intervene in the world's becoming, to contest and rework what matters and what is excluded from mattering," (827).
81. Donna J. Haraway, *When Species Meet* (Minneapolis: University of Minnesota Press), 42.
82. Barad, "On Touching," 226.

❧ The Door of No Return: An Africana Reading of Complexity

ELÍAS ORTEGA-APONTE

Me, singing their black language. I felt transported,
past shops smelling of cod to a place I had lost
in the open book of the street, and could not find.

It was another country, whose excitable
gestures I knew but could not connect with my mind,
like my mother's amnesia; untranslatable
answers accompanied these actual spirits
who had forgotten me as much as I, too, had
forgotten a continent in the narrow streets.

Now, in night's unsettling noises, what I heard
enclosed my skin with an older darkness. I stood
in a village whose fires flickered in my head

with tongues of a speech I no longer understood,
but where my flesh did not needed to be translated;
then I heard patois again, as my ears unclogged.
—DEREK WALCOTT, *Omeros*

At that moment in time and space there was a door. Those who crossed its threshold could not return: not in the flesh, not in spirit, and not even in memory. The crossing(s) of the threshold of this door, never in the singular, created other crossings; it crossed ways of knowing and being known, ways of being and belonging, ways of sensing and being sensed. Once crossed, return was not (and continues not to be) an option. Nevertheless, the closure was not total. A memorial pull fueled a longing of return in the hope of finding an unknown origin.

The crossing through the door of no return—the threshold across which enslaved Africans were forced to walk en route to the geopolitical space-time extending its dominion, birthing the New World, the Americas—was not merely a ritual act marking the fated departure. It would forever haunt the New World's future and potentially alter the form of reality itself, shattering plurality in search of oneness. These shattered and scattered fragments bearing witness to cultural memories contain important particles with which to understand reality. Perhaps the cultural memories of ancestral fragments, the memories of who remembers and what is remembered, haunt current structures of reality opening other doors.[1] It is to one such fragment, the Afro–Puerto Rican musical expression of bomba, characterized by the interactions between dancers and drummers, that this essay listens.

A turn in social theory proposes considering the ways in which experiences, often provoked by trauma and finding expression through ghosts, haunt social reality and can potentially introduce changes to its organization. Although there are different manifestations of this turn in social and affect theory (as I explain below), they exhibit a tendency to take their cue from the analysis of the literature of psychic trauma—or of the ways trauma is mobilized to raise specific justice claims. It is not clear how trauma may affect the constitution of the material world in and of itself. Therefore, I intend to locate in this turn the ways in which a deeper understanding of affect shapes the social world and may in fact be shaping the fabric of the world.

Although I agree with claims, such as Patricia Clough's, that affect studies "has introduced the infra-empirical, or . . . an empiricism of sensation, into the social sciences and the humanities" and in so doing "has intensified the difference or *différance* of subjectivity and the human body while turning attention to the sociality of the transmission of force or intensity across bodies, and not only human bodies," it is not clear to me whether such proposals go far enough.[2] For example, Clough raises the question, "What difference would it make if we thought of trauma as pointing beyond the individual subject and human sociality to an event of time in matter?"[3] Is this an invitation to consider the ways in which, say, the transatlantic slave trade or the Trail of Tears or the massive prison-warehousing of bodies of color point to transmission of forces across bodies, human and nonhuman, to "event(s) of time in matter"? A negative reply would call into question the reach of infra-empiricism. It would fail not because of an intrinsic limitation of infra-imperialism, but because interpreters have actively chosen to restrict those phenomena through colonizing tools, as I make clear below. A positive reply opens a haunting of its own, one that could potentially destabilize the study of affect.[4] Because of this possibility, Clough's understanding of "infra-empirical phenomena" plays a central role in my argument.

My proposal holds that if we grant that the picture of the world offered by complexity theory is an adequate description of material reality, then one can consider how social reality may also be given shape by the same forces and events that shape the material world.[5] If this is true, then, we should be able to glean much from practices often relegated to folklore. Fragments such as bomba should not be relegated to the realm of cultural production, but ought to be seen as vehicles of a deeper transmission of forces across bodies as events of time in matter.

My ethnographic research positioned me as a participant observer in various events where bomba was performed. Reflecting on the dynamic interactions between drummers and dancers opens a discursive space in which embodied practices reach beyond the interactions between dancer and drummer, revealing alternate constructions of reality. As a fragment of ancestral meaning, bomba participates in the construction of the world.[6] The route this essay traces moves across trauma, fragments, and complexity, moving away from the umbra cast by the door of no return and the penumbral space of the belly of the boat, in order to listen to what is communicable through shouts, *soberaos*,[7] and the beating of drums.

In the first section, I mobilize the works of Dionne Brand and Édouard Glissant to put in perspective the internal "hauntedness" of the hauntology in social theory. The second section deals with bomba as an embodied epistemology, a fragment of ancestral knowledge. The third and final section turns to an engagement of bomba with complexity theory. In this discussion, I aim to show how complexity theory's insight into the nature of the world as emerging, when added to the theorizing of Brand and Glissant, helps facilitate the move from a haunted social reality to one in which emerging properties of the world are informed by ancestral fragments.

THE DOOR, THE BOAT, AND THE SHAPE OF THE WORLD

Although crossing the threshold of the door shattered the ancestral ways of knowing, being, and sensing and cemented the root identity of coloniality, it failed to achieve full domination.[8] The cause of this failure can be traced to the ongoing refusal of ancestral ways to succumb to the imperialistic root identity. Scattered fragments remained that came and continued to know, be, and sense—but in diaspora, in the words of Glissant, "When the identity is determined by root, the immigrant is condemned . . . to being split and flattened."[9] New formations reverberated through the door; even though a physical return might not be a possibility, other forms of communication emerged. Shattered, yes, but not destroyed—they reverberate at infra-empirical levels affecting sensory empiricism, communicating from other space-times of a shifting present reality.

The resonant remains reverberate through worlds old and new. In the words of Patrick Bellegarde-Smith, these fragments are "elusive remembrances, always incomplete and altered in the passage and retelling, erected, reenacted, and re-creating spiritual systems and religions from fragmentary esoteric knowledge, the whole from the particle . . . in which patterns appear so unified that properties cannot be derived from the parts alone, and the whole cannot be divided easily into component parts that might make 'sense.'"[10] For Bellegarde-Smith, "These bits of words, uttered vibrations transported on a wing and a prayer," moved from island to island bearing witness to other ways of knowing, alternatives to the universalism proposed by Western colonial expansion. These fragments are central not only to theorizing about conditions pertaining to social realities, but also to political, economic, and cultural arrangements. Moreover, I propose that the possibility that the events unleashed by the colonial dynamics, those who joined their efforts, and—more important—those who resisted, have implications for the material condition of the world. As in Alejandro Carpentier's literary masterpiece on the Haitian Revolution *El reino de este mundo*, important events are punctuated by the sounds of drums produced by unknown players, foretelling coming events.[11] Crossing the threshold of the door of no return tore the materiality of the world itself.

As a prelude to my discussion of Dionne Brand's and Édouard Glissant's work, it is necessary for me to present the turn in social theory to *haunting*. Therefore, I now move to consider Avery Gordon's enchanting text, *Ghostly Matters: Haunting and the Sociological Imagination*.[12] In this singularly compelling and bold book, Gordon reveals that analytical categories fail to capture necessary dimensions of the world if they are not open to forms of conjuring. Without listening to ghosts, such efforts will fail to imagine alternatives to current social arrangements. Conjuring, for Gordon, is necessary to obtain a clear picture of what is going on in the world and what might yet be, because it is "a particular form of calling up and calling out the forces that make things what they are in order to fix and transform a troubling situation." Gordon therefore encourages her readers to contemplate an alternative "mode of apprehension and reformation . . . [that] merges the analytical, the procedural, the imaginative, and the effervescent."[13] Such a move pushes social theoretical analysis beyond the zones that can be explored with quantitative and qualitative methodologies. The task consists in considering something more visceral: matters of affect, ghostly apparitions, what Clough terms "infra-empirical."

Ghostly apparitions attest to alternative possibilities. As in Walter Mignolo's "logic of coloniality" analysis, they bear witness to colonized ways of knowing, being, and sensing the world.[14] As such, these ways are not readily intelligible via methodologies developed under the colonial sign.[15] It is Gordon's estimation that such an "otherwise" constitutes an indispensable lens through which to understand social suffering, particularly in societies marked by state repression and racism.

I read Gordon with deep curiosity, indeed with an affinity for the theoretical contours and ethical risks the author takes in advancing such proposals: The ghosts *are* real; their effects on the world of the living reveal social arrangements in need of transformation; and alternative methodologies are required for such explorations. I also experienced intense moments of "Oh, here we go again"—for example, when Don DeLillo's *White Noise* is used to frame the haunting of Ralph Ellison's *Invisible Man*. Or when Raymond Williams's notion of "structure of feeling" ("sense" in Gordon's usage) is used to explain the "feeling sense" of the depth of the haunting in authors such as Toni Morrison and Luisa Valenzuela.[16] The break that reading the works of Ellison, Morrison, and Valenzuela is supposed to create in current reality never materializes—the haunting is superseded by the white gaze authorizing the voices of Ellison, Morrison, and Valenzuela. Such a strategy raises suspicions of colonization at work. They have to be "translated" so that their ghosts can speak—a gesture that has the ironic effect of counter-conjuring the ghosts Gordon finds in Ellison, Morrison, and Valenzuela.[17] This irony is evident in Gordon's statement on the task that remains: "to follow the ghosts and spells of power in order to tame this sorcerer and conjure otherwise," that is, to tame the sorcerer whose conjuring possibilities, repressed by "the coloniality of power," threaten to destabilize the prevailing order.[18] To be a counter-conjurer who does not commune and learn from the ghostly wisdoms whose hauntings affect and stir our reality surfaces as the goal of her analysis. The precondition of conjuring-otherwise is the taming of the sorcerer. In other words, the fragments of ancestral knowledge are extracted from their source, from their entanglement in alternative worlds. What is left of their infra-empirical nature are the translated echoes speaking to the world in a white-colonizing vernacular. And thus reading DeLillo focuses the gaze so as to perceive Ellison without seeing him, as the reading of Williams tunes the ears to listen to Morrison's and Valenzuela's melodies without hearing the color of their voices.

Nonetheless, I affirm Gordon's intention, attunement, and intuition that ghosts are a "crucible for political mediation and historical memory" and that the power of ghost stories is their refusal to submit to the "logic of the unreconstructed spectacle." Such a refusal affirms the imperative for an alternate diagnosis that attempts to "link the politics of accounting in all its intricate political-economic, institutional, and affective dimensions, to a potent imagination of what has been done and what is to be done otherwise."[19] If I am more interested in communion with sorcerers than in "conjuring otherwise," it is because I affectively experience counter-conjuring as another attempt to make ghosts of de-colonized forms of knowing, being, and sensing instead of freeing them. I just don't hear strongly enough in Gordon, Dani Nabudere's call to an open-endedness that "allows processes of 'fusion of horizons' to take place . . . [that] can lead to a new pooling of knowledge drawn from all cultures and which is accessible to all users."[20]

In my estimation, a more suitable framing of affect theory can be found in the work of Patricia Clough, who explains that affect "subsists in matter as incorporeal potential. As soon as it begins to inform, it dissolves back into complexity across all scales of matter, like quantum effects feeding the indeterminacy appropriate to each scale of matter: the subatomic, the physical, the biological and the cultural."[21] In this alternative, the sorcerer and the ghost remain as "incorporeal potentiality," informing as they dissolve, adding complexity across matter, whether physical, biological, or cultural. It is at this juncture that I find that thinkers such as Brand and Glissant can offer much. In fact, I suggest that affectivity and materialist perspectives may be "catching up" to the insights of Afro-diasporic experience.[22]

The works of Brand and Glissant disrupt and trouble the line of discourse gaining purchase within the current sociological imagination, that is, analyses that turn to "haunting" to reflect the contemporary situation and increased complexity of life in "late modernity." What Brand's and Glissant's works reveal is that this discourse may *itself* be haunted by the specter of the transatlantic slave trade and its aftermath. Such haunting should be affirmed, indeed conjured. If haunting reveals repressed possibilities and alternate arrangements, then we should let ghosts loose to haunt every life, institution, and social arrangement, as well as the carefully planned theoretical thinking that seeks to tame the sorcerer and counter-conjure the ghosts coming out of the Africana experience. Following Nobel Prize–winning poet Derek Walcott, we ought to let the world be haunted by those "singing their black language" and by "night's unsettling noises." Walcott wants to be in a place where his "flesh [does] not needed to be translated," to hear patois with ears unclogged.[23] I want to affirm this haunting, the sorcerer, as well as the Deleuzian empiricism (infra-empiricism) that brings them forth, but I do not want to counter-conjure in order to tame.[24] With this in mind, let me proceed to the works of Brand and Glissant.

Dionne Brand's *A Map to the Door of No Return* and Édouard Glissant's *Poetics of Relation* share deep resonances. Emanating from the Black Diaspora, both works open with crucial moments at the beginning of the journey of Black bodies to the unknown. For Brand, the door of no return "is no mere physicality. It is a spiritual location. It is also a psychic destination."[25] It is a condition that, in the rupture created by the slave trade, thrives in remembering the un-rememberable, bringing it back to haunt the world. Brand narrates her own personal disappointment at being unable to get her grandfather to remember the names of his ancestors. This disappointment also pointed to a larger dimension. "It was a rupture in history," Brand says, "a rupture in the quality of being. It was also a physical rupture, a rupture in geography."[26] Such ruptures extend not only to the social lives of individuals and cultural realities but also to the fabric of the world.[27]

For Brand, the door of no return is both real and imaginary. Real as an actual place and historical event, imaginary because its symbolic power transforms doors as such. "[It] is a door," she writes, "which makes the word door impossible and dangerous."[28] The effect of the impossibilities and dangers of doors after the door of no return is the flipping of ways of knowing and being from what is present and given to what is absent. Brand's contention is that although the door exists in its absence and may be a thing and a place we don't know, "it exists as the ground we walk. Every gesture we make somehow gestures towards this door."[29] In Brand's conceptualization, the door of no return points to an experience, particularly for those of African descent, of always "being touched by or glimpsed from this door," a touch saturated by ambivalence, discomfort, anger, and unknowable memories.

Nevertheless, memories of the door—and this is central—are more than remembrances; they seep into reality.[30] And the way they seep into reality is through the haunting of all cognitive schemas by captivity itself; in fact, for Brand, "our cognitive schema is captivity." In the Diaspora, "you are constantly overwhelmed by the persistence of the specter of captivity."[31] If this is the case, then, counter-conjuring proves ineffective. From this, however, it should not be taken that Brand succumbs to the dread of captivity without hope. This is not the case. Unlike Gordon, she is not interested in offering theoretical poultices to ease our particular woundedness and tragedy resulting from captivity and its aftermath. My attention to Brand has located a tear in the fabric of historical and geographical reality—a tear that began at the accursed moment in which the threshold of the door of no return was crossed for the first time. It is also through this rupture that ancestral fragments seep through to haunt the world. Glissant's work will aid in the development of this theme.

Glissant also speaks of "the experience of deportation to the Americas" as having a shattering effect on cognition, particularly for diasporic subjects.[32] Through this deportation, according to Glissant, the terrifying, the unknown, and the abyss partake of each other. The abyss is experienced in three ways: as womb (the belly of the boat), as the depth of the sea, and as a reverse image of all, particularly the earth. Nevertheless, the experience of these three aspects of the abyss "lies inside and outside the abyss." For Glissant, those who survive the ordeal of the first two abysses, the "belly of the boat" and "the depth of the sea," and lived to experience the third, "the reverse image of all," were left with a world turned upside down: "the panic of the new land, the haunting of the former land, finally, the alliance with the imposed land, suffered and redeemed." Those who lived to experience the third aspect of the abyss did not put down roots in the soil in this world, this new location; no, the people who came forth "despite having forgotten the chasm, despite being unable to imagine the passion of those who foundered there, nonetheless wove this sail" and travel through the world, forming communities through relation but never putting down roots.[33] This sail, in Glissant's conceptualization, was not put to use on a return ship; it became a channel of knowledge.

The knowledge communicated through this channel originates from the unknown of the abyss in its three dimensions. It is not just about the particularity of the experience of enslavement. It cannot be bounded by slavery's chains nor tamed by the master's whip. Instead, it is "knowledge of the Whole, greater from having been at the abyss and freeing knowledge of Relation within the Whole."[34] This knowledge interrupts the settled, sedentary ruling form of knowledge that marked "root identity."

Glissant's understanding of relation is complex and perplexing. It requires that ways of knowing that rely on categorization and compartmentalization be suspended. Instead, cognitive surplus is the mark of knowing in relation. For Glissant, relation brings us to "that-there that cannot be split up into original elements. We are scarcely at liberty to approach the complete interaction, as much for the elements set in relation as for the relay mode relentlessly evolving." Relation is constantly diversifying "forms of humanity according to infinite strings of models infinitely brought into contact and relayed." Furthermore, unlike the root which seeks to anchor itself in being, relation's source lies "in these contacts and not in itself." Relation, therefore, "is a product that produces. What it produces does not partake in Being."[35]

Like the abyss that is divided in three parts, relation also has three parts, and they are connected to a form of poetics. These three poetics are of depth, "of language-in-itself," and of structure.[36] Thus Glissant conceives of relation as a poetics. Such a poetics intends to free ways of knowing that remain utterly open even as they reach toward abstraction and universality.[37] Nevertheless, just as Brand presents the door of no return as marking all subsequent knowing with captivity, Glissant presents the plantation, with the forms of discourses that developed there (Afro-musical, literary, and dance expressions), as a possible site for a universal explanation of reality. For Glissant, these expressions are the cry of the plantation, "transfigured into the speech of the world." Furthermore, he concludes, "This is the only form of universality there is: when, from a specific form of enclosure, the deepest voice cries out."[38] Therefore, to access this knowledge, or to be affected by it (known in relation to it), means to experience the haunting of enslavement, to be open to the reconstruction of reality in ways that take stock of its ongoing effects. Even after the plantation has closed, Glissant reminds us, "the word derived from it remains open. This is one part, a limited part, of the lesson of the world."[39]

My hope in this discussion of Brand and Glissant is to present an alternative haunting to the one developed by Gordon. Gordon's precision in describing the sociological manifestation of the effect of ghostly disruptions ought neither to displace nor to supersede Brand's and Glissant's analyses of the world-destroying and world-forming power of the deportation of Africans to the Americas and its effect on their descendants. In their experience of the abyss is the best element of exchange in relation.[40]

BOMBA AS EMBODIED EPISTEMOLOGY OF ANCESTRAL FRAGMENTS

As one visits the Museo de Las Americas in Puerto Rico, a museum dedicated to preserving the history of culture of the Americas "from Alaska to Patagonia," as stated on their website, one cannot help but wonder about the choice of location. The museum is located in El Cuartel Ballajá in Old San Juan.[41] The route will take you through narrow streets, various plazas, a view of the Atlantic Ocean, El Morro (a colonial-era Spanish fort), and La Perla. One may forget that the site chosen for preserving historical and cultural traditions in the Americas is located inside a former colonial army barracks and training site. Such a choice of space should give one pause: When walking through this space and its various exhibits, I cannot help but wonder whether the colonial gaze is still at play here. Are there ways in which local memories, in the face of broken memory, are mobilized to fit an idealized paradigm?[42] My intuition (perhaps informed by infra-empirical data) leads me to believe that this is the case. If serious consideration is given to the contributions of affect theory to thinking about trauma and ontologies that affirm the agential relationality of embodied material interactions, and, furthermore, if these insights are framed by the particular experiences of trauma and open to learning from ancestral fragments, then we can have a high degree of confidence that the collection of affect surrounding such a space, in this instance, El Cuartel Ballajá, carries the particular weight of its colonial past. Its impact on how the histories and cultures of the Americas are memorialized should receive attention.

I want to call to your attention to two of the permanent exhibits, one simply named *The African Heritage*, which pales in size compared to another, *The Indian in America: Twenty-Two Ethnic Groups Who Have Survived Colonization*. The differences between these two, especially the difference in the space they occupy, reach beyond the titles to expose what is in my estimation a colonial time-trap. Whereas *The Indian in America* covers the terrain from before and after the conquest, *The African Heritage* is not only smaller but also caught in a timeline set by the colonial dynamics of power. This is evident in the presentation of the African heritage in Puerto Rico. After the customary drums and masks, the bulk of the exhibit is composed of chains, manacles, instruments of discipline and punishment, announcements of auctions, newspaper clippings, and a few photos. Center stage is a time mosaic of racial mixture that is finished by a commentary on today's reach of Afro-culture, mostly as it pertains to food and literary impact but framed as remnants of the past with ongoing influence.

There is a significant gap here, not only in the scope of these exhibits but also between the elements included in *The African Heritage* and those that could have been included from a larger perspective. For example, in addition to the impact of Africanization on the culinary tradition and language of the island, and the overall influence on the literary and cultural scene, the exhibit could have highlighted the political and social contributions of communities of the descendants of Africans. Instead, these elements are moved into folklore.[43] Such a move in my reading claims that root identity (as identified by Glissant) relies on hierarchical notions of being and is, therefore, predicated on violence, the violence of colonization.

In the particular case of Puerto Rico, Halbert Barton observes:

> The African presence in Puerto Rico has been denied for so long and in so many different ways that to bring it center stage in any significant way was to be, and continues to be, a controversial, if not revolutionary, act of history-making proportions. And on the few occasions where the African presence is acknowledged, it is usually reduced to dance music, in the form of a color-coded folkloric dyad.[44]

Barton points out that a revival of the Africanized aspects of Puerto Rican culture is under way, primarily led by bomba players. This raises the challenge of how to use bomba without reinscribing it as a "color-coded folkloric remnant" of African culture. The acceptance and legitimacy of bomba as an evolving cultural expression beyond folklore faces challenges from the political, cultural, and aesthetic domains.[45] Nevertheless, it is well positioned to meet these challenges. Barton concludes that to "the extent that bomba may have been used as a vehicle to announce rebellions in the past . . . bomba keeps its irrepressible spirit by keeping a certain *cimarron* consciousness alive." This consciousness of the escaped slaves empowers the ongoing refusal to "allow identity, social mobility, and cultural spaces to be defined by outsiders."[46]

Whereas for the dominant white sectors of society, bomba may be tolerated as folklore, for those communities embracing its African roots and historical legacy, it represents a tool of resistance

Bomba refers to Afro-diasporic forms of musical performance in coastal zones where sugar was cultivated and processed by slave labor. Traditionally, the main form of instrumentation is made up of two *barriles* (drums), one called *el buleador* or *segundo* (second) and the other *primo* or *subidor* (first or riser), two wooden sticks known as *cua*, and maracas. Singers and dancers are integral parts of the performance of bomba. Typically, the themes of bomba songs are everyday realities of the island's legacy of slavery. Common scenes include the realities of sugar mills and the social suffering caused by them. The songs focus on personal and communal affirmation and celebration. Additionally, most of the rhythms associated with bomba are named after a particular geographical area. But the central aspect of bomba is the conversation between drummers and dancers, particularly between the drummer playing *el subidor* and the dancers. In the exchange the dancers communicate through movements and gestures to which the drummer replies with various *toques*. In this conversation the drummer follows the dancers, and the exchange continues until either the dancer or drummer stops. For Barton and Juan Cartagena, the dancing of bomba expresses a deeper reality: the hope for a future connected to political, economic, aesthetic, and historical possibilities.

Drawing on the work of José Manuel Argüelles, Juan Cartagena traces the etymology of the word "bomba" to Akan and Bantú, in which it "has a common meaning that encompasses a spiritual connotation of gathering."[47] This spiritual connection brings into communion not only the living but also the ancestors, and it connects the worlds of the ancestors and the living with the rest of nature. During my research, I observed how an alternate spiritual space materialized in the exchange between drummer and dancer, and during a *primo* drum solo. In such moments it is usual to hear members of the crowd shouting, "Habla!" "Speak!" Typically, in a bomba performance the drummers and singers take the stage. Throughout the performances, dancers will come onstage either individually or in pairs and dance before the drum, presenting two possible scenarios: one of a dialogue between a dancer and a drummer, and a second one that takes place primarily between dancers. To a person unfamiliar with bomba, the call to "speak" may be perplexing. Neither the dancer nor the drummer communicates verbally. They "speak" to each other rhythmically, in a call and response based on movements and *toques de tambor* (the playing of drums).

When a dancer "takes the floor," he or she walks up to the *primo* drum player; they acknowledge each other and perform a synchronistic dance-drumming exchange, a call and response. The dance is an exchange in which the drummer responds to the movements he or she sees the dancer performing. Although there are traditional steps in bomba, the dance is open to the innovation, creativity, and expression of each individual. Similarly, if two dancers decide to take the floor and dance with each other, they both greet the drum player first and then dance; on this occasion, the *primo* "follows" both dancers. At times, dances between two dancers take the form of *retos*, akin to dance battles, or they may instead be jovial exchanges. Other times, they may take the form of courtship. Of particular interest to me as a participant observer were the thematic discourses of such events. The theme of Africanness is constant and unabashed in bomba performances.. For example, one can hear members of the audience shouting phrases like, "Baila *Cimarron/Cimarrona!*" (Dance, Cimmaron!). At other times, singers narrate scenarios of slavery's past and make correlations with present conditions. Slavery restrictions preventing slaves from becoming literate are compared with government policies to defund public education, measures that will strongly affect poor communities. Furthermore, critiques of the social, political, and economic realities in Puerto Rico and of unequal treatment because of race are prominent and punctuated by calls to action. These calls for action are often framed as acts of *cimarronaje* (to escape slavery). Listeners are encouraged to openly defy law and order through acts of civil disobedience. To be called a *cimarrón* in the memory activated by bomba is to call forth an alternate reality—one in which reality is redefined through fragments of Africanness.[48]

These exchanges are full of memories that burst forth and do so, compellingly enough, through movement. I see in this sense the possibilities of an embodied epistemology in line with the analyses of Brand and Glissant, while posing questions like the following to Gordon: What if we were to engage DeLillo's *White Noise* in a public plaza where bomba is being performed? In that case, what ghostly apparitions might haunt the reality of the text—and the textuality of the world? For Bellegarde-Smith, "Memory is rooted in experience rather than biology. But within the concept of the transmission of knowledge we operate from . . . these memories are embodied, inscribed in our flesh transgenerationally through the notion of 'technique.'" If we accept this claim, then we would have to relationally open our horizons to the reality that "when I dance, my ancestors dance with me." And if the knowledge of reality is shaped, as Brand argues, by crossing the threshold of the door of no return, and, as Glissant writes, given voice by the cries and shouts of the plantation, then, in the words of Bellegarde-Smith: "To the European diatonic scale, we oppose the African pentatonic and hexatonic scales. It is an accretion, a layering process, complex conditions rendered into bare-bones essence."[49]

OF COMPLEXITY, GHOSTS, AND AFRICANA WAYS

In this final section I offer some brief thoughts on a potential point of interaction between complexity theory in the natural sciences and the Africana perspective I have offered. A turn to complexity theory allows a theoretical opening to the construction of a world in which multiple realities are mutually constitutive of each other. In light of my argument, such mutual constitution of reality, from a perspective that, following Brand and Glissant, perceives the world through remembrance of the unrememberable and in relation, informed by infra-empirical data, reveals the necessity of continuing the explorations of how reality is haunted by the legacy of the transatlantic slave trade. This is possible because, as Peter Stewart explains, complexity "is a matter of perspective or framing (which in our case relates to human intention and interests), level of detail (fine or coarse graining), and the result of perceiving through observation."[50] This means that, in the words of Kenneth Mossman, "Complexity forces us to think about natural phenomena in a holistic rather than elementalist way. Problems in complex systems are best resolved by looking at the whole picture rather than individual parts."[51] Such a perspective implies that reality is an ongoing process through, as Clough described it, physical, biological, and cultural phenomena interacting and shaping each other at the quantum level as well as at system-wide levels. John Urry strengthens this perspective. For Urry, a turn to complexity requires engagement with an emergent structure of feeling that "involves a greater sense of contingent openness to people, corporations and societies, of the unpredictability of outcomes in time-space, of a charity towards objects and nature, of the diverse and non-linear changes in relationships," including multiple types of human relations and the ongoing complexity added to the human by technological innovations.[52]

For the present discussion, I focus on the ways in which time and space are conceptualized through complexity theory. These understandings help elucidate the ways in which ancestral fragments, like a bomba performance, through their haunting contribute to shaping reality via counterpossibilities. Here I depend on the joint work of David Byrne and Gill Callaghan. These theorists propose an ambitious project that puts complexity theory in conversation with the social sciences and vice versa.[53] Of interest to me is their view that science, as a field, should be understood "not as the contemplative observation of the world but as a set of social practices with the most profound implications for the construction of the social world."[54] For these theorists, such an orientation is the necessary outcome of taking seriously the processes of reflexivity and emergence in the world. Byrne and Callaghan argue that once we understand the world as situated and reality as embodied, we have to dispense with any frame that makes strong claims for predicting regularities. We have to settle instead for how a reflexive understanding of reality can provide us with information about tendencies without predicting future actualizations.[55] Such a position, furthermore, is open to emergent realities.

For my purposes here, I zero in on Byrne and Callaghan's conceptualization of space in their development of complexity theory. In this conceptualization, Byrne and Callaghan reject idiographic and nomothetic explanations of space.[56] Instead, they propose that we should consider space not only geographically but also socially constructed. Such a shift is important in order to tease apart the social implications of the uses of space.[57] Their intent is to conceive of space's nature as an open ontology, which is to say that it is more than territory, place, and scale; it is also and intrinsically *network*. Space should not be conceived as an empty container; space should be "reconceived as interacting levels and networks in a loosely nested open system in which the outcomes of the processes are never, in a mechanical sense, determined, but are always actively achieved."[58] Such a change more accurately describes "the reflexive relations between "observed reality" and the way concepts and social policies that shape the construction of the world are produced.[59] In other words, spatial processes are actively related to causal configurations. Such understanding brings spatial configurations (including territory, place, scale—but primarily network) to bear in the causal relations at play in the social world.

If such an understanding of space holds true (one that is relational, unfolding as a network, and with continually emerging properties), and if we add to this frame Brand's and Glissant's theorizations, then the door of no return, the abyss, and the plantation are more than sociocultural happenings that shape history, economics, and cultural worlds. They are deeper, implicating the nature of reality itself. Through their coming into existence, new properties and effects emerge in reality—some manifested in the cries of the plantations or in the embodied epistemology of the bomba. These emergences appear in the sense of being touched and glimpsed when no one is behind us to touch or look at us or in the ghosts that haunt a world in need of transformation. But how does this take place in the performance of bomba?

As I mentioned in the second section of this essay, the performance of bomba in this time and place is undergoing a renaissance. This renaissance is not solely a matter of its gaining popularity among a larger following. If this were the case, one would be hard-pressed to differentiate bomba from some element of folklore, possibly trapped within the bounds circumscribed by the coloniality of power. But instead, as Glissant observed, folklore can transcend itself and point to resonant possibilities, demonstrating potential to transform reality, to materialize newness in systems declared rooted and closed to change.[60]

Furthermore, bomba also calls for a break in time as linear. In the playing and dancing of bomba, the resonant possibilities of a folklore that extends beyond itself also call us to reconceptualize time as the extended temporality of ancestral fragments. As Urry states, openness to the insights of complexity requires an openness to experience time, space, and the relations between humans and the networks of materiality they participate in as nonlinear; thus, it requires engagement with "emergent structures of feeling." Bomba haunts a reality in which the legacy of the African Diaspora is continually threatened by counter-conjuring and the taming of its sorcerers. Bomba, through dance and *toques de tambor*, communicates—indeed, embodies—the possibility of a different and more complex reality. I'll give the last word to Glissant: "We know ourselves as part and as crowd, in an unknown that does not terrify. We cry our cry of poetry. Our boats are open, and we sail them for everyone."[61]

NOTES

1. The notion of fragments will be explained in the first section of this essay. My use of "cultural memory" derives from the work of Jeanette Rodriguez and Ted Fortier. In their study of cultural memory, they arrive at the following conclusion: "Whether cultural memory fulfills the need to transcend certain events or to maintain a corporate identity, it passes from generation to generation through oral traditions, written accounts, images, rituals, and dramas. It is evoked around image, symbol, affect, or event precisely because it keeps alive and transforms those events of the past. They are not bound or limited to the past, but continue to give meaning to the present" (Jeanette Rodriguez and Ted Fortier, *Cultural Memory: Resistance, Faith, and Identity* [Austin: University of Texas Press, 2007], 13).
2. Patricia Ticineto Clough, "Afterword: The Future of Affect Studies," *Body & Society* 16, no. 1 (2010): doi:10.1177/1357034X09355302.
3. Ibid., 227.
4. This is a strong statement and one that I do not wholly affirm, as of now, but one that I find deeply intriguing. For example, how would a mostly white affect theorist react to Dani Nabudere's thesis in his *Afrikology, Philosophy and Wholeness: An Epistemology* (Pretoria: African Books Collective, 2011), an interdisciplinary proposal that frames the contribution of science within African ways of knowing? Or would there be an attempt to hold fast to the hermeneutical key? Either possibility would be to expose limits and the possibility of affect theorizing. In the first section of this essay, I engage this move in relation to the haunting of social theory, as presented by Avery Gordon in her *Ghostly Matters Haunting and the Sociological Imagination* (Minneapolis: University of Minnesota Press, 2008).
5. In this essay, I do not enter into a full explanation of complexity or quantum theory. Instead, I highlight particular insights.
6. I use the term "ancestral fragments" to refer to what I take to be "infra-empirical" phenomena/data that interrupt reality communicating, borrowing from Mignolo, Afro-diasporic ways of knowing, of being, and sensing.

7. "Soberao" is the term used to refer to the dance/dancing of bomba.
8. For my usage of root identity see, Édouard Glissant, *Poetics of Relation*, trans. Betsy Wing (Ann Arbor: University of Michigan Press, 1997), 141–46. I follow Walter Mignolo's understanding of coloniality. See Walter D. Mignolo, "Delinking," *Cultural Studies* 2, no. 2–3 (2007): 449–514.
9. Glissant, *Poetics of Relation*, 143.
10. Patrick Bellegarde-Smith, *Fragments of Bone: Neo-African Religions in a New World* (Urbana: University of Illinois press, 2005), 2.
11. Alejo Carpentier, *The Kingdom of This World* (New York: Noonday Press, 1989). and Alejo Carpentier, *El reino de este mundo* (New York: Rayo: Planeta, 2009).
12. Gordon, *Ghostly Matters*.
13. Ibid., 22.
14. Walter D. Mignolo, "Geopolitics of Sensing and Knowing: On (de) Coloniality, Border Thinking and Epistemic Disobedience," *Postcolonial Studies* 14, no. 3 (2011): 129–50.
15. See Mignolo, "Delinking."
16. Gordon, *Ghostly Matters*, 18.
17. Ibid., 195–201.
18. Ibid., 28.
19. Ibid., 18.
20. Nabudere, *Afrikology, Philosophy and Wholeness*, 162.
21. Furthermore, she continues, "If a method attending to affect necessarily becomes entangled with or assemblaged with affect's enactable capacities for activation or informing, it also will be subject to a causality that necessarily is a quasi-causality, where future effects—what will be activated—causes activation in the present." Patricia Ticineto Clough, "The New Empiricism Affect and Sociological Method," *European Journal of Social Theory* 12, no. 1 (2009): 43–61.
22. See, for example, Nabudere, "Afrikology as a Universal Emancipatory Epistemology," in *Afrikology, Philosophy and Wholeness*, 159ff.; Christina Elizabeth Sharpe, *Monstrous Intimacies: Making Post-slavery Subjects* (Durham, N.C.: Duke University Press, 2010), chap. 2; Teshale Tibetu, *Hegel and the Third World: The Making of Eurocentrism in World History* (Syracuse, N.Y.: Syracuse University Press, 2011), chap. 6; Peter K. J. Park, *Africa, Asia, and the History of Philosophy: Racism in the Formation of the Philosophical Canon, 1780–1830* (Albany: State University of New York Press, 2013), chap. 5.
23. Derek Walcott, *Omeros* (New York: Farrar, Straus and Giroux, 1999), 167.
24. Gordon, *Ghostly Matters*, and Clough, "New Empiricism."
25. Dionne Brand, *A Map to the Door of No Return: Notes to Belonging* (Toronto: Vintage Canada, 2002), 1.
26. Ibid., 5. This theme will be picked up again in the last section.
27. I argue later that, when complexity theory's understanding of space is incorporated into frames of thoughts like Brand's, such possibility is viable.
28. Brand, *Map*, 5.

29. Ibid.
30. Ibid.
31. Ibid.
32. Glissant, *Poetics of Relation*.
33. Ibid.
34. Ibid. Also see 143–46 for a discussion of root/relation identity.
35. Ibid., 160.
36. Ibid.
37. Ibid.
38. Ibid., 74.
39. Ibid., 75.
40. Ibid.
41. http://www.museolasamericas.org.
42. Brand, *Map*.
43. See Édouard Glissant, *Caribbean Discourse: Selected Essays* (Charlottesville: University Press of Virginia, 1999), 195–220, for his treatment of folklore.
44. Halbert Barton, "A Challenge for Puerto Rican Music : How to Build a *'soberao'* for *'bomba,'*" *Centro: Journal of the Center for Puerto Rican Studies* 16, no. 1 (2004): 68–89.
45. Ibid, 74.
46. Ibid., 85–86.
47. Juan Cartagena, "When 'bomba' Becomes the National Music of the Puerto Rican Nation," *Centro: Journal of the Center for Puerto Rican Studies* 16, no. 1 (2004): 14–35.
48. See the following sources for discussion of cimarrons. Ronald Cummings, "(Trans)nationalisms, Marronage, and Queer Caribbean Subjectivities," *Transforming Anthropology* 18, no. 2 (2010): 169–80; Yolanda Arroyo Pizarro, *Tongas, palenques y quilombos: Ensayos y columnas de afroresistencia* (CreateSpace Independent Publishing Platform, 2013); Neil Roberts, *Freedom as Marronage* (Chicago: University of Chicago Press, 2015).
49. Bellegarde-Smith also says that "dance is movement, movement is life, dance is life" (*Fragments of Bone*, 6).
50. Peter Stewart, "Complexity Theories, Social Theory, and the Question of Social Complexity," *Philosophy of the Social Sciences* 31, no. 3 (2001): 324.
51. Kenneth L Mossman, *The Complexity Paradox: The More Answers We Find, the More Questions We Have* (New York: Oxford University Press, 2014), 3.
52. John Urry, "Complexity," *Theory, Culture & Society* 23, no. 2–3 (2006): 111–15.
53. This is a direction that Stewart also affirms. He says, "Social processes and phenomena are far too complex for complexity theory to deal with, or profoundly elucidate, without the aid of the resources of the better of existing social theories and studies. Furthermore, complexity theories do not provide a particularly effective metatheory of social processes" (Stewart, "Complexity Theories," 354).
54. David Byrne and Gill Callaghan, *Complexity Theory and the Social Sciences: The State of the Art* (New York: Routledge, 2014).
55. Ibid.

56. Ibid., 130. This framework also applies to time.
57. Ibid., 131.
58. Ibid., 136.
59. Ibid., 136–37.
60. Glissant, *Caribbean Discourse*, 197–99.
61. Glissant, *Poetics of Relation*, 9.

✢ The Trouble with Commonality: Theology, Evolutionary Theory, and Creaturely Kinship

BEATRICE MAROVICH

We have to envisage the existence of "living creatures," whose plurality cannot be assembled within the single figure of an animality that is opposed to humanity.
—JACQUES DERRIDA, The Animal That Therefore I Am

Every creature, no matter how small, which we can see with our eyes or conceive of in our minds, has in itself such an infinity of parts or rather of entire creatures that they cannot be counted. . . . In every creature, whether spirit or body, there is an infinity of creatures, each of which contains an infinity in itself, and so on to infinity.
—ANNE CONWAY, The Principles of the Most Ancient and Modern Philosophy

As humans, we share the planet with an uncountable number of other entities: animals, insects, bacteria, flowers, trees, and rocks. An infinity of others, you might say. To follow the early modern philosopher Anne Conway, our own bodies are moving units of an infinite number of other bodies. And yet there are few words in the English language that gather us up into a grouping or category that evokes this confederation. More often than not we make recourse—as the philosopher Jacques Derrida notes—to descriptors such as "the animal" that blunt or mask the complexities of life. For many reasons (some commendable and some deplorable), we humans tend to guard our uniqueness. But from time to time, as language and sentiment allow, we concede—as both Derrida and Conway do—to sharing the earth with other *creatures*. This is one of the few words we have in the English language to express the fact that humans are earthlings, bound in a kind of kinship with other mortals on the planet.

One thing that creaturely life shares, however, is a kind of predicament. Creatureliness is among the most radically generic subjective states or conditions: It is defined most succinctly by what it is *not*. Creatures are not The

Creator. Creatures are bound, as if by an intangible umbilical cord, to the superhuman, the supernatural, to realms of extreme and mysterious power. In the wake of this power, creatures seem to cringe in a kind of exposed vulnerability and dependence: They are the fragile finite to the potent infinite. It is, perhaps, little wonder that many humans—whose relations with the divine or the immortal have become decidedly more complex, particularly over the course of modernity—might resist identification with creaturely life. To be a part of this generic form of life seems to demand a subjection, of sorts, to this power that makes us vulnerable and dependent. Basic survival, as well as our communal and political life, seems to demand that we avoid getting too deeply buried or lost in our inevitable vulnerabilities and dependencies.

The case might be made that the problem is driven by the conflict between religious and scientific frames of reference. If we detach the bodies of earthlings from the great metaphysical figures of the divine or the immortal, we can finally see these bodies empirically, with more accuracy. We can see, with more veracity, the powers that really do reside in the bodies of earthlings: Bacteria ravage and sustain, dogs bite, lions roar, trees push their roots through pavement and concrete. If we detach from theological metaphysics, in other words, earthlings appear to have a different kind of agency—a new way of thinking about life on earth (a more accurate form of knowledge) becomes possible. Percy Bysshe Shelley, for instance, argued that theology itself was superstition—one that served as a kind of intellectual "tyranny over the understandings of men." Citing the work of Francis Bacon, Shelley claimed that only by getting rid of the divine itself—only through atheism—could a clear-sighted knowledge emerge.[1] "If ignorance of nature gave birth to gods, knowledge of nature is made for their destruction," writes Shelley.[2] To follow Shelley, then, it might be said that creaturely kinship itself—roping us back into the tyranny of theology—fails to illuminate the nature of life on earth with real accuracy. It is the tyranny of theology, and divine power, that makes earthlings vulnerable to these radically asymmetrical forms of power and blinds us to the nature of things.

Or, perhaps not. As I aim to illuminate here, creaturely kinship may not be something that merely *distinguishes* theological perspectives on planetary life from theoretical perspectives in the sciences. Instead, creaturely kinship might actually be something that keeps theology and science strangely *entangled*. Ultimately, the ability to see the powers and capacities of creaturely or mortal bodies has less to do with severing these earthlings from theology and its metaphysics, and more to do with the refusal to lose sight of the radical plurality and diversity of creaturely life—something that a theological perspective does not necessarily hinder. In this essay I argue that figures such as Anne Conway—whose work is deeply theological—may even do a better job of emphasizing the powers and capacities of creaturely life than antitheistic scientists like Richard Dawkins, whose work resonates in certain regards with that of orthodox theologians like Andrew Linzey. The crucial move, I argue, is to emphasize the importance of creaturely *difference* (the plurality of creaturely life that Derrida underscores)—not to evacuate the cosmos of traces of divinity. Working with the "diffractive methodology" of feminist theorist and theoretical physicist Karen Barad, I argue that creaturely kinship can be read metaphorically as diffractive—as a relational bond that emphasizes what I call a "connective distinction," a differentiation that also binds.

THEOLOGY, EVOLUTIONARY THEORY, AND COMMONALITY

The theologian Andrew Linzey's project is an "animal theology." Its animating interest is in evoking the connective tissues between humans and animals, and making these connections into an ethical and theological issue. Linzey argues that, if Christianity holds—as one of its central values—a moral priority for those with less power, then Christians have been going about their animal ethics in the wrong way. Arguing that human civilization has thrown animal life into a position of vulnerability, Linzey argues that animals themselves should now "constitute a special category of moral obligation, a category to which the best, perhaps only, analogy is that of parents to children."[3] The weakness of animal life, in the face of a ravenous human industry and development, should thus give animals not only an *equal* but a "greater consideration"[4] than other humans, in a manner similar to the way that parents attend disproportionately to the needs of their own children.

If such a decentralization of the human sounds a bit unorthodox for a Christian theologian, Linzey's justification for this decentralization is a great deal more orthodox. For Linzey, shifting the moral priority away from human life (alone) is justified on the grounds of *common origin*. It is invoked, in other words, on the grounds of what I will call a "commonality." As Linzey puts it, "The common origin of all creatures is a doctrine that carries with it implications and consequences which so far only a few in the Christian tradition have fully appreciated."[5] According to this doctrine of common origin, "All creatures are God's creatures. All things proceed from the Creator's hands." Common origin, then, becomes the form of commonality that links and bonds creaturely bodies together. By this logic, of course, God is "the Creator" and therefore "sovereign in power, however we may understand that sovereignty to be manifest."[6] For his part, Linzey understands this sovereignty to be manifest through God's power to hold rights over all creation. The human, in other words, does not have sovereign rights over other creatures. These rights, rather, belong to God. The human is merely the shepherd or the steward who should respect God's rights over all creatures, "by reverencing what is given."⁷ Commonality, then, expressed through God's sovereign power over (and rights to) all creation, has moral and ethical stakes. And the impetus to obey these moral and ethical stakes comes from above, from the pressure of divine power.

Despite the theological genealogies of creaturely life, we see the term "creature" cropping up in the work of the evolutionary theorist (and vehement antireligionist) Richard Dawkins, as well. Following in the footsteps of Charles Darwin (who found the term fit for use in his own work),⁸ Dawkins deploys the term "creature" without hesitation and without reference to its historical genesis or associations. We see that this is especially true in his popular publications, where he reaches out beyond the scientific community, to convert readers who might otherwise be held in the thrall of religious or theological perspectives. In his book *The Greatest Show on Earth: The Evidence for Evolution*, Dawkins cites numerous passages from Darwin's work that use the term. Additionally, Dawkins finds his own creative uses for it. "All living creatures, whether plants or animals, have approximately the same ratio of carbon-12 to carbon-14, which is the same ratio as you'll find in the atmosphere," Dawkins declares.⁹ He even utilizes the term to evoke the commonalities at work in the evolutionary process. "Although we may lack the fossils to tell us exactly what our very ancient ancestors looked like, we are in no doubt at all that living creatures are our cousins, and cousins of each other," he writes.¹⁰ Evolution throws us into relation as if we were a family. We are all "cousins"—all creatures, great and small. For Dawkins, the power of evolution is what ignites the crucial commonality among creaturely bodies. All creatures are evolved creatures—creatures of the evolutionary process (not, of course, a God).

There are certainly differences in how this theologian and this evolutionary theorist deploy these commonalities. Given that, as Dawkins has put it, all living creatures share the bond of having been shaped by natural selection, this also means—for Dawkins—that all creatures are shaped by natural selection's favoritism for "selfish" or self-interested genes. As creatures built of smaller molecular units, we are as Dawkins puts it, "survival machines—robot vehicles blindly programmed to preserve the selfish molecules known as genes."¹¹ For Dawkins, these genes (guided by natural selection's favoritism) arguably have a kind of sovereign control over creaturely bodies. These tools of natural selection dictate the function of our nervous systems and "exert ultimate power over behavior."¹² Here, creaturely commonalities are evoked in order to underscore the rather amoral "executives" or "policy-makers" that function deep within our creaturely bodies.

In a very simple sense, the theologian seems to be deploying commonalities with a moral or ethical end in mind—for Linzey, commonality gestures toward obligations to others and a fundamental benevolence at the seat of creation. The scientist, in contrast, seems to be actively eschewing such a trajectory. For him, commonalities seem to be deployed in order to denaturalize, or make strange, the notion that there is altruism or benevolence at the core of our creaturely bodies. But such a simplistic distinction between the two would be too easy. There are deeper resonances and complicities that abide between these two frameworks.

Let me return to the point I made earlier: the strange fact that Dawkins has chosen to use the term "creature" to evoke a figure bound through commonality, despite its deep and ancient resonances with creation theologies. Dawkins is merely using a term that's on offer in the English language. Theology is not his field, and he is not trained to understand the theological genealogies of contemporary terms and concepts. And yet the power relations embedded in creaturely life are not useless in Dawkins's own rhetoric. In the context of orthodox Christian theology the creature is often the finite, mortal, and vulnerable counterpart to the infinite, immortal, and omnipotent creator. Dawkins may have done away with the creator figure, but the apparition of a vulnerable and powerless creature—at the mercy of the forces of natural selection and its favored selfish genes—continues to be useful. It is as though orthodox theology still provides him with conceptual resources.

It would seem that on peeling back the commonalities we find in the work of Linzey and Dawkins, another layer presents us not with deep difference and distinction between their perspectives, but with a resonance that begins to look more like a likeness. Despite their differential positions on the ultimate aim or origin of commonalities, each seems to find creatures and creatureliness useful because they come umbilically connected to a sovereign power over life that dwarfs or dissipates the powers within creaturely bodies. Creaturely kinship helps Linzey remind humans of their own subjection to God's law, as a way of calling them back to an ethical account. Creaturely kinship helps Dawkins remind humans of their subjection to the evolutionary process, as a way of diminishing their human sense of self-importance or specialness. More than that, it is the totalizing power to which these creatures are subject that generates and drives the commonality binding creaturely bodies together. The common bond between creatures is their fragility and powerlessness in the wake of a supreme power that reduces or diminishes them.

Here we begin to see that when the radically generic nature of creaturely life is articulated as a specific sort of bond (especially one that emphasizes the powerlessness or vulnerability of creatures), the plurality of creaturely life and the differentiated nature of creaturely powers and capacities are passed over and erased. What this means, effectively, is that the kinship bonds between the human and the nonhuman world appear to be located in the most incapacitated registers of existence. And yet mutual powerlessness, weakness, and vulnerability insufficiently address the sort of complex symbioses we witness in—for instance—the odd kinship bonds between human organs and the bacterial substrata that coexist with them. This overemphasis on commonality as the kinship bond between creatures, in other words, seems to insufficiently illuminate the actual nature of interspecies relations. My argument here, however, is that this is not necessarily a problem with the figures we are exploring: the creatures, in their creatureliness. Instead, it is a problem with the way that creaturely kinship is being articulated and addressed. Is it possible to rearticulate this kinship bond *not* as a relation of commonality, but instead as a relation of distinction and differentiation? That is to say: Can creaturely kinship be understood as a tie, a relation, or a bond wherein difference itself serves as a connective tissue?

DIFFRACTING CREATURELY KINSHIP

This is, in large part, a figurative problem. The question is whether creaturely kinship can be metaphorically rearticulated—emphasizing its resident plurality, rather than conscripting it to a universal condition that effaces its particularity. A host of projects in postmodern thought, notably feminist theory, has sought to emphasize the importance of difference itself as an ally for critical thought. If a dominantly white, male, heteronormative, and exclusively humanist Western intellectual tradition once imagined identity in universal and uniform terms, difference has become a site through which the gendered, racialized, or animalized bodies excluded from this tradition have begun to claim new intellectual space. The feminist theorist Donna Haraway (trained as a biologist) has been interested in the impact that this white, male, heteronormative, and humanist intellectual tradition has made on research in the sciences. She has also had an interest in finding alternative metaphors and figures in the sciences for feminist work. Criticizing the feminist practice of reflexivity—a process of critical self-reflection[13]—Haraway argues that this is too much like the optics of reflection. It is primarily a practice of displacing the same into a new location, "setting up the worries about copy and original and the search for the authentic and really real."[14] In the optical metaphor of diffraction—an interference pattern in waves—Haraway finds a figure for thought more suitable for feminist work attuned to difference. Diffraction remains attentive to difference and distinction, where Haraway worries that reflexivity seeks reflection, sameness, or uniformity.

Following cues from Haraway, the theoretical physicist and feminist scholar Karen Barad has developed what she calls a "diffractive methodology" for interdisciplinary work in the sciences and humanities. The ultimate aim of this work is to make use of the physics of entanglement as metaphor for relational thought, across disciplines. Where Haraway wanted to move away from reflexivity as a critical practice, Barad wants for similar reasons to encourage a move away from analogical thinking. This form of thought, she argues, tends—too quickly—to seek out affirmational *likenesses* and to eviscerate important distinctions or qualifications. Entanglement, Barad argues, is not about likeness but marks a relational bind within difference and distinction. A diffractive methodology can illuminate such entanglement in interdisciplinary work by revealing the "iterative production of boundaries, the material-discursive nature of boundary-drawing practices, the constitutive exclusions that are enacted."[15] It is a method that takes *differences* as the integral elements that create the *entangled* effects to look for. Thus for Barad, to speak about the entanglement of things is to address the diffracted patterns of difference that emerge in their relationships.

For interdisciplinary scholars who seek to illuminate the entangled nature of matter and reality Barad suggests that we set up a "diffraction apparatus." To illustrate what a diffraction apparatus might be, Barad describes ocean waves, washing over a barrier with a sizable hole or gap. As the waves push through this gap, their wave forms bend and spread out—they are diffracted, but still bound through the wave form. A diffraction apparatus might also be, Barad suggests, a cardboard tube we speak into that diffracts the sound waves that pass through it.[16] What results from this sort of intervention is a set of different and distinct, but still overlapping, wave patterns. What the diffraction apparatus helps us see is the production of what Catherine Keller calls the relational ontology of "entangled difference."[17] The wave patterns are engaged in a kind of differential becoming. Moreover, from the perspective of quantum physics (which is Barad's angle of approach), there is yet another layer of difference to tease out. This diffraction is a reminder that, on a quantum level, these diffracted waves are not simply, exclusively, or uniformly waves. Rather, these diffraction patterns can serve as a reminder that on the quantum level the matter constituting these waves exhibits—paradoxically—*both* wavelike and particle behavior.

Can we, I wonder, understand creaturely kinship itself as a kind of diffraction apparatus? Can creaturely kinship be understood as a relational bond born not out of a commonality—such as a common origin—so much as it is born from an entangled difference? Must creaturely kinship be limited to a shock of recognition that depends on our acknowledgment that another creature is *like* us humans? There is another possible cognition, or affective acknowledgment: that other creatures are distinct from us and yet coexistent, possibly even codependent with us, nonetheless. Can this *connective distinction* be understood as a bond of strange kinship? Can differentiation and distinction be a bond at the deepest recesses of creaturely life?

It is possible that kinship, as a term, is too far gone—too deeply implicated in the racist, sexist, and colonialist registers of the nineteenth-century Western anthropology that built and developed it.[18] Christopher Peterson has pointed to the irony that the term "kinship" is often used—particularly to describe human bonds with the nonhuman world—as if it naturally evokes positive affects and sentiments. His critique of this optimistic use of kinship is that it disavows or ignores the deeper violence that is also resident within kinship. Kinship, he argues, is fundamentally about a violent process of exclusion. "The conventional notion of kinship as a bond of love or affection tends to disavow the violent, dialectical absorption of otherness inherent to the concept of kinship," he writes.[19] There is something about kinship that is ultimately directed against the force of difference. Kinship "presupposes that the distance between self and other might ultimately be bridged."[20] Counterintuitively, this means that the principle of exclusion remains embedded within kinship, inevitably. Many articulations of interspecies kinship seem to tend toward the idealized, ignoring that kinship is (among other things) "always implicated in the violent reduction of the other to the same."[21] What happens, however, if we turn kinship inside out? What happens if kinship is diffracted and is used to gesture not toward the originary sameness between entities, but toward their connective distinctions?

CONNECTIVE DISTINCTION, AND CREATURELY CAPACITY

I have argued that this connective distinction—this relational connection that illuminates crucial distinctions—is not inimical to work that has been done in theology. As a brief and concluding example, I want to point to the work of the seventeenth-century British philosopher Anne Conway—a Quaker convert interested in the Kabbalah, whose work anticipated that of Leibniz but was published only after her death in 1679. The philosophy she develops in her treatise *The Principles of the Most Ancient and Modern Philosophy* is predominantly a reflection on the nature of species. But this philosophy is rooted in, and derived from, a theological metaphysic. It should be noted that Conway's theological position does indeed share much with the orthodox theology of a thinker like Linzey. Keller has suggested that Conway was certainly a thinker of relational multiplicity and plurality—a thinker whose work emphasizes and illuminates the important differences between creatures. But Keller also notes that Conway's divinity was less relationally multiple and fundamentally, "a neoplatonically stabilized One, orthodox in its omnis."[22] In this respect, then, Conway's thought does presume the orthodox power structure that I have illuminated in the work of Linzey and Dawkins—a force that produces creatures with an originally similar "essence and being."[23] But, as Carol Wayne White has noted, the fact that Conway was a female philosopher when such a vocation was almost unheard of, and that she was drawn to both Kabbalistic thought and Quaker spirituality, led her to challenge "the Christian orthodoxies of her day" in an attempt to "find a more naturalistic and pluralistic approach to many influential religious doctrines."[24]

I suggest that Conway's focus on the crucial (yet still connective) distinctions between creatures begins to illuminate how creaturely kinship might become a kind of diffraction apparatus. Creatureliness in Conway's thought, in other words, provided a form of connection between various earthlings, but it also served as a device through which to illuminate productive and necessary distinctions between them. When the differences between creatures are clearer, what also becomes clearer is their mutuality—their mutual dependence on one another. Creaturely capacities become clear in a way they are not when the emphasis is—instead—on creaturely subjection to a common force or point of origin.

Conway's work was, in part, a quarrel with the Neoplatonic philosophies of her own day (she had an active correspondence with the Cambridge Platonist Henry More). Although this conversation with Neoplatonism was largely productive, Conway did want to push back against the intellectual dualism that she also found particularly in the work influenced by Descartes. Her philosophy refused to see creaturely life constituted by a fracture between realms—between, for instance creature and creator, or matter and spirit. Instead, Conway argued that the cosmos breaks down into three basic species—God, Christ, and creatures. Christ, as the incarnational element, serves as the mediator between God and creatures. This tripartite structure was, for Conway, ultimately a hierarchical spectrum—God is a higher species than Christ, which is a higher species than creatures. This structure ultimately lends support to a hierarchy among forms of creaturely life—one that holds human life to be superior to animal life. Despite this, however, creatureliness in her philosophy resists conscription into a condition of abject dependence on, or subjection to, the creator. The speciation she sets up places the three species (and, subsequently, creatures) into a kind of power-sharing agreement that begins to highlight the importance of mutuality.

Metaphysically, Conway is interested in exploring differences between species that are not simple oppositions. This begins with a reflection on the differentiated forms of infinity to be found in the species of divine and creaturely life. Creatures, for Conway, are not simply the finite, mortal entity. Rather, for Conway all creatures are actually eternal and infinite. This position is quite radical. Not only does Conway refuse to subject creatures to a state or condition of simple mortality, she also furnishes each creaturely body with a kind of infinite capacity, an endlessness. Every creature, "no matter how small, which we can see with our eyes or conceive of in our minds, has in itself such an infinity of parts or rather of entire creatures that they cannot be counted," she wrote.[25] Not only does this underscore a kind of power and capacity in creaturely life, but this also anticipates what we know to be true, today, about the uncountable numbers of minuscule cellular and bacterial creatures who constitute this thing we call the human body.

Creaturely life is ultimately a diffraction apparatus, however. So this infinity of creatures is inevitably quite different and distinct from the form of infinity that belongs to God. Creatures are not coeternal with God. Their power is not the same form of power. Thus the infinity of creatures is "an infinity in which they were and always will be without end." Creaturely infinity is one of neverending time. The infinity of God, in contrast, is a divine eternity outside of time itself.[26] It is a time without time. Both God and creatures are infinite, but this resonance between them is ultimately also a distinction.

This crucial distinction Conway has highlighted between creature and creator also highlights their need for one another. Creatures, in an obvious way, needed the creator in order to exist in the first place. But God also needed creatures, Conway suggests. Existing on his own, outside of time, God "wished to create living beings with whom he could communicate."[27] God desired creaturely companionship and another species to share the cosmos with—God desired distinctive and different entities to connect with. God, perhaps, wanted difference (rather than a reflection)—hence, the origin of creaturely life. The figure of an omnipotent, sovereign, creator is certainly not absent—we know that it constitutes a foundation of Conway's metaphysic. And yet it is also decentralized here. The figure of the divine is, in its way, rendered vulnerable to a kind of desire—the need for companionship. The difference between God and creatures is productive, and good. Creaturely capacity and powers bubble to the surface.

This pattern of connective distinction continues in Conway's reflections on the relationships between different forms of creatures—suggesting that creaturely life retains its ability to serve as a diffraction apparatus beyond its primordial connection with the divine. It seems, on one level, that the connective distinction between creatures is almost derived from the pattern of connection and differentiation that emerges between God and creatures. Conway does argue that there is more mutability between creatures than there is between species (between God and creatures). It is clear, she acknowledges, that hard stone can change "back into softer and more pliant earth."[28] Processes such as decomposition make a lie of infinite differences and irreconcilable distinctions between creatures. But Conway is careful not to efface the real and observable differences between creatures, either. It is for this reason that Conway argues that creaturely life is made up of *finite* differences. Creatures, themselves (even animals such as cows, horses, and dogs, she argues) are "potentially infinite." They are all "always capable of greater perfection without end."[29] But their differences from other creatures are finite. Each individual creature is "endowed with different degrees of perfection," according to its material and spiritual constitution.[30]

The glory of God shines forth differently in each creature. But this difference or distinction between creatures is not (as in the difference and distinction between God and creatures) ultimately alienating or fracturing. Instead, it underscores the differentiated capacities of each creature—it highlights what creatures can do for one another and how they need one another. There is a degree of mutuality in creaturely life, Conway argues, such that "one cannot live without the other. What creature in the entire universe can be found that does not need its fellow creatures? Certainly none."[31] Creatures here are dependent but also have the power to sustain, and this sustenance comes in large part from their distinctions from one another. Here, bound by a kind of mutual dependence, creatures are powerful and sufficient enough to sustain one another in a kind of power-sharing relation.

Again, Conway shares certain presumptions with the thinkers of creaturely commonality—Linzey and Dawkins—that I highlighted in opening this essay. All three of these thinkers set creaturely life into relief against the backdrop of an arguably sovereign force (whether this force be a god or an earthly process). Creaturely life, in each case, is engaged in asymmetrical power dynamics. And yet although Conway may presume this sort of environmental condition for creaturely life, when she reflects on it in greater detail, her focus seem to shift. She does not evoke a creaturely life that is bound in kinship with other entities because of its sheer vulnerability or its subjection to the force of this divine power. Instead, she becomes attuned to differences that matter—distinctions between creature and creator and between creatures themselves. The kinship bonds that emerge between creature and creator and between creatures are not predicated on a sameness. Instead, their relational bonds of mutuality are strengthened because they are different from one another—these distinctions allow them to provide others with aid or companionship.

• • •

Given that kinship has historically been used to emphasize the bonds constituted through something shared—particularly the sameness of bloodline—it is far from logical to articulate kinship as a bond or connection that emerges through distinctions. And perhaps this stretches kinship too far. Even if we might object to the quasi-biological reductionism inherent in the notion of kinship as shared blood, there is still a power in holding things in common. It can, for instance, be a powerful thing to share a table with companions (human and nonhuman). So it may be that my emphases here have stressed differences and distinctions at the expense of things held in common, or at the expense of the commons. This argument was intended to investigate the troubles with commonality, rather than to suggest that commonality is an inherently troubled thing.

My argument that creaturely kinship might be read as a kind of diffraction apparatus was not designed to devalue things held in common, or even commonalities, as such. Rather, this argument was intended as a more pointed critique and a suggestion that the felicity of having something in common can often be used to disempowering ends or to mask a less felicitous agenda. The celebration of creaturely kinship, for example, as a shared or mutual vulnerability seems to suggest that the common bond between creatures can be reduced to their fragility and powerlessness in the wake of a supreme power that diminishes them. I intended, here, to make the point that the bonding relations between creatures—even in early modern theological articulations, such as Conway's—have never been so easily reducible. There has long been room for creaturely capacities and I suspect that thinking differently about creaturely kinship can underscore and illuminate this.

Moreover, there are many things we fail to see when we are on the lookout for common characteristics, or forms of sameness. When we are disproportionately attuned to what we share, for instance, with a bat there are an almost infinite number of things about the bat that we will fail to see. It may be that we will be better companions to the bats of the world if we are attuned to the way that our distinctions equip us to live in this common world. The benefit of creaturely life as a figure for thought is that it offers us a radically generic subjective category—one that brings an infinite number of entities into a kind of relational bind. If we seek to reduce this extended form of relation into a bond predicated on one universally shared characteristic, or quality, then we evacuate creaturely life of its power, its beauty, and its complexity from the outset. The glory of creaturely life, perhaps, is the absurd and extraordinary fact that it can name a federation in the first place.

NOTES

1. Percy Bysshe Shelley, *The Necessity of Atheism and Other Essays* (New York: Prometheus Books, 1993), 36. Accessed via Google Books on 2/24/15.
2. Ibid., 38.
3. Andrew Linzey, *Animal Theology* (Urbana: University of Illinois Press, 1994), 36.
4. Ibid., 28.
5. Ibid., 11.
6. Ibid., 22.
7. Ibid., 23.

8. "Natural Selection acts exclusively by the preservation and accumulation of variations, which are beneficial under the organic and inorganic conditions to which each *creature* is exposed at all periods of life," writes Charles Darwin. See Charles Darwin, *The Origin of Species* (New York: P. F. Collier, 1909), 134 (italics mine).
9. Richard Dawkins, *The Greatest Show on Earth: The Evidence for Evolution* (New York: Free Press, 2009), 104. He is, here, speaking about the ratio between carbon 12 and carbon 14 in relation to carbon dating. After the death of a living creature, the carbon 14 decays while the carbon 12 does not. Comparing the ratio between the two can give insight into the approximate age of an organic material.
10. Richard Dawkins, *The Magic of Reality: How We Know What's Really True* (New York: Free Press, 2011), 51.
11. Richard Dawkins, *The Selfish Gene: 30th Anniversary Edition* (1976, 1989; Oxford: Oxford University Press, 2006), xxi.
12. Ibid., 60.
13. Haraway is not raising questions about self-reflection per se as much as she is raising questions about a particular feminist methodology.
14. Donna Haraway, *Modest_Witness@Second_Millenium. FemaleMan©_Meets_OncoMouse™* (New York: Routledge, 1997), 16.
15. Karen Barad, *Meeting the Universe Halfway: Quantum Physics and the Entanglement of Matter and Meaning* (Durham, N.C.: Duke University Press, 2007), 93.
16. Ibid., 74–75.
17. See Catherine Keller, *Cloud of the Impossible: Negative Theology and Planetary Entanglement* (New York: Columbia University Press, 2014).
18. Since the early 1980s, the discourse of kinship within anthropology has come under fire for a number of reasons. The anthropologist David Schneider's critique of kinship in the 1980s argued that the discourse of kinship was reductionist—that it presented kinship in biologistic terms. This critique was incisive and deeply influential. In *A Critique of the Study of Kinship*, Schneider questioned the basic animating presumptions that had shaped the study of kinship in anthropology from its beginnings. He pondered why it was that kinship had always been seen as the most *primitive* form of social engagement and why it had traditionally been analyzed as something entirely separate from elements such as a culture's economics. He argued that kinship had, methodologically, always been about the "the primary assumption that 'Blood Is Thicker Than Water.'" That biological laws of inheritance, in other words, dictated our social lives. Given that kinship was often presented, anthropologically, as something that "they " (the non-West) had, whereas "we" (the West) had politics, culture, and civilization, this reductionist discourse of kinship helped perpetuate problematic racist assumptions in Western scholarship. See David Murray Schneider, *A Critique of the Study of Kinship* (Ann Arbor: University of Michigan Press, 1984).
19. Christopher Peterson, *Kindred Specters: Death, Mourning, and American Affinity* (Minneapolis: University of Minnesota Press, 2007), 16.
20. Ibid., 137.

21. Ibid., ix.
22. Catherine Keller, "Be a Multiplicity: Ancestral Anticipations," in *Polydoxy: Theologies of Multiplicity and Relation*, ed. Catherine Keller and Laurel Schneider (London: Routledge, 2010), 93.
23. Anne Conway, *The Principles of the Most Ancient and Modern Philosophy*, ed. Allison P. Coudert and Taylor Corse (Cambridge: Cambridge University Press, 1996), 38.
24. Carol Wayne White, *The Legacy of Anne Conway (1631–1679): Reverberations from a Mystical Naturalism* (Albany: State University of New York Press, 2008), ix.
25. Conway, *Principles of the Most Ancient and Modern Philosophy*, 17.
26. Ibid., 13.
27. Ibid., 10. This is also, arguably, among the most kabbalistically influenced passage in Conway's text. In the next two lines of text, Conway speaks about God withdrawing to make space for creatures, evoking the concept of *Tzimtzum* (God's contraction, or withdrawal, that allows for the emergence of creatures) from the kabbalistic philosophy of Isaac Luria, whom she was likely reading at the time.
28. Ibid., 29.
29. Ibid., 33.
30. Ibid., 32.
31. Ibid., 55.

CONTRIBUTORS

KAREN BARAD is a professor of feminist studies, philosophy, and history of consciousness at the University of California, Santa Cruz. Barad is the author of *Meeting the Universe Halfway: Quantum Physics and the Entanglement of Matter and Meaning* (Duke, 2007).

JANE BENNETT is a professor of political science at Johns Hopkins University and the editor of the journal *Political Theory*. Her most recent book is *Vibrant Matter* (Duke, 2010). She is working on a book on Walt Whitman.

LORILIAI BIERNACKI is an associate professor in the Department of Religious Studies at the University of Colorado at Boulder. Her first book, *Renowned Goddess of Desire: Women, Sex, and Speech in Tantra* (Oxford, 2007) won the Kayden Award in 2008. She is co-editor of *God's Body: Panentheism across the World's Religious Traditions* (Oxford, 2013).

PHILIP CLAYTON holds the Ingraham Chair at Claremont School of Theology in Claremont, California. He is the author of *The Predicament of Belief* and *In Quest of Freedom: The Emergence of Spirit in the Natural World*, among other works.

JACOB J. ERICKSON is an assistant professor of theological ethics at Trinity College Dublin. His writing and research focus on contemporary constructive ecotheologies, climate ethics, and ecological resilience. He is a contributor, most recently, to *Living Traditions and Universal Conviviality: Prospects and Challenges for Peace in Multireligious Communities* (Lexington Books, 2016), *For Our Common Home: Process-Relational Responses to Laudato Si'* (Process Century Press, 2015), and *Divinanimality: Animal Theory, Creaturely Theology* (Fordham University Press, 2014).

CATHERINE KELLER is George T. Cobb Professor of Constructive Theology in the Theological School and Graduate Division of Religion of Drew University. Books she has authored include *Apocalypse Now and Then, God and Power; Face of the Deep: A Theology of Becoming,* and *Cloud of the Impossible: Negative Theology and Planetary Entanglement.* She has co-edited numerous volumes of the Drew Transdisciplinary Theological Colloquium.

BEATRICE MAROVICH is an assistant professor of theological studies at Hanover College in Indiana. She is working on a book-length project on creaturely life.

ELÍAS ORTEGA-APONTE is an assistant professor of Afro-Latino/a religions and cultural studies at Drew University Theological School. He is the author of "Democratic Futures in the Shadow of Mass Incarceration: Towards a Political Theology of Prison Abolition," in *Common Goods: Economy, Ecology, and Political Theology,* edited by Melanie Johnson-Debaufre, Catherine Keller, and Elías Ortega-Aponte (Fordham University Press, 2015).

TERRA S. ROWE has a PhD in theological and philosophical studies from Drew University's Graduate Division of Religion. She teaches writing and theology at Marist College and Wartburg Theological Seminary and is the author of a forthcoming book on ecology, economy, and the Protestant tradition (Fortress Press).

MARY-JANE RUBENSTEIN is a professor of religion at Wesleyan University, where she is also core faculty in feminist, gender, and sexuality studies and affiliated faculty in the Science in Society Program. She is the author of *Strange Wonder: The Closure of Metaphysics and the Opening of Awe* (Columbia, 2009) and *Worlds without End: The Many Lives of the Multiverse* (Columbia, 2014).

ELIZABETH SINGLETON is a PhD student of religion, ethics, and society at Claremont School of Theology. She has taught religion and philosophy at Augustana University and Minnesota State Community and Technical College.

MANUEL A. VÁSQUEZ is the author of *More Than Belief: A Materialist Theory of Religion* (Oxford, 2011). He is working on *The Wiley-Blackwell Companion to Religion and Materiality.*

THEODORE WALKER JR. is an associate professor of theology of ethics and society at the Perkins School of Theology at Southern Methodist University, where he teaches Christian ethics, metaphysics of nature, metaphysics of morals, Black theology, and process cosmology. He is the author of *Empower the People: Social Ethics for the African-American Church* (Orbis Books, 1991) and *Mothership Connections: A Black Atlantic Synthesis of Neoclassical Metaphysics and Black Theology* (State University of New York Press, 2004). He is co-editor and contributor with Mihály Tóth of *Whiteheadian Ethics: Abstracts and Papers from the Ethics Section of the Philosophy Group at the 6th International Whitehead Conference at the University of Salzburg, July 2006* (Cambridge Scholars Publishing, 2008), and co-author with Chandra Wickramasinghe of *The Big Bang and God: An Astro-Theology* (Palgrave Macmillan, 2015).

CAROL WAYNE WHITE is a professor of philosophy of religion at Bucknell University. Her most recent books are *Black Lives and Sacred Humanity: Toward an African American Religious Naturalism* (Fordham University Press, 2016) and *The Legacy of Anne Conway (1631–70): Reverberations from a Mystical Naturalism* (State University of New York Press, 2009).

TRANSDISCIPLINARY THEOLOGICAL COLLOQUIA

Laurel Kearns and Catherine Keller, eds., *Ecospirit: Religions and Philosophies for the Earth.*

Virginia Burrus and Catherine Keller, eds., *Toward a Theology of Eros: Transfiguring Passion at the Limits of Discipline.*

Ada María Isasi-Díaz and Eduardo Mendieta, eds., *Decolonizing Epistemologies: Latina/o Theology and Philosophy.*

Stephen D. Moore and Mayra Rivera, eds., *Planetary Loves: Spivak, Postcoloniality, and Theology.*

Chris Boesel and Catherine Keller, eds., *Apophatic Bodies: Negative Theology, Incarnation, and Relationality.*

Chris Boesel and S. Wesley Ariarajah, eds., *Divine Multiplicity: Trinities, Diversities, and the Nature of Relation.*

Stephen D. Moore, ed., *Divinanimality: Animal Theory, Creaturely Theology.* Foreword by Laurel Kearns.

Melanie Johnson-DeBaufre, Catherine Keller, and Elias Ortega-Aponte, eds., *Common Goods: Economy, Ecology, and Political Theology.*

Catherine Keller and Mary-Jane Rubenstein, eds., *Entangled Worlds: Religion, Science, and New Materialisms.*

www.ingramcontent.com/pod-product-compliance
Lightning Source LLC
Chambersburg PA
CBHW030432300426
44112CB00009B/961